CPSS Power Electronics Series

This series comprises advanced textbooks, research monographs, professional books, and reference works covering different aspects of power electronics, such as Variable Frequency Power Supply, DC Power Supply, Magnetic Technology, New Energy Power Conversion, Electromagnetic Compatibility as well as Wireless Power Transfer Technology and Equipment. The series features leading Chinese scholars and researchers and publishes authored books as well as edited compilations. It aims to provide critical reviews of important subjects in the field, publish new discoveries and significant progress that has been made in development of applications and the advancement of principles, theories, and designs, and report cutting-edge research and relevant technologies. The CPSS Power Electronics series has an editorial board with members from the China Power Supply Society and a consulting editor from Springer.

Readership: Research scientists in universities, research institutions and the industry, graduate students, and senior undergraduates.

More information about this series at http://www.springer.com/series/15422

Bo Zhang · Xujian Shu

Fractional-Order Electrical Circuit Theory

Bo Zhang
School of Electric Power
South China University of Technology
Guangzhou, China

Xujian Shu
School of Electric Power
South China University of Technology
Guangzhou, China

ISSN 2520-8853 ISSN 2520-8861 (electronic)
CPSS Power Electronics Series
ISBN 978-981-16-2824-5 ISBN 978-981-16-2822-1 (eBook)
https://doi.org/10.1007/978-981-16-2822-1

This Springer imprint is published by the registered company Springer Nature Singapore Pte Ltd.
The registered company address is: 152 Beach Road, #21-01/04 Gateway East, Singapore 189721, Singapore

Preface

Fractional-order circuit is an unfamiliar concept to most researchers and engineers in the field of electrical engineering. In traditional electrical circuit theory, inductors and capacitors are regarded as integer-order components, and their voltage-current relationship is described by a first-order differential or integral equation. However, it is a fact that inductors and capacitors in engineering applications have fractional-order characteristics, which cannot be explained by the concepts of integer-order electric circuits. It is more available to adopt the fractional-order calculus to analyze inductors, capacitors, and their circuits. More importantly, some novel features that cannot be reflected in the integer-order system will be discovered and applied.

Inspired by the concept of fractional-order systems, we proposed a novel wireless power transmission mechanism and system based on fractional-order inductors and capacitors in 2013, which was immediately funded by the key project of the National Natural Science Foundation of China due to its unique innovation. Relying on the fractional-order wireless power transmission technology, we have solved the problem of stabilizing transmission efficiency and power in the existing wireless power transmission system.

However, when we were conducting research on the fractional-order wireless power transmission, we found that there was no systematic theory of fractional-order circuits as a reference. Therefore, while the fractional-order wireless power transmission was explored, we also developed the theory of fractional-order circuits and extended the basic theorems of integer-order circuits to fractional-order circuits. This monograph is the result of our research on fractional-order circuits; we hope that it is helpful for more graduate students and researchers involved in the area of electrical engineering to study fractional-order circuits and systems.

This monograph consists of nine chapters. Chapter 1 briefly introduces several special functions (such as Gamma function, Beta function, and Mittag-Leffler function), the basic definitions and properties of Grünwald-Letnikov, Riemann-Liouville, and Caputo fractional calculus, the methods of processing fractional differential equation, and some software tools for solving fractional calculus. Chapter 2 gives the fundamental concepts and properties of fractional-order components, and expands them to fractional-order circuits composed of series, parallel, and series-parallel combinations of fractional-order components. Chapter 3 summarizes several network theorems of fractional-order circuits based on the characteristics of fractional-order components and Kirchhoff's laws. Chapter 4 is devoted to the time-domain analysis of fractional-order circuits. Taking RC_α, RL_β, $RL_\beta C_\alpha$ series and parallel fractional-order circuits as examples, the dynamic behaviors under different excitations are analyzed, including zero-input response, zero-state response, and total response. Chapter 5 provides the sinusoidal steady-state analysis of fractional-order circuits. Through the phasor representations of voltage, current, impedance, active and reactive power of fractional-order circuits, the sinusoidal steady-state response of several typical fractional-order circuits such as series and parallel resonant circuits, mutual inductance coupling circuits, and filter circuits are discussed. Chapter 6 focuses on the sinusoidal steady-state analysis and calculation of fractional-order symmetrical and asymmetrical three-phase circuits. Chapter 7 investigates the steady-state response of fractional-order circuits under non-sinusoidal periodic excitation. The non-sinusoidal periodic steady-state analysis of fractional-order symmetrical three-phase circuits is given. Chapter 8 elaborates the parameters, equations, and properties of fractional-order two-port networks. Chapter 9 deals with the state equations of fractional-order circuits and their solutions.

Guangzhou, China
March 2021

Bo Zhang
Xujian Shu

Acknowledgements

For the completion of this monograph, the authors would like to acknowledge the support of the Key Program of the National Natural Science Foundation of China (No. 51437005) to the research. We would also like to express our gratitude to Prof. Dongyuan Qiu, Prof. Yanfeng Chen, and Dr. Yanwei Jiang for carefully reading the earlier version of this monograph and offering many helpful comments, and the effort of Mr. Min Li, Mr. Jianguo Li, Mr. Yuhu Qu, Mr. Yadong Wei, and Ms. Shuyun Guo for sorting out the relevant information and references. In addition, Dr. Chao Rong, Dr. Lihao Wu, Dr. Yunhua Liu, and Dr. Wenjie Ma read various portions of the manuscript, which we greatly appreciate. Finally, we would like to sincerely thank all the editors of the CPSS Power Electronics Series of Springer. Without their invitation, encouragement, recommendations, and comments, this monograph would not have been possible.

Guangzhou, China
March 2021

Bo Zhang
Xujian Shu

Contents

About the Authors

Bo Zhang (M'03-SM'15) was born in Shanghai, China, in 1962. He received the B.S. degree in electrical engineering from Zhejiang University, Hangzhou, China, in 1982, the M.S. degree in power electronics from Southwest Jiaotong University, Chengdu, China, in 1988, and the Ph.D. degree in power electronics from Nanjing University of Aeronautics and Astronautics, Nanjing, China, in 1994.

He is currently a Professor of the School of Electric Power, South China University of Technology, Guangzhou, China. He has authored or co-authored over 600 papers and held 130 patents. He has authored nine monographs. His research interests include nonlinear analysis and control of power electronics, wireless power transfer technology, and ac drives.

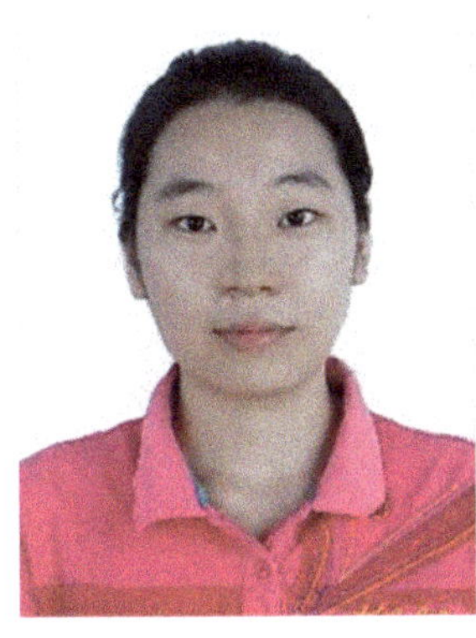

Xujian Shu was born in Anhui, China, in 1993. She received the B.S. degree in Electrical Engineering and Automation from China University of Mining and Technology, Xuzhou, China, in 2015. She is currently working toward the Ph.D. degree in power electronics at School of Electric Power, South China University of Technology, Guangzhou, China.

Her research interests include wireless power transfer applications and power electronics converters.

Chapter 1
Introduction to Fractional Calculus

Fractional calculus is the basis of the fractional-order circuit, which was born almost simultaneously as the classical integer-order calculus [1–3]. It can be traced back to September 30, 1695, by Leibniz's letter to L'Hospital, in which the meaning of the differentiation of order 1/2 is discussed [4–6]. The special day is regarded as the birth date of fractional calculus. However, due to the lack of accurate physical meaning and geometric interpretation, fractional calculus has been considered a purely mathematical theory studied only by mathematicians [7–9]. Its research has been stuck at the level of mathematics without real applications for a long time [10]. Until the 1970s, the significant progress in fractal theory and fractional-order physical models' emergence promoted fractional calculus's widespread application [11–14]. In recent years, with the further development of science and technology and the deepening of human understanding of nature, the classical integer-order calculus can no longer satisfy the human pursuit of science and technology and the exploration of nature [15–17]. Therefore, fractional calculus has gradually become a hot topic in domestic and international research [18]. Nowadays, fractional calculus has been successfully applied in electrochemistry [13, 19, 20], biology [21, 22], electromagnetics [23–25], signal processing [26, 27], viscoelastic and hereditary mechanics [28–31], etc. [32–35]. In the field of electrical engineering, existing research shows that many devices and systems have fractional-order properties. For example, actual capacitors and inductors are non-integral [36–38], and the skin effect of conductors at high frequencies can be described by fractional calculus and so on [39]. For researchers in the engineering field, fractional calculus is just a mathematical tool for solving engineering problems. Therefore, this chapter briefly introduces the definition and calculation of several special functions of fractional calculus and three different definitions of fractional calculus, several methods for solving fractional differential equations, and common software tools for solving fractional calculus.

B. Zhang and X. Shu, *Fractional-Order Electrical Circuit Theory*, CPSS Power Electronics Series, https://doi.org/10.1007/978-981-16-2822-1_1

1.1 Special Functions of Fractional Calculus

1.1.1 Gamma Function

Gamma function is one of the essential functions of fractional calculus, which generalizes the factorial $n!$ and allows n to take non-integer and even complex values [5, 40]. The Gamma function $\Gamma(z)$ is defined by

$$\Gamma(z) = \int_0^\infty e^{-t} t^{z-1} dt \tag{1.1}$$

where the integral converges in the right half of the complex plane $\mathrm{Re}(z) > 0$.

Besides, the Gamma function can also be represented by

$$\Gamma(z) = \lim_{n\to\infty} \frac{n! n^z}{z(z+1)\cdots(z+n)} = \frac{1}{z}\prod_{k=1}^{\infty} \frac{\left(1+\frac{1}{k}\right)^z}{1+\frac{z}{k}} \tag{1.2}$$

where $z \neq 0, -1, -2, \ldots, n$.

The Gamma function has several basic properties as follows:

$$\Gamma(z+1) = z\Gamma(z) \tag{1.3}$$

$$\Gamma(1) = \Gamma(2) = 1, \Gamma\left(\frac{1}{2}\right) = \sqrt{\pi}, \Gamma(0) = \pm\infty \tag{1.4}$$

$$\Gamma(n) = (n-1)! \tag{1.5}$$

where (1.3) can be easily proved by integration by parts

$$\Gamma(z+1) = \int_0^\infty e^{-t} t^{(z+1)-1} dt = \left[-e^{-t} t^z\right]_{t=0}^{t=\infty} + z\int_0^\infty e^{-t} t^{z-1} dt = z\Gamma(z) \tag{1.6}$$

1.1.2 Beta Function

The Beta function $B(z, m)$ is a specific combination of values of the Gamma function [5, 16, 41], which is defined as

$$B(z,m) = \int_0^1 \tau^{z-1}(1-\tau)^{m-1}\mathrm{d}\tau \tag{1.7}$$

where $\mathrm{Re}(z) > 0$ and $\mathrm{Re}(m) > 0$.

The relationship between the Beta function and the Gamma function is

$$B(z,m) = \frac{\Gamma(z)\Gamma(m)}{\Gamma(z+m)} \tag{1.8}$$

With the help of the Beta function, some crucial relationships of the Gamma function can be obtained as follows:

$$\Gamma(z)\Gamma(1-z) = \frac{\pi}{\sin(\pi z)},\ z \neq 0, \pm 1, \pm 2, \cdots \tag{1.9}$$

$$\Gamma(z)\Gamma\left(z+\frac{1}{2}\right) = \sqrt{\pi} z^{1-2z}\Gamma(2z),\ 2z \neq 0, -1, -2, \cdots \tag{1.10}$$

$$\Gamma(z)\Gamma\left(z+\frac{1}{2}\right)\Gamma\left(z+\frac{2}{3}\right) = \sqrt{\pi} z^{\frac{1}{2}-3z}\Gamma(3z),\ 3z \neq 0, -1, -2, \cdots \tag{1.11}$$

$$\begin{aligned}&\Gamma(z)\Gamma\left(z+\frac{1}{k}\right)\cdots\Gamma\left(z+\frac{k-1}{k}\right)\\ &= (2k)^{\frac{k-1}{2}} k^{-kz+1}\Gamma(kz),\ kz \neq 0, -1, -2, \cdots\end{aligned} \tag{1.12}$$

$$\Gamma\left(n+\frac{1}{2}\right) = \frac{\sqrt{\pi}(2n)!}{2^{2n}n!} \tag{1.13}$$

1.1.3 Mittag-Leffler Function

The Mittag-Leffler function is an extension of the exponential function. The exponential function e^z occupies a significant role in integer-order calculus equations [5, 16, 18]. Similarly, the Mittag–Leffler function also plays an important role in fractional calculus.

The one-parameter Mittag–Leffler function $E_\alpha(z)$ is the most straightforward Mittag–Leffler function, which is denoted by

$$E_\alpha(z) = \sum_{k=0}^{\infty} \frac{z^k}{\Gamma(\alpha k + 1)} \tag{1.14}$$

where $\alpha > 0$.

The exponential function e^z is a particular case of the single-parameter Mittag–Leffler function when $\alpha = 1$, that is

$$E_1(z) = \sum_{k=0}^{\infty} \frac{z^k}{\Gamma(k + 1)} = \sum_{k=0}^{\infty} \frac{z^k}{k!} = e^z \tag{1.15}$$

Furthermore, the case with $\alpha = 2$ yields

$$E_2(z) = \sum_{k=0}^{\infty} \frac{z^k}{\Gamma(2k + 1)} = \sum_{k=0}^{\infty} \frac{\left(\sqrt{z}\right)^{2k}}{(2k)!} = \cosh\left(\sqrt{z}\right) \tag{1.16}$$

where cosh() is the hyperbolic cosine function, and its definition is $\cosh(z) = \left(e^z + e^{-z}\right)/2$.

For the case of $\alpha = 1/2$, (1.14) can be written as

$$E_{\frac{1}{2}}(z) = \sum_{k=0}^{\infty} \frac{z^k}{\Gamma\left(\frac{1}{2}k + 1\right)} = e^{z^2} \operatorname{erfc}(-z) \tag{1.17}$$

where $\operatorname{erfc}(z)$ is the complementary error function defined by [42, 43]

$$\operatorname{erfc}(z) = \frac{2}{\sqrt{\pi}} \int_0^{\infty} e^{-t^2} dt \tag{1.18}$$

The two-parameter Mittag–Leffler function $E_{\alpha,\beta}(z)$ is defined as

$$E_{\alpha,\beta}(z) = \sum_{k=0}^{\infty} \frac{z^k}{\Gamma(\alpha k + \beta)} \tag{1.19}$$

where $\alpha > 0$ and $\beta > 0$.

In the case that $\beta = 1$, the two-parameter Mittag–Leffler function can be simplified to a one-parameter Mittag–Leffler function, which is

$$E_{\alpha,1}(z) = \sum_{k=0}^{\infty} \frac{z^k}{\Gamma(\alpha k + 1)} \equiv E_\alpha(z) \tag{1.20}$$

The one-parameter Mittag–Leffler function is a particular case of the two-parameter Mittag–Leffler function.

For particular values of α and β, the two-parameter Mittag–Leffler function of general functions can be expressed as follows:

$$\begin{aligned} E_{1,2}(z) &= \sum_{k=0}^{\infty} \frac{z^k}{\Gamma(k+2)} = \sum_{k=0}^{\infty} \frac{z^k}{(k+1)!} \\ &= \frac{1}{z} \sum_{k=0}^{\infty} \frac{z^{k+1}}{(k+1)!} = \frac{\mathrm{e}^z - 1}{z} \end{aligned} \tag{1.21}$$

$$\begin{aligned} E_{1,3}(z) &= \sum_{k=0}^{\infty} \frac{z^k}{\Gamma(k+3)} = \sum_{k=0}^{\infty} \frac{z^k}{(k+2)!} \\ &= \frac{1}{z^2} \sum_{k=0}^{\infty} \frac{z^{k+2}}{(k+2)!} = \frac{\mathrm{e}^z - 1 - z}{z^2} \end{aligned} \tag{1.22}$$

More generally, it can be described as

$$\begin{aligned} E_{1,m}(z) &= \sum_{k=0}^{\infty} \frac{z^k}{\Gamma(k+m)} = \frac{1}{z^{m-1}} \sum_{k=0}^{\infty} \frac{z^{k+m-1}}{(k+m-1)!} \\ &= \frac{1}{z^{m-1}} \left(e^z - \sum_{k=0}^{m-2} \frac{z^k}{k!} \right) \end{aligned} \tag{1.23}$$

Besides, the hyperbolic functions can also be expressed by two-parameter Mittag–Leffler functions as

$$E_{2,1}\left(z^2\right) = \sum_{k=0}^{\infty} \frac{z^{2k}}{\Gamma(2k+1)} = \sum_{k=0}^{\infty} \frac{z^{2k}}{(2k)!} = \cosh(z) \tag{1.24}$$

$$E_{2,2}\left(z^2\right) = \sum_{k=0}^{\infty} \frac{z^{2k}}{\Gamma(2k+2)} = \frac{1}{z} \sum_{k=0}^{\infty} \frac{z^{2k+1}}{(2k+1)!} = \frac{\sinh(z)}{z} \tag{1.25}$$

where sinh() is the hyperbolic sine function, and its definition is $\sinh(z) = \left(\mathrm{e}^z - \mathrm{e}^{-z}\right)/2$.

The Laplace transform of the two-parameter Mittag–Leffler function plays a vital part in the process of solving fractional differential equations [44], which can be written as

$$\mathcal{L}\left[t^{\alpha k+\beta-1} E_{\alpha,\beta}^{(k)}(\pm a t^{\alpha})\right] = \int_0^{\infty} \mathrm{e}^{-st} t^{\alpha k+\beta-1} E_{\alpha,\beta}^{(k)}(\pm a t^{\alpha}) \mathrm{d}t$$

$$= \frac{k! s^{\alpha-\beta}}{(s^\alpha \mp a)^{k+1}} \tag{1.26}$$

where $\mathcal{L}$ is Laplace operator, and $\mathrm{Re}(s) > |a|^{1/\alpha}$.

1.2 Definitions and Basic Properties of Fractional Calculus

Fractional calculus has not made any substantial progress in early research. Until the middle of the nineteenth century, some well-known mathematicians such as Liouville, Riemann, Grünwald, and Letnikov successively proposed a significant definition of fractional calculus, which facilitates the establishment of the theory of fractional calculus [45–48]. Subsequently, the Liouville and Riemann definitions of fractional calculus were unified into the Riemann–Liouville (R-L) definition, the Grünwald and Letnikov definitions of fractional calculus were unified into the Grünwald-Letnikov (G-L) definition as well. Both of the two definitions are widely used in scientific research and engineering practice. When the R-L and G-L definitions are used to solve fractional differential equations, it is necessary to know the values of the fractional derivatives of the function at the initial condition [5, 49]. However, the physical meaning of the fractional derivative is not clear yet, so there will be enormous obstacles in measuring the required initial values. To solve this problem, Caputo proposed the Caputo definition of fractional calculus in 1967 [5, 50]. In this definition, only the value of the function and the value of its integer-order derivative at the initial conditions are required, which can be easily obtained in the actual applications when solving fractional differential equations.

In addition to the above three definitions of fractional calculus, there are other definitions of fractional calculus. Whereas this book does not introduce other definitions except for the three most widely applicable definitions of fractional calculus (G-L, R-L, and Caputo definitions).

To more accurately describe the three different definitions of fractional calculus, a unified fractional differential and integral operator is introduced

$${}_{t_0}D_t^\alpha f(t) = \begin{cases} \dfrac{\mathrm{d}^\alpha}{\mathrm{d}t^\alpha} f(t), & \alpha > 0 \\ f(t), & \alpha = 0 \\ \displaystyle\int_{t_0}^{t} f(\tau)\mathrm{d}\tau^{-\alpha}, & \alpha < 0 \end{cases} \tag{1.27}$$

where t is the independent variable, t_0 is the lower boundary of the variable, α is the order of fractional calculus, and α can be an arbitrary real number. $\alpha > 0$ represents fractional differential, while $\alpha < 0$ denotes fractional integral.

1.2.1 Grünwald-Letnikov Fractional Calculus

For random positive real number $\alpha > 0$, $[\alpha]$ represents the integer part of α. In other words, $[\alpha]$ is the largest integer less than α. Supposing that $f(t)$ has $m + 1$ continuous derivatives on the interval $[t_0, t]$, and m is at least equal to $[\alpha]$ when $\alpha > 0$, the G-L definition of the fractional derivative of order α can be represented by

$$
\begin{aligned}
{}_{t_0}^{\mathrm{GL}}D_t^{\alpha} f(t) &= \lim_{h\to 0,\, nh=t-t_0} f_h^{(\alpha)}(t) \\
&= \lim_{h\to 0,\, nh=t-t_0} h^{-\alpha} \sum_{r=0}^{n} \begin{bmatrix} -\alpha \\ r \end{bmatrix} f(t-rh)
\end{aligned} \tag{1.28}
$$

where $\begin{bmatrix} -\alpha \\ r \end{bmatrix} = \frac{(-\alpha)(-\alpha+1)(-\alpha+2)\cdots(-\alpha+r-1)}{r!}$.

Using integration by parts, the G-L definition of fractional differentiation can also be expressed in another form as

$$
\begin{aligned}
{}_{t_0}^{\mathrm{GL}}D_t^{\alpha} f(t) &= \lim_{\substack{h\to 0 \\ nh=t-t_0}} f_h^{(\alpha)}(t) \\
&= \sum_{k=0}^{m} \frac{f^{(k)}(t_0)(t-t_0)^{-\alpha+k}}{\Gamma(-\alpha+k+1)} \\
&\quad + \frac{1}{\Gamma(-\alpha+m+1)} \int_{t_0}^{t} (t-\tau)^{-\alpha+m} f^{(m+1)}(\tau)\mathrm{d}\tau
\end{aligned} \tag{1.29}
$$

From (1.29), it is easily seen that the G-L definition of the fractional differential is obtained from the limit of finite difference, which is widely used in solving practical engineering problems. Nevertheless, due to its complicated expression, it may be challenging to apply it to actual problems.

For arbitrary positive real number α, assuming that $f(t)$ has $m + 1$ continuous derivative in $[t_0, t]$, and m takes at least $[\alpha]$ when $\alpha > 0$, the G-L definition of the fractional integral is defined as

$$
\begin{aligned}
{}_{t_0}^{\mathrm{GL}}D_t^{-\alpha} f(t) &= \lim_{h\to 0,\, nh=t-t_0} f_h^{(-\alpha)}(t) \\
&= \lim_{h\to 0,\, nh=t-t_0} h^{\alpha} \sum_{r=0}^{n} \begin{bmatrix} \alpha \\ r \end{bmatrix} f(t-rh) \\
&= \sum_{k=0}^{m} \frac{f^{(k)}(t_0)(t-t_0)^{\alpha+k}}{\Gamma(\alpha+k+1)} + \frac{1}{\Gamma(\alpha+m+1)} \int_{t_0}^{t} (t-\tau)^{\alpha+m} f^{(m+1)}(\tau)\mathrm{d}\tau
\end{aligned} \tag{1.30}
$$

The definition formulas of G-L fractional differential and integral can be expressed in a unified form, as shown in (1.30), which represents the fractional differentiation when $\alpha > 0$ and fractional integral when $\alpha < 0$.

G-L fractional calculus operator ${}_{t_0}^{\mathrm{GL}}D_t^{\alpha}$ is a linear operator, for arbitrary constant λ_1 and λ_2, it follows

$$
{}_{t_0}^{\mathrm{GL}}D_t^{\alpha}(\lambda_1 f_1(t) + \lambda_2 f_2(t)) = \lambda_1 {}_{t_0}^{\mathrm{GL}}D_t^{\alpha} f_1(t) + \lambda_2 {}_{t_0}^{\mathrm{GL}}D_t^{\alpha} f_2(t) \tag{1.31}
$$

The composition operation of integer-order differential operator and G-L fractional differential operator satisfies the superposition relation, that is

$$
\frac{\mathrm{d}^n}{\mathrm{d}t^n}\left[{}_{t_0}^{\mathrm{GL}}D_t^{\alpha} f(t)\right] = {}_{t_0}^{\mathrm{GL}}D_t^{\alpha}\left[\frac{\mathrm{d}^n f(t)}{\mathrm{d}t^n}\right] + \sum_{k=0}^{n-1} \frac{f^{(k)}(t_0)(t-t_0)^{-\alpha-n+k}}{\Gamma(-\alpha-n+k+1)} \tag{1.32}
$$

In particular, if the fractional differentiation at the lower terminal $t = t_0$ satisfies $f^{(k)}(t_0) = 0$ $(k = 0, 1, \ldots, n-1)$, the integer-order differential operator $\frac{\mathrm{d}^n}{\mathrm{d}t^n}$ and the G-L fractional differential operator ${}_{t_0}^{\mathrm{GL}}D_t^{\alpha}$ is commutative, namely

$$
\frac{\mathrm{d}^n}{\mathrm{d}t^n}\left[{}_{t_0}^{\mathrm{GL}}D_t^{\alpha} f(t)\right] = {}_{t_0}^{\mathrm{GL}}D_t^{\alpha}\left[\frac{\mathrm{d}^n f(t)}{\mathrm{d}t^n}\right] = {}_{t_0}^{\mathrm{GL}}D_t^{\alpha+n} f(t) \tag{1.33}
$$

The composition operation between G-L fractional calculus operators ${}_{t_0}^{\mathrm{GL}}D_t^{p}$ and ${}_{t_0}^{\mathrm{GL}}D_t^{q}$ can be divided into three cases as follows:

1. In the case of $p < 0$ and arbitrary real q, it can be written as

$$
{}_{t_0}^{\mathrm{GL}}D_t^{q}\left[{}_{t_0}^{\mathrm{GL}}D_t^{p} f(t)\right] = {}_{t_0}^{\mathrm{GL}}D_t^{p+q} f(t) \tag{1.34}
$$

2. For the case that $0 \le m < p < m+1$, if $f^{(k)}(t_0) = 0 (k = 0, 1, \ldots, m-1)$, (1.34) holds also for any real q.
3. If $0 \le m < p < m+1, 0 \le n < q < n+1$ and $f^{(k)}(t_0) = 0 (k = 0, 1, \ldots r-1)$ where $r = \max\{m, n\}$, it follows

$$
{}_{t_0}^{\mathrm{GL}}D_t^{q}\left[{}_{t_0}^{\mathrm{GL}}D_t^{p} f(t)\right] = {}_{t_0}^{\mathrm{GL}}D_t^{p}\left[{}_{t_0}^{\mathrm{GL}}D_t^{q} f(t)\right] = {}_{t_0}^{\mathrm{GL}}D_t^{p+q} f(t) \tag{1.35}
$$

1.2.2 Riemann-Liouville Fractional Calculus

For arbitrary positive real $\alpha > 0$, the R-L fractional differential of a given function $f(t)$ is defined as

$$ {}_{t_0}^{\mathrm{RL}}D_t^{\alpha} f(t) = \frac{1}{\Gamma(n-\alpha)} \frac{\mathrm{d}^n}{\mathrm{d}t^n} \int_{t_0}^{t} (t-\tau)^{n-1-\alpha} f(\tau)\mathrm{d}\tau \tag{1.36} $$

where $n - 1 \le \alpha < n$.

Under the same condition, the R-L fractional integral of order α can be written as

$$ {}_{t_0}^{\mathrm{RL}}D_t^{-\alpha} f(t) = \frac{1}{\Gamma(\alpha)} \int_{t_0}^{t} (t-\tau)^{\alpha-1} f(\tau)\mathrm{d}\tau \tag{1.37} $$

R-L fractional differential and integral operators are linear operators. According to this fundamental property, it can be obtained that

$$ {}_{t_0}^{\mathrm{RL}}D_t^{\alpha}(\lambda_1 f_1(t) + \lambda_2 f_2(t)) = \lambda_1 {}_{t_0}^{\mathrm{RL}}D_t^{\alpha} f_1(t) + \lambda_2 {}_{t_0}^{\mathrm{RL}}D_t^{\alpha} f_2(t) \tag{1.38} $$

$$ {}_{t_0}^{\mathrm{RL}}D_t^{-\alpha}(\lambda_1 f_1(t) + \lambda_2 f_2(t)) = \lambda_1 {}_{t_0}^{\mathrm{RL}}D_t^{-\alpha} f_1(t) + \lambda_2 {}_{t_0}^{\mathrm{RL}}D_t^{-\alpha} f_2(t) \tag{1.39} $$

When the given function $f(t)$ is continuous on the interval $t > t_0$, the R-L fractional integral operators ${}_{t_0}^{\mathrm{RL}}D_t^{-p}$ and ${}_{t_0}^{\mathrm{RL}}D_t^{-q}$ satisfy the commutative law

$$ {}_{t_0}^{\mathrm{RL}}D_t^{-p}\left[{}_{t_0}^{\mathrm{RL}}D_t^{-q} f(t)\right] = {}_{t_0}^{\mathrm{RL}}D_t^{-q}\left[{}_{t_0}^{\mathrm{RL}}D_t^{-p} f(t)\right] = {}_{t_0}^{\mathrm{RL}}D_t^{-p-q} f(t) \tag{1.40} $$

where $p > 0$ and $q > 0$.

Unlike conventional integer-order differentiation and integration, R-L fractional differentiation and integration do not commute. If $0 \le m - 1 < p < m$ and $0 \le n - 1 < q < n$, the compound operation of R-L fractional differentiation and integration satisfies

$$ {}_{t_0}^{\mathrm{RL}}D_t^{p}\left[{}_{t_0}^{\mathrm{RL}}D_t^{-q} f(t)\right] = {}_{t_0}^{\mathrm{RL}}D_t^{p-q} f(t) \tag{1.41} $$

where $f(t)$ is continuous, and the derivative ${}_{t_0}^{\mathrm{RL}}D_t^{p-q}$ exists.

Interchanging R-L fractional differential operation and integral operation, (1.41) becomes

$$ \begin{aligned} {}_{t_0}^{\mathrm{RL}}D_t^{-p}\left[{}_{t_0}^{\mathrm{RL}}D_t^{q} f(t)\right] &= {}_{t_0}^{\mathrm{RL}}D_t^{q-p} f(t) \\ &\quad - \sum_{k=1}^{n} \left[{}_{t_0}^{\mathrm{RL}}D_t^{q-k} f(t)\right]_{t=t_0} \frac{(t-t_0)^{p-k}}{\Gamma(p-k+1)} \end{aligned} \tag{1.42} $$

The n-th derivative of the R-L fractional differentiation of order α is

$$ \frac{\mathrm{d}^n}{\mathrm{d}t^n}\left[{}_{t_0}^{\mathrm{RL}}D_t^{\alpha} f(t)\right] = {}_{t_0}^{\mathrm{RL}}D_t^{n+\alpha} f(t) \tag{1.43} $$

but the R-L fractional differentiation of the integer-order derivative can be deduced as

$$ {}_{t_0}^{\mathrm{RL}}D_t^{\alpha}\left[\frac{\mathrm{d}^n f(t)}{\mathrm{d}t^n}\right] = {}_{t_0}^{\mathrm{RL}}D_t^{n+\alpha} f(t) - \sum_{k=0}^{n-1}\frac{f^{(k)}(t_0)(t-t_0)^{-\alpha-n+k}}{\Gamma(-\alpha-n+k+1)} \tag{1.44} $$

If the fractional differentiation at the lower terminal $t = t_0$ satisfies $f^{(k)}(t_0) = 0$ ($k = 0, 1, \ldots, n-1$), the integer-order differential operator $\frac{\mathrm{d}^n}{\mathrm{d}t^n}$ commutes with the G-L fractional differential operator ${}_{t_0}^{\mathrm{GL}}D_t^{\alpha}$, that is

$$ \frac{\mathrm{d}^n}{\mathrm{d}t^n}\left[{}_{t_0}^{\mathrm{RL}}D_t^{\alpha} f(t)\right] = {}_{t_0}^{\mathrm{RL}}D_t^{\alpha}\left[\frac{\mathrm{d}^n f(t)}{\mathrm{d}t^n}\right] = {}_{t_0}^{\mathrm{RL}}D_t^{\alpha+n} f(t) \tag{1.45} $$

The composition operation of two R-L fractional differential operators ${}_{t_0}^{\mathrm{GL}}D_t^{p}$ and ${}_{t_0}^{\mathrm{GL}}D_t^{q}$ can be expressed as follows:

$$ {}_{t_0}^{\mathrm{RL}}D_t^{p}\left[{}_{t_0}^{\mathrm{RL}}D_t^{q} f(t)\right] = {}_{t_0}^{\mathrm{RL}}D_t^{p+q} f(t) - \sum_{k=1}^{n}\left[{}_{t_0}^{\mathrm{RL}}D_t^{q-k} f(t)\right]_{t=t_0}\frac{(t-t_0)^{-p-k}}{\Gamma(1-p-k)} \tag{1.46} $$

$$ {}_{t_0}^{\mathrm{RL}}D_t^{q}\left[{}_{t_0}^{\mathrm{RL}}D_t^{p} f(t)\right] = {}_{t_0}^{\mathrm{RL}}D_t^{p+q} f(t) - \sum_{k=1}^{n}\left[{}_{t_0}^{\mathrm{RL}}D_t^{p-k} f(t)\right]_{t=t_0}\frac{(t-t_0)^{-q-k}}{\Gamma(1-q-k)} \tag{1.47} $$

Generally, the composition operation of R-L fractional differential operators ${}_{t_0}^{\mathrm{GL}}D_t^{p}$ and ${}_{t_0}^{\mathrm{GL}}D_t^{q}$ does not satisfy the commutative law. Only when the function $f(t)$ satisfies the following two conditions simultaneously:

$$ \left[{}_{t_0}^{\mathrm{RL}}D_t^{p-k} f(t)\right]_{t=t_0} = 0,\ (k = 1, 2, \ldots, m) \tag{1.48} $$

$$ \left[{}_{t_0}^{\mathrm{RL}}D_t^{q-k} f(t)\right]_{t=t_0} = 0,\ (k = 1, 2, \ldots, n) \tag{1.49} $$

it can be obtained that

$$ {}_{t_0}^{\mathrm{RL}}D_t^{p}\left[{}_{t_0}^{\mathrm{RL}}D_t^{q} f(t)\right] = {}_{t_0}^{\mathrm{RL}}D_t^{q}\left[{}_{t_0}^{\mathrm{RL}}D_t^{p} f(t)\right] = {}_{t_0}^{\mathrm{RL}}D_t^{p+q} f(t) \tag{1.50} $$

The R-L definitions of fractional derivatives of some general functions are given in Tables 1.1 and 1.2.

Table 1.1 Fractional derivatives of order q in R-L definition with the lower terminal $t_0 = 0$

$x(t)$	${}_0D_t^q x(t), (t > 0, q \in \mathbb{R})$
$\varepsilon(t)$	$\frac{t^{-q}}{\Gamma(1-q)}$
$\varepsilon(t-a)$	$\begin{cases} \frac{(t-a)^{-q}}{\Gamma(1-q)}, (t > a) \\ 0, (0 \le t \le a) \end{cases}$
$\varepsilon(t-a)x(t)$	$\begin{cases} {}_0D_t^q x(t), (t > a) \\ 0, (0 \le t \le a) \end{cases}$
$\delta(t)$	$\frac{t^{-q-1}}{\Gamma(-q)}$
$\delta^{(n)}(t)$	$\frac{t^{-q-n-1}}{\Gamma(-q-n)}, (n \in \mathbb{N})$
$\delta^{(n)}(t-a)$	$\begin{cases} \frac{(t-a)^{-n-q-1}}{\Gamma(-n-q)}, (t > a, n \in \mathbb{N}) \\ 0, (0 \le t \le a), (n \in \mathbb{N}) \end{cases}$
t^{υ}	$\frac{\Gamma(\upsilon+1)}{\Gamma(\upsilon+1-q)} t^{\upsilon+q}, (\upsilon > -1)$
$\mathrm{e}^{\lambda t}$	$t^{-q} E_{1,1-q}(\lambda t)$
$\cosh\left(\sqrt{\lambda t}\right)$	$t^{-q} E_{2,1-q}\left(\lambda t^2\right)$
$\frac{\sinh\left(\sqrt{\lambda t}\right)}{\sqrt{\lambda t}}$	$t^{1-q} E_{2,2-q}\left(\lambda t^2\right)$
$\ln(t)$	$\frac{t^{-q}}{\Gamma(1-q)}[\ln(t) + \psi(1) - \psi(1-q)]$
$t^{\beta-1}\ln(t)$	$\frac{\Gamma(\beta)t^{\beta-q-1}}{\Gamma(\beta-q)}[ln(t) + \psi(\beta) - \psi(\beta-q)]$, $(\mathrm{Re}(\beta) > 0)$
$t^{\beta-1}E_{\mu,\beta}(\lambda t^{\mu})$	$t^{\beta-q-1}E_{\mu,\beta-q}(\lambda t^{\mu}), (\beta > 0, \mu > 0)$

1.2.3 Caputo Fractional Calculus

The Caputo definition of fractional derivative is

$$ {}_{t_0}^{\mathrm{C}}D_t^{\alpha} f(t) = \frac{1}{\Gamma(n-\alpha)} \int_{t_0}^{t} \frac{f^{(n)}(\tau)}{(t-\tau)^{\alpha-n+1}} \mathrm{d}\tau \tag{1.51} $$

where $n - 1 < \alpha < n$.

The expressions of the fractional integral defined by G-L, R-L, and Caputo definitions can be unified as

$$ {}_{t_0}D_t^{-\alpha} f(t) = \frac{1}{\Gamma(\alpha)} \int_{t_0}^{t} (t-\tau)^{\alpha-1} f(\tau) \mathrm{d}\tau \tag{1.52} $$

Table 1.2 Fractional derivatives of order q in R-L definition with the lower terminal $t_0 \to -\infty$

$x(t)$	${}_{-\infty}D_t^q x(t)$, $(t > 0, q \in \mathbb{R})$
$\varepsilon(t-a)$	$\begin{cases} \frac{(t-a)^{-q}}{\Gamma(1-q)}, (t > a) \\ 0, (t \le a) \end{cases}$
$\varepsilon(t-a)x(t)$	$\begin{cases} {}_{-\infty}D_t^q x(t), (t > a) \\ 0, (t \le a) \end{cases}$
$e^{\lambda t}$	$\lambda^q e^{\lambda t}$
$e^{\lambda t+\mu}$	$\lambda^q e^{\lambda t+\mu}$
$\sin \lambda t$	$\lambda^q \sin\left(\lambda t + \frac{\pi q}{2}\right)$, $(\lambda > 0, q > -1)$
$\cos \lambda t$	$\lambda^q \cos\left(\lambda t + \frac{\pi q}{2}\right)$, $(\lambda > 0, q > -1)$
$e^{\lambda t} \sin \mu t$	$r^q e^{\lambda t} \sin(\mu t + q\varphi)$, $\left(r = \sqrt{\lambda^2 + \mu^2}, \tan\varphi = \frac{\mu}{\lambda}, \lambda > 0, \mu > 0\right)$
$e^{\lambda t} \cos \mu t$	$r^q e^{\lambda t} \cos(\mu t + q\varphi)$, $\left(r = \sqrt{\lambda^2 + \mu^2}, \tan\varphi = \frac{\mu}{\lambda}, \lambda > 0, \mu > 0\right)$

For the Caputo definition of fractional calculus, the fractional derivative of arbitrary constant C is zero, which is

$$ {}_{t_0}^{\mathrm{C}}D_t^{\alpha} C = 0 \tag{1.53}$$

whereas the R-L fractional derivative of a constant C on a finite lower terminal t_0 is not equal to zero but is

$$ {}_{t_0}^{\mathrm{RL}}D_t^{\alpha} C = \frac{Ct^{-\alpha}}{\Gamma(1-\alpha)} \tag{1.54}$$

Supposing $T > 0, f(t) \in C^1[0, T]$, $\alpha, \beta \in \mathbb{R}^+$ and $\alpha + \beta \le 1$, it can be obtained that

$$ {}_{t_0}^{\mathrm{C}}D^{\alpha}\,{}_{t_0}^{\mathrm{C}}D^{\beta} f(t) = {}_{t_0}^{\mathrm{C}}D^{\beta}\,{}_{t_0}^{\mathrm{C}}D^{\alpha} f(t) = {}_{t_0}^{\mathrm{C}}D^{\alpha+\beta} f(t) \tag{1.55}$$

For the case that $T > 0, f(t) \in C^m[0, T]$, $\alpha = \sum_{i=1}^{n} \alpha_i$, $\alpha_i \in (0, 1]$, $m - 1 \le \alpha < m$, $m \in \mathbb{Z}^+$, and $i_k < n$ exists so that $\sum_{j=1}^{i_k} \alpha_j = k$ $(k = 1, 2, \ldots, m-1)$, the composition operation of fractional calculus in Caputo definition can be expressed as

$$ {}_{t_0}^{\mathrm{C}}D^{\alpha} f(t) = {}_{t_0}^{\mathrm{C}}D^{\alpha_1}\,{}_{t_0}^{\mathrm{C}}D^{\alpha_2} \ldots {}_{t_0}^{\mathrm{C}}D^{\alpha_{n-1}}\,{}_{t_0}^{\mathrm{C}}D^{\alpha_n} f(t) \tag{1.56}$$

The fractional derivatives in Caputo definition of some common functions are shown in Tables 1.3 and 1.4.

Table 1.3 Fractional derivatives of order q in Caputo definition with the lower terminal $t_0 = 0$

$x(t)$	${}_0D_t^q x(t),\ (t > 0, q \in \mathbb{R})$
$\varepsilon(t)$	$\frac{t^{-q}}{\Gamma(1-q)} - \sum\limits_{k=0}^{m-1} \frac{\varepsilon^{(k)}(t)\vert_{t=0}}{\Gamma(k-q+1)} t^{k-q}$
$\varepsilon(t-a)$	$\begin{cases} \frac{(t-a)^{-q}}{\Gamma(1-q)} - \sum\limits_{k=0}^{m-1} \frac{\varepsilon^{(k)}(t-a)\vert_{t=0}}{\Gamma(k-q+1)} (t-a)^{k-q}, (t > a) \\ 0, (0 \le t \le a) \end{cases}$
$\varepsilon(t-a)x(t)$	$\begin{cases} {}_0D_t^q x(t) - \sum\limits_{k=0}^{m-1} \frac{[\varepsilon(t-a)x(t)]^{(k)}\vert_{t=0}}{\Gamma(k-q+1)} (t-a)^{k-q}, (t > a) \\ 0, (0 \le t \le a) \end{cases}$
$\delta(t)$	$\frac{t^{-q-1}}{\Gamma(-q)} - \sum\limits_{k=0}^{m-1} \frac{\delta^{(k)}(t)\vert_{t=0}}{\Gamma(k-q+1)} t^{k-q}$
$\delta^{(n)}(t)$	$\frac{t^{-q-n-1}}{\Gamma(-q-n)} - \sum\limits_{k=0}^{m-1} \frac{\delta^{(n+k)}(t)\vert_{t=0}}{\Gamma(k-q+1)} t^{k-q}, (n \in \mathbb{N})$
$\delta^{(n)}(t-a)$	$\begin{cases} \frac{(t-a)^{-n-q-1}}{\Gamma(-n-q)} - \sum\limits_{k=0}^{m-1} \frac{\delta^{(n+k)}(t-a)\vert_{t=0}}{\Gamma(k-q+1)} t^{k-q}, (t > a, n \in \mathbb{N}) \\ 0, (0 \le t \le a, n \in \mathbb{N}) \end{cases}$
t^{υ}	$\begin{cases} 0, \quad \upsilon \in \mathbb{N}, n > \upsilon) \\ \dfrac{\Gamma(\upsilon+1)}{\Gamma(\upsilon+1-q)} t^{\upsilon-q}, \text{ others} \end{cases}$
$e^{\lambda t}$	$\lambda^m t^{m-q} E_{1,1+m-q}(\lambda t)$
$(t+C)^{\upsilon}$	$\frac{\Gamma(\upsilon+1)}{\Gamma(\upsilon+1-m)} \frac{C^{\upsilon-m-1} t^{m-q}}{\Gamma(1+m-\alpha)} E_{1,1+m-q}^{m-\upsilon}\left(\frac{-t}{C}\right), (C > 0)$
$\sin \lambda t$	$\frac{\upsilon^m t^{m-q}}{2j} \left[E_{1,1+m-q}(j\upsilon t) + (-1)^m E_{1,1+m-q}(-j\upsilon t)\right]$
$\cos \lambda t$	$\frac{\upsilon^m t^{m-q}}{2} \left[E_{1,1+m-q}(j\upsilon t) + (-1)^m E_{1,1+m-q}(-j\upsilon t)\right]$

1.2.4 Comparison of Three Definitions of Fractional Calculus

The three different definitions of fractional calculus (G-L, R-L, and Caputo) are related under certain conditions. If the function $f(t)$ is $(n-1)$ - th continuously differentiation and its n-th derivative is integrable in the interval $[t_0, T]$, the G-L definition is equivalent to the R-L definition, that is

$$ {}_{t_0}^{\mathrm{GL}}D_t^{\alpha} f(t) = {}_{t_0}^{\mathrm{RL}}D_t^{\alpha} f(t) \tag{1.57} $$

where $0 \le m - 1 < \alpha < m \le n$.

If the function $f(t)$ is continuously differentiable of order m, and all its derivatives on the lower terminal t_0 are equal to 0, the fractional derivatives in the above three

Table 1.4 Fractional derivatives of order q in Caputo definition with the lower terminal $t_0 \to -\infty$

$x(t)$	${}_{-\infty}D_t^q x(t),\ (t > 0, q \in \mathbb{R})$
$\varepsilon(t-a)$	$\begin{cases} \frac{(t-a)^{-q}}{\Gamma(1-q)}, & (t > a) \\ 0, & (t \le a) \end{cases}$
$\varepsilon(t-a)x(t)$	$\begin{cases} {}_{-\infty}D_t^q x(t), & (t > a) \\ 0, & (t \le a) \end{cases}$
$e^{\lambda t}$	$\lambda^q e^{\lambda t}$
$e^{\lambda t+\mu}$	$\lambda^q e^{\lambda t+\mu}$
$\sin \lambda t$	$\lambda^q \sin\left(\lambda t + \frac{\pi q}{2}\right),\ (\lambda > 0, q > -1)$
$\cos \lambda t$	$\lambda^q \cos\left(\lambda t + \frac{\pi q}{2}\right),\ (\lambda > 0, q > -1)$
$e^{\lambda t} \sin \mu t$	$r^q e^{\lambda t} \sin(\mu t + q\varphi)$, $\left(r = \sqrt{\lambda^2 + \mu^2}, \tan \varphi = \frac{\mu}{\lambda}, \lambda > 0, \mu > 0\right)$
$e^{\lambda t} \cos \mu t$	$r^q e^{\lambda t} \cos(\mu t + q\varphi)$, $\left(r = \sqrt{\lambda^2 + \mu^2}, \tan \varphi = \frac{\mu}{\lambda}, \lambda > 0, \mu > 0\right)$

definitions can be expressed in the same form as

$$ {}_{t_0}^{\ C}D_t^{\alpha} f(t) = {}_{t_0}^{GL}D_t^{\alpha} f(t) = {}_{t_0}^{RL}D_t^{\alpha} f(t) = {}_{t_0}D^{\alpha} f(t) \tag{1.58} $$

where ${}_{t_0}D^q$ denotes any fractional differential operator under the three definitions mentioned above.

In practical applications, the basic properties of fractional calculus are most frequently used. According to the three definition expressions of fractional calculus, there are several common properties as follows:

1. Linearity

$$ {}_{t_0}D^q(\lambda x(t) + \mu y(t)) = \lambda {}_{t_0}D^q x(t) + \mu {}_{t_0}D^q y(t) \tag{1.59} $$

The linearity of fractional calculus can be derived from the corresponding definition of fractional differentiation.

2. Leibniz rule for fractional derivatives

Given two functions $x(t)$ and $u(t)$, by using the Leibniz rule, the n-th derivative of the product of the functions $x(t)$ and $u(t)$ can be obtained as

$$ \frac{d^n}{dt^n}(x(t) \cdot u(t)) = \sum_{j=0}^{n} \binom{n}{j} x^{(k)}(t) u^{(n-j)}(t) \tag{1.60} $$

Similarly, extending it to fractional calculus, (1.60) becomes

$$ {}_{t_0}D_t^q(x(t)\cdot u(t)) = \sum_{j=0}^{n}\binom{q}{j}x^{(k)}(t){}_{t_0}D_t^{q-j}u(t) \tag{1.61} $$

where $x(t)$ and $u(t)$ and all their derivatives are continuous in the interval $[t_0, t]$.

3. The fractional derivative of a composite function

Supposing that $x(t)$ is a composite function, and $x(t) = f(h(t))$. Then, the fractional derivative of the composite function can be written as

$$ \begin{aligned} {}_{t_0}D_t^q x(t) &= {}_{t_0}D_t^q f(h(t)) \\ &= \frac{(t-t_0)^{-q}}{\Gamma(1-q)}x(t) \\ &\quad + \sum_{j=1}^{\infty}\binom{q}{j}\frac{j!(t-t_0)^{j-q}}{\Gamma(j-q+1)}\sum_{n=1}^{j}f^{(n)}(h(t))\sum\prod_{r=1}^{j}\frac{1}{a_r!}\left(\frac{h^{(r)}(t)}{r!}\right)^{a_r} \end{aligned} \tag{1.62} $$

where $\sum_{r=1}^{j} r\alpha_r = j$ and $\sum_{r=1}^{j}\alpha_r = n$.

There are also some distinctions among the G-L, R-L, and Caputo definitions of fractional calculus. The definition of G-L fractional differentiation is consistent with integer-order differentiation. Both are obtained from the limits of difference and are widely used to solve practical problems. Yet, due to its complicated expression, the description of actual engineering problems is limited. The mathematical expression of the G-L definition is not brief, and its restrictions are not strict enough, which causes that it can only be used for theoretical calculations of some simple functions. Consequently, this book mainly discusses the Caputo definition and R-L definition, which are widely used in the theoretical derivation.

The difference between Caputo and R-L definitions of fractional differentiation can be explained from three aspects [5, 51]. First, the application condition of the Caputo definition of the fractional differentiation is that the function $f(t)$ is k-times continuous and differentiable in the integration interval. In contrast, the R-L definition of fractional differentiation only requires the function $f(t)$ to be continuous in the integration interval. Secondly, the fractional differential of the constant in Caputo definition is equal to 0, while the fractional differential of the constant in the R-L fractional definition is not equal to zero. Finally, when solving a fractional differential equation, Caputo definition requires the integral-order derivatives of the function at the initial condition. However, the R-L definition requires the fractional differential of the function at the initial condition, which is difficult to determine because of the lack of physical meaning in actual systems.

1.2.5 Geometric and Physical Interpretation of Fractional Calculus

The geometric and physical meaning of fractional calculus has always been a major obstacle to the application of fractional calculus. In 2001, Professor Podlubny gave a new explanation to this challenging problem [52].

Considering the Riemann–Liouville fractional integral of order α

$$ {}_0^{\mathrm{RL}}D_t^{-\alpha} f(t) = \frac{1}{\Gamma(\alpha)} \int_0^t \frac{f(\tau)}{(t-\tau)^{1-\alpha}} \mathrm{d}\tau \tag{1.63} $$

By introducing the following auxiliary function

$$ g_t(\tau) = \frac{1}{\Gamma(\alpha+1)} \{t^\alpha - (t-\tau)^\alpha\} \tag{1.64} $$

(1.63) can be rewritten as

$$ {}_0^{\mathrm{RL}}D_t^{-\alpha} f(t) = \int_0^t f(\tau) \mathrm{d}g_t(\tau) \tag{1.65} $$

here, the function $g_t(\tau)$ has the property of scaling transformation. By supposing that $t_1 = kt$ and $\tau_1 = k\tau$, it can be obtained that

$$ g_{t_1}(\tau_1) = g_{kt}(k\tau) = k^\alpha g_t(\tau) \tag{1.66} $$

If τ, g, and f are regarded as the axes, the function $g_t(\tau)$ for $0 \le \tau \le t$ in the plane (τ, g) can be drawn. Along the obtained curve of $g_t(\tau)$, the varying height $f(\tau)$ can be plotted as a fence, then the upper boundary of the fence forms a three-dimensional curve, as shown in Fig. 1.1. This fence can be projected onto two planes, where the projection of the fence onto the plane (τ, f) corresponds to the integral of ${}_0^{\mathrm{RL}}D_t^{-1} f(t)$, and the projection of the fence onto the plane (g, f) corresponds to the integral of ${}_0^{\mathrm{RL}}D_t^{-\alpha} f(t)$. The two projection areas form the geometric interpretation of integer-order integral and fractional integral for a fixed t, respectively.

Obviously, if $\alpha = 1$ and $g_t(\tau) = \tau$, the two projection areas are equal. This indicates that, even from the geometric point of view, the classical integer-order definite integration is a particular case of the R-L fractional integration. When t changes, the length and shape of the fence change simultaneously, as shown in Figs. 1.2 and Fig. 1.3. Therefore, a dynamical geometric interpretation of the fractional integration (1.63) is built.

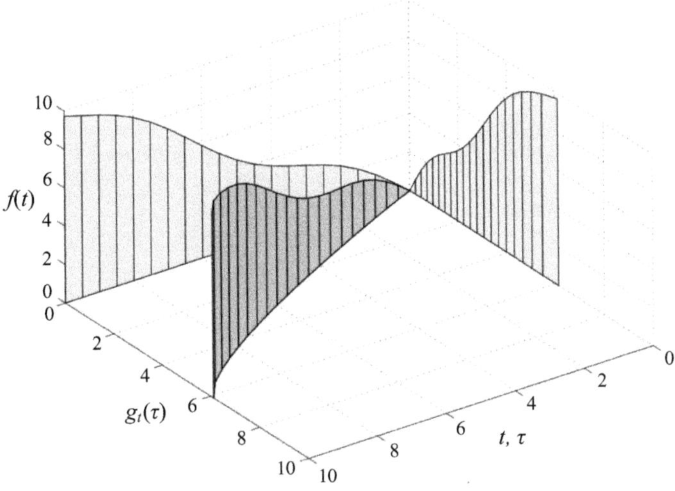

Fig. 1.1 The fence ${}_0^{\mathrm{RL}}D_t^{-1} f(t)$ and its projection ${}_0^{\mathrm{RL}}D_t^{-\alpha} f(t)$ for $\alpha = 0.75, 0 \le t \le 10, f(t) = t + 0.5\sin(t)$, and the auxiliary function $g_t(\tau) = [t^\alpha - (t-\tau)^\alpha]/\Gamma(\alpha+1)$, where $0 \le \tau \le t$

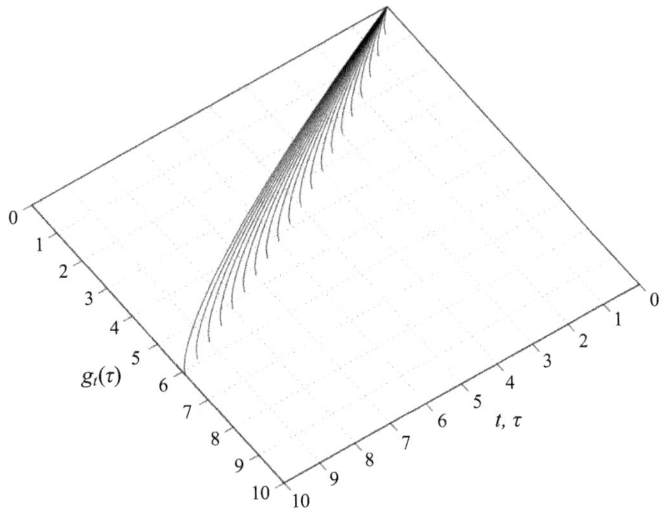

Fig. 1.2 The changing process of the projection length of the fence for $\alpha = 0.75$, $0 \le t \le 10$, the auxiliary function $g_t(\tau) = [t^\alpha - (t-\tau)^\alpha]/\Gamma(\alpha+1)$, where $0 \le \tau \le t$

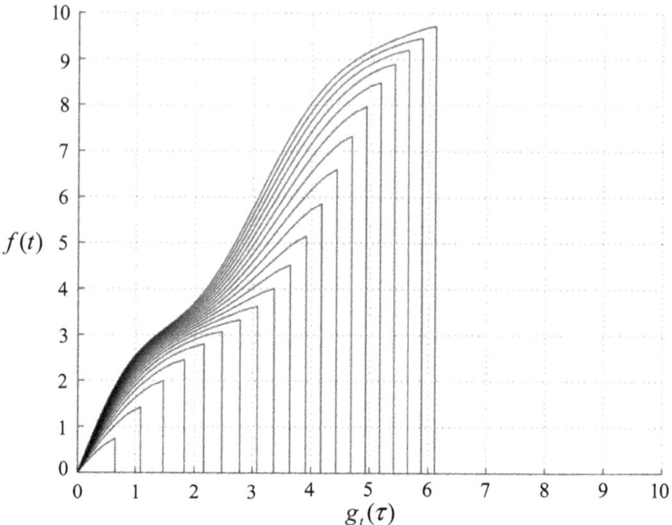

Fig. 1.3 The changing process of the projection area of the fence, where the time interval Δt is 0.5, $f(t) = t + 0.5\sin(t)$ and the auxiliary function $g_t(\tau) = [t^{\alpha} - (t - \tau)^{\alpha}]/\Gamma(\alpha + 1)$

Although the geometric interpretation of fractional integrals has been preliminarily described, it still needs more effort to get more easily understood geometric and physical interpretations of various fractional calculus definitions.

1.3 Methods of Solving Fractional Differential Equations

1.3.1 Laplace Transform Method

Laplace transform can be used to solve fractional differential equations, and it is the Fourier transform of multiplying the time-domain function $f(t)$ by $e^{-\sigma t}$ (where σ is a sufficiently large positive number) [5, 17, 53]. The Laplace transform $F(s)$ of a function $f(t)$ defined for $-\infty \leq t < +\infty$ is

$$F(s) = \int_{-\infty}^{\infty} f(t) e^{-\sigma t} e^{-j\omega t} dt = \int_{-\infty}^{\infty} f(t) e^{-st} dt \tag{1.67}$$

where $s = \sigma + j\omega$ is a complex variable.

If $\lim\limits_{s\to\infty} F(s) = 0$, the function $f(t)$ can be obtained by applying the inverse Laplace transform to $F(s)$, which is represented as

$$f(t) = \mathcal{L}^{-1}[F(s)] = \frac{1}{2\pi \mathrm{j}} \int_{\sigma - j\infty}^{\sigma + j\infty} F(s)\mathrm{e}^{st}\mathrm{d}s \tag{1.68}$$

where $j = \sqrt{-1}$ and $\sigma \in \mathbb{R}$. It is generally complicated to directly calculate the inverse Laplace transform of function $F(s)$ by using (1.68). In practice, the original function $f(t)$ of $F(s)$ is obtained directly by viewing the Laplace transform relationship table.

Rigorous mathematical proof shows that Laplace transform and inverse Laplace transform are linear and have a one-to-one correspondence. Hence, the Laplace transform $F(s)$ of the function $f(t)$ can be used to analyze the transition process of the linear time-invariant circuit.

When analyzing the transition process of a circuit, it is necessary to discuss the transition process when switching occurs at $t = 0$. For instance, the function $f(t)$, which is used to analyze the transition process of a circuit, can be generally expressed as

$$f(t) = f(t) \cdot \varepsilon(t) = \begin{cases} 0, & t < 0 \\ f(t), & t > 0 \end{cases} \tag{1.69}$$

The Laplace transform of the function (1.69) is called the unilateral Laplace transform, which is mainly discussed in this book. The definition of the unilateral Laplace transform is

$$F(s) = \mathcal{L}[f(t)] = \int_{0}^{\infty} f(t)\mathrm{e}^{-st}\mathrm{d}t \tag{1.70}$$

where $f(t)$ is defined for $t \geq 0$.

The precondition for the existence of the unilateral Laplace transform shown in (1.70) is that $f(t)$ is piecewise continuous in any finite interval $t \geq 0$. And when $t \to \infty$, $f(t)$ always satisfies the condition $|f(t)| \leq M\mathrm{e}^{\sigma_0 t}$ (where M and σ_0 are sufficiently large positive numbers).

There are the Laplace transforms of several standard functions as follows:

1. Convolution

$$f(t) * g(t) = \int_{0}^{t} f(t - \tau)g(\tau)\mathrm{d}\tau = \int_{0}^{t} f(\tau)g(t - \tau)\mathrm{d}\tau \tag{1.71}$$

2. Laplace transform of a convolution

$$\mathcal{L}[f(t) * g(t)] = F(s)G(s) \tag{1.72}$$

3. Laplace transform of integer-order differentiation $f^{(n)}(t)$

$$\mathcal{L}\left[f^{n}(t)\right]=s^{n}F(s)-\sum_{k=0}^{n-1}s^{n-k-1}f^{(k)}(0)=s^{n}F(s)-\sum_{k=0}^{n-1}s^{k}f^{(n-k-1)}(0) \tag{1.73}$$

4. Laplace transform of a power function

$$\mathcal{L}\left[t^{n}\right]=\int_{0}^{+\infty}t^{n}s^{-st}\mathrm{d}t=\Gamma(n+1)s^{-n-1} \tag{1.74}$$

Laplace transform can be utilized for the simplified calculation of fractional calculus. However, there are some differences between the three definitions of fractional calculus.

Since R-L, Caputo, and G-L fractional integrals have the same definition (1.52), the Laplace transform of the fractional integrals in three different definitions can be written in the same expressions as

$${}_{0}D_{t}^{-\alpha}f(t)=\frac{1}{\Gamma(\alpha)}\int_{0}^{t}(t-\tau)^{\alpha-1}f(\tau)\mathrm{d}\tau \tag{1.75}$$

Under the condition of $\alpha>0$, according to the convolution (1.71), the definition of fractional integral can be written as the convolution form of functions $g(t)=t^{\alpha-1}$ and $f(t)$, which is

$${}_{0}D_{t}^{-\alpha}f(t)=\frac{1}{\Gamma(\alpha)}t^{\alpha-1}*f(t) \tag{1.76}$$

The Laplace transform of $t^{\alpha-1}$ yields

$$G(s)=\mathcal{L}\left[t^{\alpha-1}\right]=\Gamma(\alpha)s^{-\alpha} \tag{1.77}$$

By applying the Laplace transform of the convolution to (1.76) and substituting (1.77) into it, the Laplace transform of the fractional integral can be described as

$$\mathcal{L}\left[{}_{0}D_{t}^{-\alpha}f(t)\right]=\frac{1}{\Gamma(\alpha)}G(s)F(s)=s^{-\alpha}F(s) \tag{1.78}$$

When $0\leq\alpha<1$, if the function $f(t)$ is continuous on the closed interval $[0, t]$, and its derivative $f'(t)$ is continuous on the interval $[0, b]$, the G-L definition of fractional differentiation can be written as

$$ {}_{0}^{\mathrm{GL}}D_t^{\alpha} f(t) = \frac{f(0)t^{-\alpha}}{\Gamma(1-\alpha)} + \frac{1}{\Gamma(1-\alpha)} \int_0^t (t-\tau)^{-\alpha} f'(\tau)\mathrm{d}\tau \tag{1.79} $$

Based on (1.71), rewriting the integral term of (1.79) into the convolution of $t^{-\alpha}$ and $f'(t)$ as

$$ \int_0^t (t-\tau)^{-\alpha} f'(\tau)\mathrm{d}\tau = t^{-\alpha} * f'(t) \tag{1.80} $$

The Laplace transform of $t^{-\alpha}$ can be calculated by

$$ \mathcal{L}\left[t^{-\alpha}\right] = \Gamma(1-\alpha)s^{\alpha-1} \tag{1.81} $$

By applying (1.73), the Laplace transform of $f'(t)$ can be represented as

$$ \mathcal{L}\left[f'(t)\right] = sF(s) - f(0) \tag{1.82} $$

Last, by applying the Laplace transform of convolution to (1.79) and substituting, (1.80)–(1.82) into it, the Laplace transform of G-L fractional differential can be deduced as

$$ \mathcal{L}\left[{}_{0}^{\mathrm{GL}}D_t^{\alpha} f(t)\right] = \frac{f(0)}{s^{1-\alpha}} + \frac{1}{s^{1-\alpha}}(sF(s) - f(0)) = s^{\alpha}F(s) \tag{1.83} $$

where $0 \le \alpha < 1$.

In the classical sense, the Laplace transform of G-L fractional differential of order $\alpha > 1$ does not exist, because in such a case, there are non-integrable terms in the definition of G-L fractional differential, and the Laplace transform of these terms can only be given by divergent integrals. However, the generalized function approach can be used to treat the Laplace transform of the power function as an analytical extension of the parameter α and calculate the finite-part integral to obtain the Laplace transform of these terms, thereby obtaining the same Laplace transform formula as (1.83). The boundary condition of the formula is extended from $0 \le \alpha < 1$ to $n-1 < \alpha \le n$. It is worth noting that the formula (1.83) holds in the classical sense only in the condition of $0 \le \alpha < 1$, and it holds in the sense of generalized function definition when $\alpha > 1$.

The fractional differential can be written in the form of a fractional integral of order n-α followed by an n-th integer-order differential as

$$ {}_{0}^{\mathrm{RL}}D_t^{\alpha} f(t) = D^n {}_0D_t^{-(n-\alpha)} f(t) \tag{1.84} $$

Replacing the fractional integral part with function $g(t)$, becomes

$$g(t) = {}_0D_t^{-(n-\alpha)} f(t) = \frac{1}{\Gamma(n-\alpha)} \int_0^t (t-\tau)^{n-\alpha-1} f(\tau)\mathrm{d}\tau \tag{1.85}$$

where $n - 1 \leq \alpha < n$.

Thus, (1.84) can be rewritten as

$${}_0^{\mathrm{RL}}D_t^{\alpha} f(t) = g^{(n)}(t) \tag{1.86}$$

Applying the Laplace transform of integer-order derivatives, (1.86) becomes

$$\mathcal{L}\left[{}_0^{\mathrm{RL}}D_t^{\alpha} f(t)\right] = \mathcal{L}\left[g^{(n)}(t)\right] = s^n G(s) - \sum_{k=0}^{n-1} s^k g^{(n-k-1)}(0) \tag{1.87}$$

By using (1.78), the Laplace transform of the function $g(t)$ yields

$$G(s) = s^{-(n-\alpha)} F(s) \tag{1.88}$$

and based on the R-L definition of fractional differential, it can be obtained that

$$g^{(n-k-1)}(t) = \frac{\mathrm{d}^{n-k-1}}{\mathrm{d}t^{n-k-1}} {}_0D_t^{-(n-\alpha)} f(t) = {}_0D_t^{\alpha-k-1} f(t) \tag{1.89}$$

Then, substituting (1.88) and (1.89) into (1.87), the Laplace transform of R-L fractional differential of order $\alpha > 0$ can be written as

$$\mathcal{L}\left[{}_0^{\mathrm{RL}}D_t^{\alpha} f(t)\right] = s^{\alpha} F(s) - \sum_{k=0}^{n-1} s^k {}_0D_t^{\alpha-k-1} f(0) \tag{1.90}$$

Since the Laplace transform of the R-L fractional differential (1.90) requires the initial values of the fractional derivatives at the lower terminal $t = 0$, these initial values are difficult to determine and lack a reasonable physical interpretation, which hinders its application in practical engineering.

As for the Caputo definition of fractional differential, it can be written in the form of $(n - \alpha)$-fold fractional integral of an n-th integer-order derivative as

$${}_0^{\mathrm{C}}D_t^{\alpha} f(t) = {}_0D_t^{-(n-\alpha)} f^{(n)}(t) \tag{1.91}$$

Defining a function $g(t) = f^{(n)}(t)$, then (1.91) can be rewritten as

$${}_0^{\mathrm{C}}D_t^{\alpha} f(t) = {}_0D_t^{-(n-\alpha)} g(t) \tag{1.92}$$

where $n - 1 < \alpha \leq n$.

Applying the Laplace transform of fractional integral shown in (1.78) to (1.92), there is

$$\mathcal{L}\left[{}_0^C D_t^\alpha f(t)\right] = \mathcal{L}\left[{}_0 D_t^{-(n-\alpha)} g(t)\right] = s^{-(n-\alpha)} G(s) \tag{1.93}$$

By using (1.73), the Laplace transform of $g(t) = f^{(n)}(t)$ is

$$G(s) = s^n F(s) - \sum_{k=0}^{n-1} s^{n-k-1} f^{(k)}(0) = s^n F(s) - \sum_{k=0}^{n-1} s^k f^{(n-k-1)}(0) \tag{1.94}$$

Substituting (1.94) into (1.93), the Laplace transform of Caputo fractional differential can be derived as

$$\mathcal{L}\left[{}_0^C D_t^\alpha f(t)\right] = s^\alpha F(s) - \sum_{k=0}^{n-1} s^{\alpha-k-1} f^{(k)}(0) \tag{1.95}$$

The Laplace transform of Caputo fractional differential as shown in (1.95) contains the initial values of the function $f(t)$ and its integer-order derivatives at the lower terminal $t = 0$, all of these have particular physical meanings. Taking the dynamic system as an example, $f(0)$ represents the initial position, $f'(0)$ denotes the initial velocity, and $f''(0)$ is the initial acceleration. Therefore, the initial conditions required for the Laplace transform of the fractional differential in Caputo definition can be given in the form of integer-order derivatives, which is very useful for solving practical problems described by linear fractional differential equations. Consequently, the Caputo fractional definition is widely used in modeling practical issues.

The Laplace transform of fractional calculus defined by different definitions has some common properties [54, 55], as follows:

1. Linearity

$$\begin{aligned}\mathcal{L}\left[k{}_0 D_t^\alpha (f_1(t) + f_2(t))\right] &= \mathcal{L}\left[k{}_0 D_t^\alpha f_1(t) + k{}_0 D_t^\alpha f_2(t)\right] \\ &= k\mathcal{L}\left[{}_0 D_t^\alpha f_1(t)\right] + k\mathcal{L}\left[{}_0 D_t^\alpha f_2(t)\right]\end{aligned} \tag{1.96}$$

2. Composition with fractional differential

The Laplace transform of the composition of the fractional differential and fractional integral operators is

$$\mathcal{L}\left\{{}_0 D_t^p \left[{}_0 D_t^{-q} f(t)\right]\right\} = \mathcal{L}\left[{}_0 D_t^{p-q} f(t)\right] \tag{1.97}$$

where $p > 0$ and $q > 0$.

The Laplace transform of the composition of two fractional differential operators can be given as

$$\mathcal{L}\{{}_0D_t^p[{}_0D_t^q f(t)]\} = \mathcal{L}\left[{}_0D_t^{p+q} f(t)\right] - \mathcal{L}\left\{\sum_{j=1}^{k}\left[{}_0D_t^{q-j} f(t)\right]_{t=0}\frac{(t-t_0)^{-p-j}}{\Gamma(1-p-j)}\right\} \tag{1.98}$$

where $0 \leq k-1 \leq p < k$ and $k-1 \leq q < k$.

3. Composition with fractional integral

The Laplace transform of the compound operation of two fractional integral operators is

$$\mathcal{L}\left\{{}_{t_0}D_t^{-p}\left[{}_{t_0}D_t^{-q} f(t)\right]\right\} = \mathcal{L}\left[{}_{t_0}D_t^{-p-q} f(t)\right] \tag{1.99}$$

where $p > 0$ and $q > 0$.

The Laplace transform of the compound operation of the fractional integral and fractional differential operators can be denoted as

$$\mathcal{L}\left\{{}_{t_0}D_t^{-p}[{}_{t_0}D_t^q f(t)]\right\} = \mathcal{L}\left[{}_{t_0}D_t^{q-p} f(t)\right] - \mathcal{L}\left\{\sum_{j=1}^{k}\left[{}_{t_0}D_t^{q-j} f(t)\right]_{t=t_0}\frac{(t-t_0)^{p-j}}{\Gamma(1+p-j)}\right\} \tag{1.100}$$

where $0 \leq k-1 \leq q < k$ and $p > 0$.

4. Shifting property

Supposing

$$\mathcal{L}[{}_0D_t^\alpha f(t)] = s^\alpha F(s) - \sum_{j=0}^{n-1} s^k {}_0D_t^{\alpha-j-1} f(0) \tag{1.101}$$

and $n-1 < \alpha < n$, there is

$$\mathcal{L}[\mathrm{e}^{\lambda t} {}_0D_t^\alpha f(t)] = (s-\lambda)^\alpha F(s-\lambda) - \sum_{j=0}^{n-1} (s-\lambda)^k {}_0D_t^{\alpha-j-1} f(0) \tag{1.102}$$

5. Time delay

Assuming (1.101) and $n-1 < \alpha < n$, $L[{}_0D_t^\alpha f(t)]$ has the property of time delay

$$\mathcal{L}[{}_0D_t^\alpha f(t-\tau)\cdot\varepsilon(t-\tau)] = \mathrm{e}^{-st}\left[s^\alpha F(s) - \sum_{j=0}^{n-1} s^k {}_0D_t^{\alpha-j-1} f(0)\right] \tag{1.103}$$

6. Scaling property

Under the assumption of (1.101) and $n - 1 < \alpha < n$, the scaling property can be described as

$$\mathcal{L}\left[{}_0D_t^{\alpha} f(at)\right] = \frac{1}{a}\left[\left(\frac{s}{a}\right)^{\alpha} F\left(\frac{s}{a}\right) - \sum_{j=0}^{n-1}\left(\frac{s}{a}\right)^{k} {}_0D_t^{\alpha-j-1} f(0)\right] \tag{1.104}$$

7. Initial value theorem

Supposing (1.101) and $n - 1 < \alpha < n$, there is

$${}_0D_t^{\alpha} f(0_+) = \lim_{s\to\infty}\left[s^{\alpha+1}F(s) - \sum_{j=0}^{n-1} s^{k+1}{}_0D_t^{\alpha-j-1} f(0)\right] \tag{1.105}$$

8. Final value theorem

Under the assumptions of (1.101) and $n - 1 < \alpha < n$, if $\lim_{t\to\infty} f(t)$ exists as a constant, the limit of ${}_0D_t^{\alpha} f(t)$ at infinity can be calculated by

$${}_0D_t^{\alpha} f(\infty) = \lim_{s\to 0}\left[s^{\alpha+1}F(s) - \sum_{j=0}^{n-1} s^{k+1}{}_0D_t^{\alpha-j-1} f(0)\right] \tag{1.106}$$

Table 1.5 gives a brief summary of some useful Laplace transform pairs, and these functions play an important role in solving fractional differential equations [54, 56, 57].

Here, it is worth noting that erf(z) is the error function, $J_v(z)$ is the Bessel function, $I_v(z)$ is the modified Bessel function, $H_n(t)$ is the Hermite polynomial, and daw(z) is Dawson function [58], which are defined as follows:

$$\operatorname{erf}(z) = \frac{2}{\sqrt{\pi}}\int_0^z \mathrm{e}^{-t^2}\,\mathrm{d}t = \frac{1}{\sqrt{\pi}}\int_{-z}^{z} \mathrm{e}^{-t^2}\,\mathrm{d}t \tag{1.107}$$

$$J_v(z) = \sum_{k=0}^{\infty}\frac{(-1)^k (z/2)^{2k+v}}{k!\Gamma(v+k+1)} \tag{1.108}$$

$$I_v(z) = \sum_{k=0}^{\infty}\frac{(z/2)^{2k+v}}{k!\Gamma(v+k+1)} \tag{1.109}$$

$$H_n(t) = \mathrm{e}^{t^2}\frac{\mathrm{d}^n}{\mathrm{d}t^n}\mathrm{e}^{-t^2} \tag{1.110}$$

Table 1.5 Laplace transform pairs

$F(s)$	$f(t)=\mathcal{L}^{-1}[F(s)]$
$\frac{1}{s^\alpha}$	$\frac{t^{\alpha-1}}{\Gamma(\alpha)}$
$\frac{1}{\sqrt{s+a}\sqrt{s+b}}$	$\mathrm{e}^{-(a+b)t/2}I_0\left(\frac{a-b}{2}t\right)$
$\frac{1}{\left(s+\sqrt{s^2+a^2}\right)^N}$	$\frac{NJ_N(at)}{at}$, $N>0$
$\frac{1}{s\sqrt{s}}\mathrm{e}^{-\sqrt{s}}$	$2\sqrt{\frac{t}{\pi}}\mathrm{e}^{-1/(4t)}-\mathrm{erfc}\left(\frac{1}{2\sqrt{t}}\right)$
$\frac{\mathrm{e}^{-\sqrt{s}}}{\sqrt{s}(\sqrt{s}+1)}$	$\mathrm{e}^{t+1}\mathrm{erfc}\left(\sqrt{t}+\frac{1}{2\sqrt{t}}\right)$
$\frac{s^\alpha}{s-a}$	$-t^\alpha E_{1,1-\alpha}(at)$, $0<\alpha<1$
$\frac{b^2-a^2}{(s-a^2)(\sqrt{s}+b)}$	$\mathrm{e}^{a^2t}\left[b-a\,\mathrm{erf}(a\sqrt{t})\right]-b\mathrm{e}^{b^2t}\mathrm{erfc}(b\sqrt{t})$
$\frac{1}{s^v}\mathrm{e}^{k/s}$	$\left(\frac{t}{k}\right)^{(v-1)/2}I_{v-1}\left(2\sqrt{kt}\right)$
$\frac{1}{s^v}\mathrm{e}^{-k/s}$	$\left(\frac{t}{k}\right)^{(v-1)/2}J_{v-1}\left(2\sqrt{kt}\right)$, $v>0$
$\sqrt{s-a}-\sqrt{s-b}$	$\frac{1}{2\sqrt{\pi t^3}}(\mathrm{e}^{bt}-\mathrm{e}^{at})$
$\frac{\sqrt{s+2a}-\sqrt{s}}{\sqrt{s}}$	$a\mathrm{e}^{-at}[I_1(at)+I_0(at)]$
$\frac{\sqrt{s}}{s+1}$	$\frac{1}{\sqrt{\pi t}}-\frac{2}{\sqrt{\pi}}\mathrm{daw}(\sqrt{t})$
$\frac{s}{(s-a)\sqrt{s-a}}$	$\frac{1}{\sqrt{\pi t}}\mathrm{e}^{at}(1+2at)$
$\frac{1}{\sqrt{s}+a}$	$\frac{1}{\sqrt{\pi t}}-a\mathrm{e}^{a^2t}\mathrm{erfc}(a\sqrt{t})$
$\frac{\sqrt{s}}{s-a^2}$	$\frac{1}{\sqrt{\pi t}}+a\mathrm{e}^{a^2t}\mathrm{erf}(a\sqrt{t})$
$\frac{1}{\sqrt{s}(s+a^2)}$	$\frac{2}{a\sqrt{\pi}}\mathrm{e}^{-a^2t}\int_0^{a\sqrt{t}}\mathrm{e}^{\tau^2}\mathrm{d}\tau$
$\frac{s\sqrt{s}}{s+1}$	$2\sqrt{\frac{t}{\pi}}-\frac{2}{\sqrt{\pi}}\mathrm{daw}(\sqrt{t})$
$\frac{k!}{\sqrt{s}\pm\lambda}$	$t^{(k-1)/2}E_{1/2,1/2}^{(k)}(\mp\lambda\sqrt{t})$, $\mathrm{Re}(s)>\lambda^2$
$\frac{s^{\alpha-1}}{s^\alpha\pm\lambda}$	$E_\alpha(\mp\lambda t^\alpha)$, $\mathrm{Re}(s)>\lvert\lambda\rvert^{1/\alpha}$
$\frac{s^{\alpha\gamma-\beta}}{(s^\alpha+a)^\gamma}$	$t^{\beta-1}E_{\alpha,\beta}^\gamma(-at^\alpha)$
$\arctan\frac{k}{s}$	$\frac{1}{t}\sin kt$
$\ln\frac{s^2-a^2}{s^2}$	$\frac{2}{t}(1-\cosh at)$
$\ln\frac{s^2+a^2}{s^2}$	$\frac{2}{t}(1-\cos at)$
$\frac{(1-s)^n}{s^{n+1/2}}$	$\frac{n!}{(2n)!\sqrt{\pi t}}H_{2n}\left(\sqrt{t}\right)$
$\frac{1}{\sqrt{s^2+a^2}}$	$J_0(at)$

(continued)

Table 1.5 (continued)

$\frac{1}{\sqrt{s^2-a^2}}$	$I_0(at)$
$\frac{1}{(s+a)^\alpha}$	$\frac{t^{\alpha-1}}{\Gamma(\alpha)}\mathrm{e}^{-at}$
$\frac{1}{s^\alpha+a}$	$t^{\alpha-1}E_{\alpha,\alpha}(-at^\alpha)$
$\frac{1}{s}e^{-k/s}$	$J_0\left(2\sqrt{kt}\right)$
$\frac{1}{s\sqrt{s}}e^{-k/s}$	$\frac{1}{\sqrt{\pi k}}\sin 2\sqrt{kt}$
$\frac{1}{\sqrt{s}}e^{k/s}$	$\frac{1}{\sqrt{\pi k}}\cosh 2\sqrt{kt}$
$\ln\frac{s-a}{s-b}$	$\frac{1}{t}\left(\mathrm{e}^{bt}-\mathrm{e}^{at}\right)$
$\frac{a}{s(s^\alpha+a)}$	$1-E_\alpha(-at^\alpha)$
$\frac{1}{s^n\sqrt{s}}, n=1,2,\ldots$	$\frac{2^n t^{n-1/2}}{1\cdot 3\cdot 5\cdots(2n-1)\sqrt{\pi}}$
$\frac{k}{s^2+k^2}\coth\left(\frac{\pi s}{2k}\right)$	$\lvert\sin kt\rvert$
$\frac{1}{s\sqrt{s}}\mathrm{e}^{-k/\sqrt{s}}$	$2\sqrt{\frac{t}{\pi}}\mathrm{e}^{-k^2/(4t)}-k\operatorname{erfc}\left(\frac{k}{2\sqrt{t}}\right)$
$\frac{\mathrm{e}^{-k/\sqrt{s}}}{\sqrt{s}(a+\sqrt{s})}$	$\mathrm{e}^{ak}\mathrm{e}^{a^2t}\operatorname{erfc}\left(a\sqrt{t}+\frac{k}{2\sqrt{t}}\right)$
$\frac{1}{\sqrt{s+b}(s+a)}$	$\frac{1}{\sqrt{b-a}}\mathrm{e}^{-at}\operatorname{erf}\left[\sqrt{(b-a)t}\right]$
$\frac{(1-s)^n}{s^{n+3/2}}$	$-\frac{n!}{(2n+1)!\sqrt{\pi}}H_{2n+1}\left(\sqrt{t}\right)$
$\frac{(a-b)^k}{\left(\sqrt{s+a}+\sqrt{s+b}\right)^{2k}}$	$\frac{k}{t}\mathrm{e}^{-(a+b)t/2}I_k\left(\frac{a-b}{2}t\right), k>0$
$\frac{\sqrt{s+2a}-\sqrt{s}}{\sqrt{s+2a}+\sqrt{s}}$	$\frac{1}{t}\mathrm{e}^{-at}I_1(at)$
$\frac{\left(\sqrt{s^2+a^2}-s\right)^v}{\sqrt{s^2+a^2}}$	$a^vJ_v(at), v>-1$
$\frac{1}{(s^2-a^2)^k}$	$\frac{\sqrt{\pi}}{\Gamma(k)}\left(\frac{t}{2a}\right)^{k-1/2}I_{k-1/2}(at)$
$\frac{\left(\sqrt{s^2+a^2}+s\right)^v}{\sqrt{s^2-a^2}}$	$a^vI_v(at), v>-1$
$\frac{1}{\left(\sqrt{s^2+a^2}\right)^k}$	$\frac{\sqrt{\pi}}{\Gamma(k)}\left(\frac{t}{2a}\right)^{k-1/2}J_{k-1/2}(at)$
$\left(\sqrt{s^2+a^2}-s\right)^k$	$\frac{ka^k}{t}J_k(at), k>0$
$\frac{1}{s+\sqrt{s^2+a^2}}$	$\frac{J_1(at)}{at}$
$\frac{1}{\sqrt{s(s+a)}\left(\sqrt{s+a}+\sqrt{s}\right)2v}$	$\frac{1}{a^v}\mathrm{e}^{-at/2}I_v\left(\frac{a}{2}t\right), k>0$
$\frac{\Gamma(k)}{(s+a)^k(s+b)^k}$	$\sqrt{\pi}\left(\frac{t}{a-b}\right)^{k-1/2}\mathrm{e}^{-(a+b)t/2}I_{k-1/2}\left(\frac{a-b}{2}t\right)$

(continued)

Table 1.5 (continued)

$\frac{1}{\sqrt{s(s+a)}\left(\sqrt{s+a}+\sqrt{s}\right)^{2v}}$	$\frac{1}{a^v}\mathrm{e}^{-at/2}I_v\left(\frac{a}{2}t\right), k>0$
$\frac{1}{\sqrt{s^2+a^2}\left(s+\sqrt{s^2+a^2}\right)^N}$	$\frac{J_N(at)}{a^N}$
$\frac{1}{\sqrt{s^2+a^2}\left(s+\sqrt{s^2+a^2}\right)}$	$\frac{J_1(at)}{a}$
$\frac{b^2-a^2}{\sqrt{s}(s-a^2)(\sqrt{s}+b)}$	$\mathrm{e}^{a^2t}\left[\frac{b}{a}\mathrm{erf}(a\sqrt{t})-1\right]+\mathrm{e}^{b^2t}\mathrm{erfc}(b\sqrt{t})$
$\frac{a\mathrm{e}^{-k\sqrt{s}}}{s(a+\sqrt{s})}$	$-\mathrm{e}^{ak}\mathrm{e}^{a^2t}\mathrm{erfc}\left(a\sqrt{t}+\frac{k}{2\sqrt{t}}\right)+\mathrm{erfc}\left(\frac{k}{2\sqrt{t}}\right)$
$\frac{1}{\sqrt{s+a}(s+b)\sqrt{(s+b)}}$	$t\mathrm{e}^{-(a+b)t/2}\left[I_0\left(\frac{a-b}{2}t\right)+I_1\left(\frac{a-b}{2}t\right)\right]$
$\frac{\mathrm{e}^{-\sqrt{s}}}{\sqrt{s}+1}$	$\frac{\mathrm{e}^{-1/(4k)}}{\sqrt{\pi t}}-\mathrm{e}^{t+1}\mathrm{erfc}\left(\sqrt{t}+\frac{1}{2\sqrt{t}}\right)$
$\frac{\mathrm{e}^{-\sqrt{s}}}{s(\sqrt{s}+1)}$	$\mathrm{erfc}\left(\frac{1}{2\sqrt{t}}\right)-\mathrm{e}^{t+1}\mathrm{erfc}\left(\sqrt{t}+\frac{1}{2\sqrt{t}}\right)$

$$\mathrm{daw}(z)=\mathrm{e}^{-z^2}\int_0^z \mathrm{e}^{\tau^2}\mathrm{d}\tau \tag{1.111}$$

where $z\in\mathbb{C}(-\infty,0]$ and $v\in\mathbb{C}$.

1.3.2 Integer-Order Approximation of Fractional-Order Transfer Function

In addition to the Laplace transform method, one of the universal frequency-domain methods of solving fractional differential equations is to approximate the transfer function in the frequency domain of $(1/s)^\alpha$.

By utilizing frequency-domain techniques based on the Bode diagram, W. M. Ahmad et al. give the approximate linear transfer functions for the fractional integrator of order α in the frequency domain [59–61], which can be represented by the transfer function as

$$H(s)=\frac{1}{s^\alpha} \tag{1.112}$$

where α varies from 0.1 to 0.9.

If the discrepancy between the actual and approximate lines is specified as y dB over a frequency range of $\omega_{\max}$, and for a corner frequency p_{T}, then (1.112) can be approximated as

Table 1.6 Transfer function approximations of the fractional integrator of order α with maximum discrepancy $y = 2$ dB

q	$H(s) = 1/s^{\alpha}$
0.1	$H(s) = \frac{1584.8932(s+0.1668)(s+27.83)}{(s+0.1)(s+16.68)(s+2783)}$
0.2	$H(s) = \frac{79.4328(s+0.05623)(s+1)(s+17.78)}{(s+0.03162)(s+0.5623)(s+10)(s+177.8)}$
0.3	$H(s) = \frac{39.8107(s+0.0416)(s+0.3728)(s+3.34)(s+29.94)}{(s+0.02154)(s+0.1931)(s+1.73)(s+15.51)(s+138.9)}$
0.4	$H(s) = \frac{35.4813(s+0.03831)(s+0.261)(s+1.778)(s+12.12)(s+82.54)}{(s+0.01778)(s+0.1212)(s+0.8254)(s+5.623)(s+38.31)(s+261)}$
0.5	$H(s) = \frac{15.8489(s+0.03981)(s+0.2512)(s+1.585)(s+10)(s+63.1)}{(s+0.01585)(s+0.1)(s+0.631)(s+3.981)(s+25.12)(s+158.5)}$
0.6	$H(s) = \frac{10.7978(s+0.04642)(s+0.3162)(s+2.154)(s+14.68)(s+100)}{(s+0.01468)(s+0.1)(s+0.6813)(s+4.642)(s+31.62)(s+215.4)}$
0.7	$H(s) = \frac{9.3633(s+0.06449)(s+0.587)(s+5.179)(s+46.42)(s+416)}{(s+0.01389)(s+0.1245)(s+1.116)(s+10)(s+89.62)(s+803.1)}$
0.8	$H(s) = \frac{5.3088(s+0.1334)(s+2.371)(s+42.17)(s+749.9)}{(s+0.01334)(s+0.2371)(s+4.217)(s+74.99)(s+1334)}$
0.9	$H(s) = \frac{2.2675(s+1.292)(s+215.4)}{(s+0.01292)(s+2.154)(s+359.4)}$

$$H(s) = \frac{1}{s^{\alpha}} \approx \frac{1}{\left(1 + \frac{s}{p_T}\right)^{\alpha}} \approx \frac{\prod_{i=0}^{N-1}\left(1 + \frac{s}{z_i}\right)}{\prod_{i=0}^{N}\left(1 + \frac{s}{p_i}\right)} \tag{1.113}$$

where $p_0 = p_T 10^{(y/20\alpha)}$, $a = 10^{[y/10(1-\alpha)]}$, $b = 10^{(y/10\alpha)}$ and

$$N = 1 + \text{Integer}\left(\frac{\log\left(\frac{\omega_{\max}}{p_0}\right)}{\log(ab)}\right) \tag{1.114}$$

Assuming $\omega_{\max} = 100$ and $p_T = 0.01$, the resulting approximating transfer functions with maximum discrepancy of $y = 2$ dB and $y = 3$ dB are shown in Tables 1.6 and Table 1.7.

The considerable disadvantages of the integer-order approximation method are that only a few transfer functions corresponding to α can be obtained by this method. The difference between the original system and the approximate system is still unknown at present.

1.3.3 Numerical Solutions

The Laplace transform method and the integer-order approximation method of the fractional-order transfer function are both frequency-domain methods. Recent

Table 1.7 Transfer function approximations of the fractional integrator of order α with maximum discrepancy $y = 3$ dB

Q	$H(s) = 1/s^{\alpha}$
0.1	$H(s) = \frac{501.14(s+0.6811)}{(s+0.3162)(s+681.1)}$
0.2	$H(s) = \frac{141.2538(s+0.1334)(s+10)}{(s+0.05623)(s+4.217)(s+316.2)}$
0.3	$H(s) = \frac{125.8925(s+0.08483)(s+2.276)(s+61.05)}{(s+0.03162)(s+0.8483)(s+22.76)(s+610.5)}$
0.4	$H(s) = \frac{26.6073(s+0.07499)(s+1.334)(s+23.71)}{(s+0.02371)(s+0.4217)(s+7.499)(s+133.4)}$
0.5	$H(s) = \frac{50.1187(s+0.07943)(s+1.259)(s+19.95)(s+316.2)}{(s+0.01995)(s+0.3162)(s+5.012)(s+79.43)(s+1259)}$
0.6	$H(s) = \frac{28.1838(s+0.1)(s+1.778)(s+31.62)(s+562.3)}{(s+0.01778)(s+0.3162)(s+5.623)(s+100)(s+1778)}$
0.7	$H(s) = \frac{7.9433(s+0.1638)(s+4.394)(s+117.9)}{(s+0.01638)(s+0.4394)(s+11.79)(s+100)(s+316.2)}$
0.8	$H(s) = \frac{8.1752(s+0.487)(s+36.52)(s+2738)}{(s+0.0154)(s+1.155)(s+86.6)(s+6494)}$
0.9	$H(s) = \frac{4.2987(s+14.68)(s+31620)}{(s+0.01468)(s+31.62)(s+68130)}$

studies have shown that the time-domain methods are another better choice for numerically solving fractional differential equations than the frequency-domain methods. One of the time-domain methods commonly used is the Adams–Bashforth-Moulton (ABM) predictor–corrector approach [62–64]. This method is based on the Volterra integral equation, discretizes the fractional differential system by a predictor–corrector algorithm, and then gives the approximate numerical solution of the fractional differential equation, which is a more effective numerical solution.

Considering a nonlinear dynamic system, which can be described by the following fractional differential equation, that is

$$\begin{cases} \frac{\mathrm{d}^{\alpha} x(t)}{\mathrm{d}t^{\alpha}} = f(t, x(t)) \\ x^{(k)}(0) = x_0^{(k)}, k = 0, 1, \ldots, m-1 \end{cases} \tag{1.115}$$

where $\alpha > 0$, $m = [\alpha]$, and $x_0^{(k)} (k = 0, 1, 2, \ldots, m-1)$ are given real numbers.

Continuous function $x(t)$ is a solution of (1.115) if and only if it is a solution of the Volterra integral equation, which is

$$x(t) = \sum_{k=0}^{m-1} x_0^{(k)} \frac{t^k}{k!} + \frac{1}{\Gamma(\alpha)} \int_0^t (t-\tau)^{\alpha-1} f(\tau, x(\tau)) \mathrm{d}\tau \tag{1.116}$$

The analytical properties of (1.115) and (1.116) are equivalent [65] in the Caputo fractional definition. Taking $t_n = nh$, $n = 0, 1, 2, \ldots, N$, $N \in \mathbb{N}$, $h > 0$ is the step size, then (1.116) can be discretized as

$$x_h(t_{n+1}) = \sum_{k=0}^{m-1} \frac{t_{n+1}^k}{k!} x_0^{(k)} + \frac{h^{\alpha}}{\Gamma(\alpha+2)} f(t_{n+1}, x_h^{\alpha}(t_{n+1})) + \frac{h^{\alpha}}{\Gamma(\alpha+2)}$$

$$\sum_{i=0}^{n} \varepsilon_{i,n+1} f(t_i, x_h(t_i)) \tag{1.117}$$

where

$$x_h^{\alpha}(t_{n+1}) = \sum_{k=0}^{m-1} \frac{t_{n+1}^k}{k!} x_0^{(k)} + \frac{1}{\Gamma(\alpha)} \sum_{i=0}^{n} b_{i,n+1} f(t_i, x_h(t_i)) \tag{1.118}$$

$$\varepsilon_{i,n+1} = \begin{cases} n^{\alpha+1} - (n-q)(n+1)^{\alpha}, i = 0 \\ (n-i+2)^{\alpha+1} + (n-i)^{\alpha+1} - 2(n-i+1)^{\alpha+1}, 1 \le i \le n \\ 1, \ i = n+1 \end{cases} \tag{1.119}$$

$$b_{i,n+1} = \frac{h^{\alpha}}{q}\left[(n-i+1)^{\alpha} - (n-i)^{\alpha}\right] \tag{1.120}$$

The discrepancy between (1.116) and (1.117) is

$$\max_{i=0,1,2,\ldots,N} |x(t_i) - x_h(t_i)| = O(h^{\alpha}) \tag{1.121}$$

Furthermore, another time-domain method is based on the Grünwald-Letnikov fractional definition for numerical calculation. However, this method is limited by the simulation speed and storage capacity of the computer in practice.

The third time-domain method is the Adomian decomposition method (ADM) [66, 67], which provides a powerful tool for solving linear and nonlinear fractional differential equations. ADM divides the given equation into linear and nonlinear parts, inverts the highest-order derivative operator contained in the linear operator on both sides, calculates the Adomian polynomials, and finally uses the Adomian polynomials to find the successive terms of the series solution through the recursive relationship [68, 69].

For every nonlinear differential equation, it can be described by a general nonlinear equation in the following form

$$Lu + Ru + Nu = g \tag{1.122}$$

where L is the highest-order differential operator, which is assumed to be easy to invert, R is the linear differential operator whose order is lower than L, Nu represents the nonlinear terms, and g is the source term.

For ADM, the solution can be decomposed into the form of an infinite series as

$$u = \sum_{n=0}^{\infty} u_n \tag{1.123}$$

Since L is reversible, applying the inverse operation L^{-1} on both sides of (1.122), it follows

$$u = \varphi + L^{-1}g - L^{-1}(Ru) - L^{-1}(Nu) \tag{1.124}$$

where φ satisfies $L\varphi = 0$ and the initial conditions.

Using the Adomian polynomials to represent the nonlinear term Nu, yields

$$Nu = \sum_{k=0}^{\infty} A_k \tag{1.125}$$

where A_k is determined by $u_0, u_1, \ldots, u_n$ and can be given by

$$\begin{cases} A_0 = Nu_0 \\ A_1 = u_1 N^{(1)} u_0 \\ A_2 = u_2 N^{(1)} u_0 + \dfrac{1}{2!} u_1^2 N^{(2)} u_0 \\ A_3 = u_3 N^{(1)} u_0 + u_1 u_2 N^{(2)} u_0 + \dfrac{1}{3!} u_1^3 N^{(3)} u_0 \\ \vdots \\ A_n = \dfrac{1}{n!} \dfrac{d^n}{d\lambda^n} \left[N\left(\sum_{k=0}^{\infty} \lambda^k u_k \right) \right]_{\lambda=0} \end{cases} \tag{1.126}$$

In the same way, (1.122) can be rewritten as

$$\sum_{n=0}^{\infty} u_n = -L^{-1}R \sum_{n=0}^{\infty} u_n - L^{-1} \sum_{k=0}^{\infty} A_k + L^{-1}g + \varphi \tag{1.127}$$

where

$$\begin{cases} u_0 = L^{-1}g + \varphi \\ u_1 = -L^{-1}Ru_0 - L^{-1}A_0 \\ u_2 = -L^{-1}Ru_1 - L^{-1}A_1 \\ \vdots \\ u_n = -L^{-1}Ru_{n-1} - L^{-1}A_{n-1} \end{cases} \tag{1.128}$$

For a given u_0, u_n can be calculated iteratively by (1.128).

1.4 Matlab-Based Numerical Tools for Fractional Calculus

Several Matlab-based numerical tools are widely used in the numerical calculation of fractional calculus, including Ninteger Toolbox, CRONE Toolbox, FOTF Toolbox, and FOMCON Toolbox.

Ninteger Toolbox is a non-integer control toolbox in MATLAB, which intends to help develop fractional-order controllers and evaluate their performance [70, 71]. It essentially uses integer-order transfer functions to approximate the fractional-order integrator and differentiator, $C(s) = ks^{\alpha}, \alpha \in \mathbb{R}$. It offers three approximation methods as follows:

1. The CRONE method uses the recursive distribution of n poles and n zeros, which can be expressed as

$$C(s) = k' \prod_{n=1}^{N} \frac{1 + s/\omega_{zn}}{1 + s/\omega_{pn}} \tag{1.129}$$

where the gain k' is adjusted to make sure $|C(s)|=0$ at 1 rad/s when k is 1. Poles and zeros are in the range of $[\omega_l, \omega_h]$. For a positive α, poles and zeros are given by

$$\begin{cases} v = (\omega_h/\omega_l)^{\frac{\alpha}{N}}, \eta = (\omega_h/\omega_l)^{\frac{1-\alpha}{N}} \\ \omega_{z1} = \omega_l\sqrt{\eta}, \omega_{zn} = \omega_{p,n-1}\eta, n = 2, \ldots, N \\ \omega_{pn} = \omega_{z,n-1}v, n = 1, \ldots, N \end{cases} \tag{1.130}$$

where the control variables are k, α, ω_l, ω_h, and N. For a negative α, the role of poles and zeros is interchanged.

2. The Carlson method uses the Newton iterative method to solve $H^{\alpha}(s) = g\ (s)$. The iterative formula is

$$H_n(s) = H_{n-1}(s)\frac{(\alpha - 1)H_{n-1}^{\alpha}(s) + (\alpha + 1)g(s)}{(\alpha + 1)H_{n-1}^{\alpha}(s) + (\alpha - 1)g(s)} \tag{1.131}$$

where the fractional order is $1/\alpha$, and the function is $H(s) = s$.

This method requires iteration to reach or exceed a specified minimum number of poles and zeros, and the result is valid in the frequency range centered on 1 rad/s.

3. The Matsuda method, which approximates the transfer function $C(s)$ by iteration with a known gain at the frequency points ω_0, ω_1, ω_2, ..., can be described by

$$\begin{cases} C(s) = \left[d_0(\omega_0); (s - \omega_{k-1})/d_k(\omega_k)\right]_{k=1}^{+\infty} \\ d_0(\omega) = |C(\mathrm{j}\omega)| \\ d_{k+1}(\omega) = (\omega - \omega_k)/\left[d_k(\omega) - d_k(\omega_k)\right] \end{cases} \tag{1.132}$$

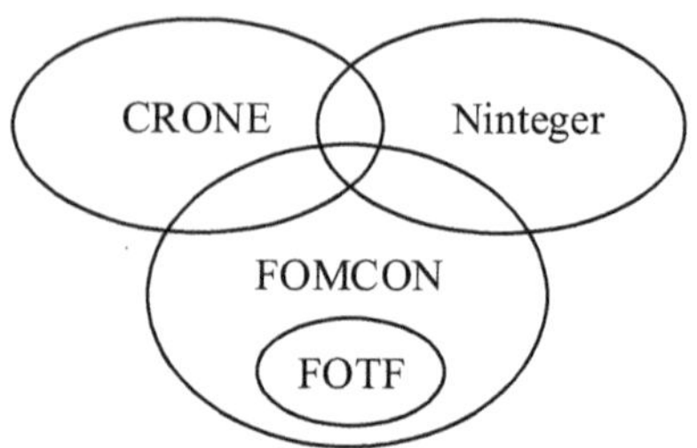

Fig. 1.4 The relationships among FOMCON, CRONE, Ninteger, and FOTF Toolboxes

The CRONE Toolbox, developed by the CRONE research group (Oustaloup et al.), is a Matlab and Simulink toolbox devoted to applications of fractional differentiation in engineering and science [72]. It is a practical tool for solving the implementation of fractional calculus algorithm, the identification of fractional-order models in the time and frequency domain, and the realization of fractional-order robust control design [73, 74]. However, the toolbox is issued in the form of Matlab pseudo-code encryption, and users cannot modify and extend any function of the toolbox.

The FOTF Toolbox is a control toolbox for fractional-order systems developed by Xue et al. [71, 75, 76], which uses the Matlab built-in functions to deal with fractional-order models. The toolbox is composed of the functions in the @fotf folder, and those functions cover the basic operations of the FOTF objects, stability test, simulation of time responses, etc. Yet, there is a disadvantage that the sampling time has a great influence on the accuracy.

The FOMCON Toolbox, developed by Tepljiakov, Petlenkov, and Belikov, is a new fractional-order modeling and control toolbox for Matlab [77, 78]. Its core is the use of the algorithms in FOTF, Ninteger, and CRONE, and it provides graphical user interfaces (GUIs) by encapsulating the main functions of these three toolboxes. The relation of the FOMCOM Toolbox with other toolboxes is depicted in Fig. 1.4. The goal of this toolbox is to extend classical control schemes for the fractional-order controller design. It is widely applicable to the analysis of the fractional-order system, system identification in time and frequency domains, and the design of the fractional-order PID controller.

In addition to the above-mentioned fundamental tools related to fractional calculus and fractional-order control, there are other numerical tools [71], but this book does not enumerate them one by one.

1.5 Summary

Fractional calculus is the basis of fractional-order circuit analysis. This chapter mainly introduces the definition and basic properties of several special functions in fractional calculus, elaborates the three definitions of fractional calculus and their basic properties, and compares the basic characteristics of fractional calculus in G-L, R-L, and Caputo definition. Furthermore, this chapter also introduces several methods for solving fractional differential equations and several software tools commonly used in the numerical analysis of fractional calculus and fractional-order controller, which builds a foundation for the characteristic analysis of complicated fractional-order circuits.

References

1. Herrmann R (2014) Fractional calculus—An introduction for physicists. World Scientific, Singapore
2. Lazarevi MP Rapai MR, Ekara TB (2014) Introduction to fractional calculus with brief historical background WSEAS Press. In: Advanced topics on applications of fractional calculus on control problems, system stability and modelling. WSEAS Press
3. Cafagna (2007) Past and present—Fractional calculus: A mathematical tool from the past for present engineers. IEEE Ind Electron Mag 1(2):35–40
4. Gerhardt (1849) Leibnizens mathematische Schriften. A. Asher Press
5. Oldham K, Spanier J (1974) The fractional calculus: theory and applications of differentiation and integration to arbitrary order. Academic Press, New York
6. Ross B (1975) A brief history and exposition of the fundamental theory of fractional calculus. In: Fractional calculus and its applications. Springer, Berlin, Heidelberg, vol 1, pp 1–36
7. Qiang D, Jan (2019) Preface to the focused issue on fractional derivatives and general nonlocal models. Commun Appl Math Comput 1(4):503–504
8. Wongsaijai B, Sukantamala N (2015) Applications of fractional q-calculus to certain subclass of analytic p-valent functions with negative coefficients. Abstr Appl Anal 2015:273236
9. Li C, Chen Y, Kurths J (2013) Fractional calculus and its applications. Philos Trans Ser A Math Phys Eng Sci 371:1–3
10. Evans MR, Edwards AD et al (2017) Applications of fractional calculus in solving Abel-type integral equations: surface-volume reaction problem. Comput Math Appl: Int J 73(6):1346–1362
11. Mandelbrot, Benoit B (1998) The fractal geometry of nature. Am J Phys 51(3):468
12. Caputo M, Mainardi F (1971) A new dissipation model based on memory mechanism. Pure Appl Geophys 91:134–147
13. Ichise M, Nagayanagi Y, Kojima T (1971) An analog simulation of non-integer order transfer functions for analysis of electrode processes. J Electroanal Chem 33(2):253–265
14. Tarasov V (2013) Review of some promising fractional physical models. Int J Mod Phys B 27:1330005
15. Kilbas AA, Srivastava HM, Trujillo JJ (2003) Fractional differential equations: a emergent field in applied and mathematical sciences. In: Factorization singular operators and related problems. Springer, Netherland, pp 151–173
16. Miller KS, Ross B (1993) An Introduction to the fractional calculus and fractional differential equations. Wiley, New York
17. Podlubny I (1999) Fractional differential equations: an introduction to fractional derivatives, fractional differential equations, to methods of their solution and some of their applications. Academic Press, San Diego
18. Lavoie JL, Tremblay TJO (1976) Fractional derivatives and special functions. SIAM Rev 18(2):240–268
19. Jesus IS, Machado JAT (2009) Development of fractional order capacitors based on electrolyte processes. Nonlinear Dyn 56(1):45–55
20. Sivarama Krishna M, Das S, Biswas K et al (2011) Fabrication of a fractional order capacitor with desired specifications: A study on process identification and characterization. IEEE Trans Electron Devices 58(11):4067–4073

21. Caponetto R, Ebrary I (2010) Fractional order systems: modeling and control applications. World Scientific, Singapore
22. Bickel DR, West BJ (1998) Multiplicative and fractal process in DNA evolution. In: Fractals-complex geometry patterns & scaling in nature & society, vol 6, no 03, pp 211–217
23. Engheta N (2000) Fractionalization methods and their applications to radiation and scattering problems. In: Proceedings of the international conference on mathematical methods in electromagnetic theory (MMET 2000), Kharkov, Ukraine
24. Aguilar JFG, Hernndez JRR, Jimnez RFE et al (2014) Fractional electromagnetic waves in plasma. In: Proceedings of the Romanian Academy—Series A: mathematics, physics, technical sciences, information science, vol 17, no 1, pp 31–38
25. Zubair M, Mughal MJ, Naqvi QA (2012) Electromagnetic fields and waves in fractional dimensional space. Springer, Berlin Heidelberg
26. Chen Y, Petras I, Xue D (2009) Fractional order control—A tutorial. In: Proceedings of the American control conference, St. Louis, MO, USA
27. Monje CA, Chen YQ, Vinagre BM et al (2010) Fractional-order systems and controls. Springer, London
28. Bagley RL, Calico RA (1991) Fractional order state equations for the control of viscoelastically damped structures. J Guid Control Dyn 14(2):304–311
29. Freed AD, Diethelm K (2007) Caputo derivatives in viscoelasticity: a nonlinear finite-deformation theory for tissue. Fract Calcul Appl Anal 10(3):219–248
30. Meral FC, Royston TJ, Magin R (2010) Fractional calculus in viscoelasticity: an experimental study. Commun Nonlinear Sci Numer Simul 15(4):939–945
31. Mainardi F (2010) Fractional calculus and waves in linear viscoelasticity. Imperial College Press, UK, p 368
32. Radwan AG (2012) Stability analysis of the fractional-order RLC circuit. J Fract Calcul Appl 3(1):1–15
33. Sprott WMAC (2003) Chaos in fractional-order autonomous nonlinear systems. Chaos Solit Fract 16(2):339–351
34. Chen X, Chen Y, Zhang B, Qiu D (2017) A modeling and analysis method for fractional-order DC-DC converters. IEEE Trans Power Electron 32(9):7034–7044
35. Wu C, Si G, Zhang Y et al (2015) The fractional-order state-space averaging modeling of the Buck-Boost DC/DC converter in discontinuous conduction mode and the performance analysis. Nonlinear Dyn 79(1):689–703
36. Westerlund S, Ekstam L (1994) Capacitor theory. IEEE Trans Dielectrics Electr Insul 1(5):826–839
37. Lazarus N, Meyer CD, Bedair SS (2013) Fractal inductors. IEEE Trans Magn 50(4):1–8
38. Sch Fer I, Krüger K (2008) Modelling of lossy coils using fractional derivatives. J Phys D Appl Phys 41(4):045001
39. Machado JAT, Galhano AMSF (2012) Fractional order inductive phenomena based on the skin effect. Nonlinear Dyn 68(1):107–115
40. Oldham KB, Spanier J (1974) The fractional calculus. Academic Press, New York
41. Baleanu D, Diethelm K, Scalas E et al (2012) Fractional calculus: models and numerical methods. World Scientific, New Jersey, United States
42. Hunter D, Regan T (1972) A Note on the evaluation of the complementary error function. Math Comput 26:539–541
43. Deaño A, Temme NM (2010) Analytical and numerical aspects of a generalization of the complementary error function. Appl Math Comput 216(12):3680–3693
44. Chamati H, Tonchev N (2005) Generalized Mittag-Leffler functions in the theory of finite-size scaling for systems with strong anisotropy and/or long-range interaction. J Phys A: Gen Phys 39(3):1–14
45. Liouville J (2021) Mémoire sur quelques questions de géométrie et de mécanique, et sur un nouveau genre de calcul pour résoudre ces questions. Journal de l'École polytechnique 13:1–69
46. Riemann B (1853) Versuch einer allgemeinen auffassung der integration und differentiation. Gesammelte Mathematische Werke 331–344

47. Grünwald AK (1867) Ueber begrenzte derivationen und deren anwendung. Z Angew Math Phys 12:441–480
48. Caputo M (1967) Linear models of dissipation whose Q is almost frequency independent-II. Geophys J Int 13(5):529–539
49. Samko SG, Kilbas AA, Marichev OI (1987) Integrals and derivatives of fractional order and some of their applications. Gordon and Breach Science Publishers
50. Mainardi F (1997) Fractional calculus. In: Fractals and fractional calculus in continuum mechanics. Springer, Vienna, pp 291–348
51. Jiang Y, Zhang B (2019) Comparative study of Riemann-Liouville and Caputo derivative definitions in time-domain analysis of fractional-order capacitor. IEEE Trans Circuits Syst II Express Briefs 67(10):2184–2188
52. Podlubny I (2001) Geometric and physical interpretation of fractional integration and fractional differentiation. Fract Calcul Appl Anal 5(4):230–237
53. Zahra WK, Hikal MM, Bahnasy TA (2017) Solutions of fractional order electrical circuits via Laplace transform and nonstandard finite difference method. J Egyptian Math Soc 25(2):252–261
54. Magin R (2006) Fractional calculus in bioengineering. Crit Rev Biomed Eng 25(2):252–261
55. Mccollum PA, Brown BF (1965) Laplace transform tables and theorems. Rinehart and Winston, New York
56. Chen Y, Petrasz I, Vinagre B (2001) A list of laplace and inverse laplace transforms related to fractional order calculus
57. Lin S-D, Lu C-H (2013) Laplace transform for solving some families of fractional differential equations and its applications. Adv Differ Equ 2013(1):137
58. Kilbas AAA, Srivastava H, Trujillo J (2006) Theory and applications of fractinal differential equations. Elsevier Science Inc., New York, United States
59. Ahmad WM, Sprott JC (2003) Chaos in fractional-order autonomous nonlinear systems. Chaos Solit Fract 16(2):339–351
60. Charef A, Sun HH, Tsao YY et al (1992) Fractal system as represented by singularity function. IEEE Trans Autom Control 37(9):1465–1470
61. Özyetkin M, Tan N (2010) Integer order approximation of fractional order systems. In: 2010 IEEE 18th signal processing and communications applications conference, Diyarbakir, Turkey
62. Diethelm K, Ford NJ, Freed AD (2002) A predictor-corrector approach for the numerical solution of fractional differential equations. Nonlinear Dyn 29(1):3–22
63. Owusu-Mensah I, Akinyemi L, Oduro B et al (2020) A fractional order approach to modeling and simulations of the novel COVID-19. Adv Differ Equ 2020(1):683
64. Douaifia R, Abdelmalek S (2019) A predictor-corrector method for fractional delay-differential system with multiple lags. Commun Nonlinear Anal 6(1):78–88
65. Kai D (1997) An algorithm for the numerical solution of differential equations of fractional order. Electron Trans Numer Anal 5:1–6
66. El-Wakil SA, Elhanbaly A, Abdou MA (2006) Adomian decomposition method for solving fractional nonlinear differential equations. Appl Math Comput 182(1):313–324
67. Guo P (2019) The Adomian decomposition method for a type of fractional differential equations. J Appl Math Phys 07(10):2459–2466
68. Adomian G (1988) A review of the decomposition method in applied mathematics. J Math Anal Appl 135(2):501–544
69. Adomian G (1994) Solution of nonlinear evolution equations. Math Comput Model 20(12):1–2
70. Valério D, Costa J (2004) Ninteger: a non-integer control toolbox for MatLab. In: Proceedings of the first IFAC workshop on fractional differentiation and applications, Portugal
71. Li Z, Liu L, Dehghan S, Chen YQ et al (2015) A review and evaluation of numerical tools for fractional calculus and fractional order control. Int J Control 90(6):1165–1181
72. Oustaloup A, Sabatier J, Moreau X (1998) From fractal robustness to the CRONE approach. In: ESAIM proceedings, vol 5, pp 177–192
73. Melchior P, Orsoni B, Lavialle O et al (2001) The CRONE toolbox for Matlab: fractional path planning design in robotics. In: Proceedings of 10th IEEE international workshop on robot and human interactive communication, Paris, France

74. Malti R, Melchior P, Lanusse P et al (2011) Towards an object oriented CRONE toolbox for fractional differential systems. IFAC Proc Vol 44(1):10830–10835
75. Moroz V, Borovets T (2017) Using of FOTF toolbox and Z-transform for fractional control systems. In: 2017 IEEE international young scientists forum on applied physics and engineering, Lviv
76. Xue D (2019) FOTF toolbox for fractional-order control systems. In: Applications in control, vol 6. De Gruyter, Boston, pp 237–266
77. Tepljakov A, Petlenkov E, Belikov J (2011) FOMCON: fractional-order modeling and control toolbox for MATLAB. In: Proceedings of the 18th international conference mixed design of integrated circuits and systems (MIXDES 2011), Gliwice, Poland
78. Bingi K, Ibrahim R, Karsiti MN et al Scilab based toolbox for fractional-order systems and PID controllers. In: Fractional-order systems and PID controllers, using scilab and curve fitting based approximation techniques. Springer, Switzerland

Chapter 2
Fractional-Order Components and Their Basic Circuits

Fractional-order circuit is a kind of circuit that contains fractional-order components, such as the fractional-order inductor, fractional-order capacitor, and fractional-order mutual inductor. Different from the integer-order component that is described by first-order calculus, the current–voltage relationship of the fractional-order components is modeled as fractional-order calculus. In fact, the ideal integer-order inductor or capacitor does not exist. Since the orders of most components are normally close to 1, they are treated as 1 with neglecting their fractional-order characteristics. However, some components with strong fractional-order properties are being found [1]. Due to the existence of the order, the characteristics of the fractional-order components and integer-order components are significant components. Moreover, fractional-order circuits have much richer and more interesting features than integer-order circuits [2–4].

First, the basic concepts and properties of fractional-order components are introduced, including the fractional-order capacitor, fractional-order inductor, and fractional-order mutual inductor. Then, the circuit of the series and parallel combinations of different fractional-order components are studied, which provides a theoretical basis for analyzing the more complicated fractional-order circuits.

2.1 Fractional-Order Components

2.1.1 Fractional-Order Capacitor

In the conventional integer-order circuit theory, the capacitor is regarded as an ideal component with absolutely no current flowing between the plates, since its internal resistance is considered to be infinite [5]. However, a practical capacitor has resistive losses and its resistance is finite, resulting in a small leakage current between the

B. Zhang and X. Shu, *Fractional-Order Electrical Circuit Theory*, CPSS Power Electronics Series, https://doi.org/10.1007/978-981-16-2822-1_2

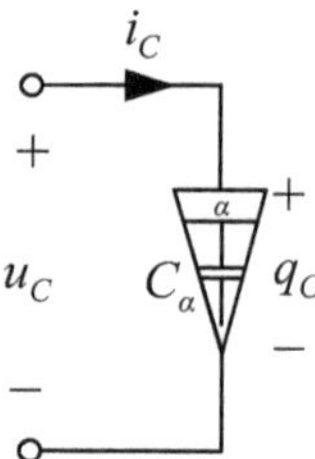

Fig. 2.1 Graphical symbol for the fractional-order capacitor

plates [6]. The ideal integer-order capacitor does not exist in the real world. Moreover, the current researches show that the models of capacitors are described more accurately by fractional-order calculus [7, 8].

The graphical symbol for a fractional-order capacitor is shown in Fig. 2.1, where α ($0 < \alpha < 2$) and C_α is the order and the pseudo-capacitance value of the fractional-order capacitor, respectively. The fractional-order capacitors can be divided into the active capacitor ($1 < \alpha < 2$) and the passive capacitor ($0 < \alpha < 1$).

The fractional-order capacitor is a two-terminal component, and its performance equation can be written as

$$q_C = f(u_C) \text{ or } u_C = f(q_C) \tag{2.1}$$

where u_C and q_C are the voltage and charge of the fractional-order capacitor, respectively.

Based on the definition of the fractional-order capacitor, its current i_C could be expressed as [9, 10]

$$i_C = \frac{\mathrm{d}q_C}{\mathrm{d}t} = C_\alpha \frac{\mathrm{d}^\alpha u_C}{\mathrm{d}t^\alpha} = \frac{C_\alpha}{\Gamma(n-\alpha)} \frac{\mathrm{d}^n}{\mathrm{d}t^n} \int_0^t (t-\tau)^{n-1-\alpha} u_C(\tau) \mathrm{d}\tau \tag{2.2}$$

where n is the largest integer less than α, $\mathrm{d}^\alpha/\mathrm{d}t^\alpha$ is termed the fractional-order derivative, the unit of the pseudo-capacitance value C_α is Farad/(second)$^{1-\alpha}$, which can also be expressed by F/s$^{1-\alpha}$.

It can be seen from (2.2) that when $t > 0$, the current i_C depends not only on the initial state of the voltage u_C, but also on its historical process. Therefore, the current of the fractional-order capacitor has memory characteristics [11, 12].

The charge is the time integral of the current, which can be derived as

$$q_C = \int i_C \,\mathrm{d}t = \int C_\alpha \frac{\mathrm{d}^\alpha u_C}{\mathrm{d}t^\alpha} \mathrm{d}t = C_\alpha \frac{\mathrm{d}^{\alpha-1} u_C}{\mathrm{d}t^{\alpha-1}} \tag{2.3}$$

When the R-L definition of fractional calculus is applied, (2.3) can be rearranged as

$$
\begin{aligned}
u_C(t) &= \frac{1}{C_\alpha} {}_0D_t^{-\alpha} i_C(t) + \sum_{j=1}^{k} \left[{}_0D_t^{-\alpha} u_C(t) \right]_{t=0} \frac{t^{\alpha-j}}{\Gamma(\alpha-j+1)} \\
&= \frac{1}{C_\alpha \Gamma(\alpha)} \int_0^t (t-\tau)^{\alpha-1} i_C(\tau) \mathrm{d}\tau + \sum_{j=1}^{k} \left[{}_0D_t^{-\alpha} u_C(t) \right]_{t=0} \frac{t^{\alpha-j}}{\Gamma(\alpha-j+1)}
\end{aligned}
\tag{2.4}
$$

where k is the largest integer less than α, its values are related to whether the fractional-order capacitor is active or passive. $k = 1$ represents that the capacitor is passive, and $k = 2$ denotes that the capacitor is active.

It can be concluded from (2.4) that when $t > 0$, the voltage $u_C(t)$ is determined by the initial state and historical process of the current $i_C(t)$. Therefore, the voltage of the fractional-order capacitor also has memory characteristics [13].

According to (2.2) and (2.4), the instantaneous power p_C absorbed by the fractional-order capacitor can be deduced as

$$
p_C = u_C i_C = u_C C_\alpha \frac{\mathrm{d}^\alpha u_C}{\mathrm{d}t^\alpha} = \frac{u_C C_\alpha}{\Gamma(n-\alpha)} \frac{\mathrm{d}^n}{\mathrm{d}t^n} \int_0^t (t-\tau)^{n-1-\alpha} u_C(\tau) \mathrm{d}\tau \tag{2.5}
$$

Since the energy is the integral of instantaneous power in time, the total energy storage of the fractional-order capacitor in the period from $-\infty$ to t can be expressed as [14]

$$
\begin{aligned}
W_C(t) &= \int_{-\infty}^{t} p_C \, \mathrm{d}\xi = \int_{-\infty}^{t} u_C i_C \, \mathrm{d}\xi = \int_{-\infty}^{t} u_C C_\alpha \frac{\mathrm{d}^\alpha u_C}{\mathrm{d}\xi^\alpha} \mathrm{d}\xi \\
&= \int_{-\infty}^{t} \left[\frac{u_C C_\alpha}{\Gamma(n-\alpha)} \frac{\mathrm{d}^n}{\mathrm{d}t^n} \int_0^t (t-\tau)^{n-1-\alpha} u_C(\tau) \mathrm{d}\tau \right] \mathrm{d}\xi
\end{aligned}
\tag{2.6}
$$

Due to the memory characteristics of the current and voltage of the fractional-order capacitor, it can be seen from (2.6) that the total energy storage $W_C(t)$ also has memory properties.

2.1.2 *Fractional-Order Inductor*

The graphical symbol of the fractional-order inductor is shown in Fig. 2.2, where β ($0 < \beta < 2$) and L_β are the order and the pseudo-inductance value of the fractional-order inductor. The fractional-order inductor can be divided into the active fractional-order inductor ($1 < \beta < 2$) and the passive fractional-order inductor ($0 < \beta < 1$) by

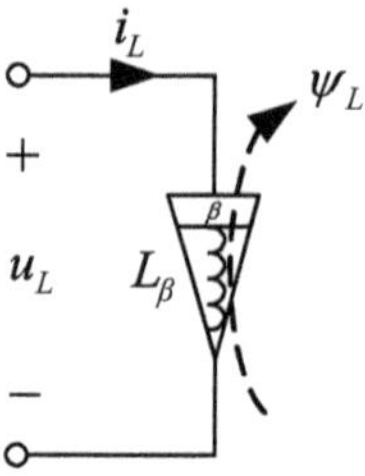

Fig. 2.2 Graphical symbol for fractional-order inductor

its order. The fractional-order inductor is a two-terminal electrical component, and its performance equation is

$$\Psi_L = f(i_L) \text{ or } i_L = f(\Psi_L) \tag{2.7}$$

where Ψ_L is the flux linkage, i_L is the current flowing through the fractional-order inductor.

According to Faraday electromagnetic laws, when the flux linkage Ψ_L changes with time, the induced electromotive force (e.m.f) of the fractional-order circuit will be generated [15].

Based on the definition of the fractional-order inductor, the relationship between the voltage and current can be described as [16]

$$u_L = \frac{\mathrm{d}^\beta \Psi_L}{\mathrm{d}t^\beta} = L_\beta \frac{\mathrm{d}^\beta i_L}{\mathrm{d}t^\beta} = \frac{L_\beta}{\Gamma(n-\alpha)} \frac{\mathrm{d}^n}{\mathrm{d}t^n} \int_0^t (t-\tau)^{n-1-\alpha} i_L(\tau)\mathrm{d}\tau \tag{2.8}$$

where n is the largest integer less than β, $\mathrm{d}^\beta/\mathrm{d}t^\beta$ is a fractional-order differential operator, the unit of the pseudo-inductance value L_β of the fractional-order inductor is Henry/(second)$^{1-\beta}$, which can also be expressed by H/s$^{1-\beta}$.

From (2.8), it can be noted that when $t > 0$, the voltage $u_L(t)$ of the fractional-order inductor not only depends on the initial state of the current $i_L(t)$ flowing through the fractional-order inductor, but also involves its historical process. Therefore, the voltage of the fractional-order inductor has memory characteristics [17].

By using the R-L definition of fractional-order calculus, (2.8) can be further presented as

$$\begin{aligned} i_L(t) &= \frac{1}{L_\beta} {}_0D_t^{-\beta} u_L(t) + \sum_{j=1}^{k} \left[{}_0D_t^{-\beta} i_L(t) \right]_{t=0} \frac{t^{\beta-j}}{\Gamma(\beta-j+1)} \\ &= \frac{1}{L_\beta \Gamma(\beta)} \int_0^t (t-\tau)^{\beta-1} u_L(\tau)\,\mathrm{d}\tau + \sum_{j=1}^{k} \left[{}_0D_t^{-\beta} i_L(t) \right]_{t=0} \frac{t^{\beta-j}}{\Gamma(\beta-j+1)} \end{aligned} \tag{2.9}$$

where k is the largest integer less than β. $k = 1$ and 2 represent the passive and active fractional-order inductors, respectively.

As can be seen from (2.9), the current $i_L(t)$ of the fractional-order inductor is determined by the initial state and the historical process of the voltage $u_L(t)$. Thus, the current $i_L(t)$ of the fractional-order inductor also has memory characteristics.

Combining (2.8) and (2.9), the instantaneous power absorbed by the fractional-order inductor is

$$p_L = u_L i_L = L_\beta i_L \frac{\mathrm{d}^\beta i_L}{\mathrm{d}t^\beta} = \frac{L_\beta i_L}{\Gamma(n-\beta)} \frac{\mathrm{d}^n}{\mathrm{d}t^n} \int_0^t (t-\tau)^{n-1-\beta} i_L(\tau)\,\mathrm{d}\tau \tag{2.10}$$

The total energy supplied to the fractional-order inductor is integral of the power from $-\infty$ to t, it can be expressed as

$$\begin{aligned} W_L(t) &= \int_{-\infty}^t p_L\,\mathrm{d}\xi = \int_{-\infty}^t L_\beta i_L \frac{\mathrm{d}^\beta i_L}{\mathrm{d}\xi^\beta}\mathrm{d}\xi \\ &= \int_{-\infty}^t \left[\frac{L_\beta i_L}{\Gamma(n-\beta)} \frac{\mathrm{d}^n}{\mathrm{d}t^n} \int_0^t (t-\tau)^{n-1-\beta} i_L(\tau)\mathrm{d}\tau \right] \mathrm{d}\xi \end{aligned} \tag{2.11}$$

Due to the memory characteristics of the current and voltage of the fractional-order inductor, it can be deduced from (2.11) that the total energy storage $W_L(t)$ also has memory characteristics.

2.1.3 Fractional-Order Mutual Inductor

The circuit model of the fractional-order mutual inductor is shown in Fig. 2.3, where $L_{\beta 1}$ and $L_{\beta 2}$ are the pseudo-inductance values of the fractional-order inductors of

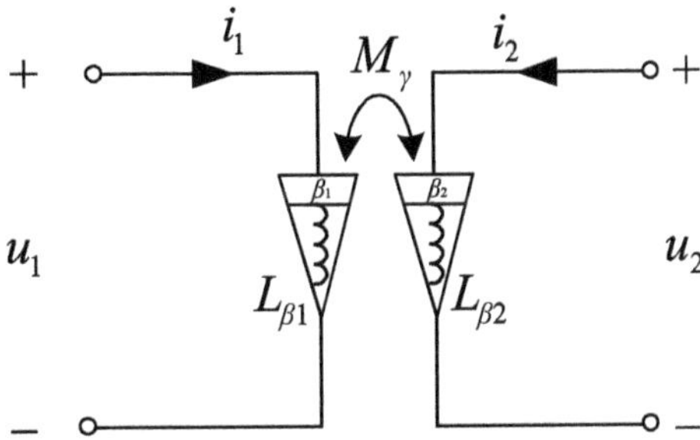

Fig. 2.3 Fractional-order mutual inductor

primary and secondary circuits, respectively. β_1 and β_2 are their orders. The mutual inductor is denoted as M_γ and the corresponding order is γ [18, 19].

Based on the KVL, the voltages at each fractional-order inductors of the primary and secondary circuits can be given by

$$\begin{cases} u_1 = L_{\beta 1}\, {}_0D_t^{\beta_1} i_{L1} + M_\gamma\, {}_0D_t^{\gamma} i_{L2} \\ u_2 = L_{\beta 2}\, {}_0D_t^{\beta_2} i_{L2} + M_\gamma\, {}_0D_t^{\gamma} i_{L1} \end{cases} \tag{2.12}$$

2.2 Fractional-Order Circuits

Fractional-order circuit is a specific form of the fractional-order electromagnetic field. It is generated by connecting the power supply (or signal source) and electrical equipment through intermediate links based on certain requirements. The power supply refers to the generator of electrical energy or electric signal, and the electrical equipment such as motors, light bulbs, etc., are called loads. Intermediate links include wires, fractional-order components, switches, measuring equipment, controllers, etc. Besides, the power supply can also be called excitation or input, and the voltage and current generated by the excitation are called the response or output.

Fractional-order circuit theory focuses on studying the electromagnetic phenomena in the circuit, so some physical quantities are required to described or measured corresponding properties. The physical quantities commonly used in fractional-order circuit theory are current, charge, voltage, electromotive force, magnetic flux, magnetic linkage, power, and energy. It is worth noting that the fractional-order circuits discussed in this book are not actual fractional-order circuits but their circuit models. The circuit models of actual fractional-order circuits are made up of ideal fractional-order circuit components connected with each other, and these ideal circuit components, which have precise mathematical definitions, are the smallest units of circuit models.

In addition, the fractional-order circuit is also called the fractional-order network. The first term is used to discuss some specific circuit issues, and the second term is often used in the study of general principles. In this book, fractional-order circuits and fractional-order networks are equivalent.

2.2.1 *Circuits of Fractional-Order Capacitors*

The series connection of fractional-order capacitors can be elaborated from two aspects: the same order and different orders [20].

The series connection of fractional-order capacitors with the same order is shown in Fig. 2.4, where u_{C1}, u_{C2}, …, and u_{Cn} are the voltages across the fractional-order

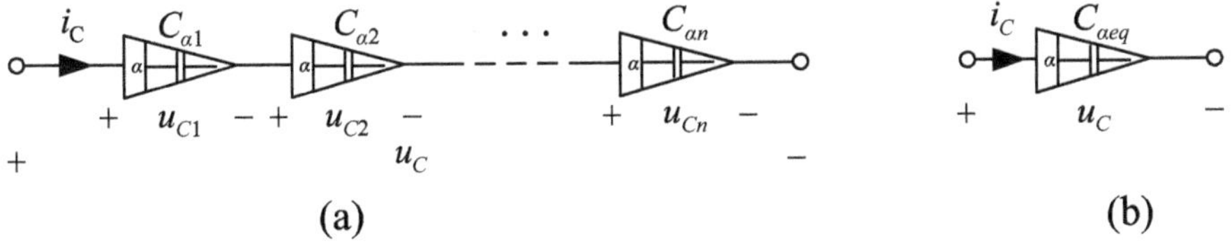

Fig. 2.4 Fractional-order capacitors in series: **a** series connection of fractional-order capacitors with the same order, **b** equivalent fractional-order capacitor

capacitors $C_{\alpha 1}$, $C_{\alpha 2}$, ..., $C_{\alpha n}$, respectively, i_C is the current flowing through the series branch, u_C is the total voltage across the series branch, $C_{\alpha eq}$ is the equivalent fractional-order capacitor of the series branch and the order of all fractional-order capacitors is α.

Since all the fractional-order capacitors are in series, the currents flowing through the capacitors are equal, $i_C = i_{C1} = i_{C2} = \cdots = i_{Cn}$. Therefore, according to the KVL, the total voltage can be deduced as

$$\begin{aligned} u_C &= u_{C1} + u_{C2} + \cdots + u_{Cn} \\ &= \frac{1}{C_{\alpha 1}} {}_0D_t^{-\alpha} i_{C1} + \frac{1}{C_{\alpha 2}} {}_0D_t^{-\alpha} i_{C2} + \cdots + \frac{1}{C_{\alpha n}} {}_0D_t^{-\alpha} i_{Cn} \\ &= \left(\frac{1}{C_{\alpha 1}} + \frac{1}{C_{\alpha 2}} + \cdots + \frac{1}{C_{\alpha n}} \right) {}_0D_t^{-\alpha} i_C = \frac{1}{C_{\alpha eq}} {}_0D_t^{-\alpha} i_C \end{aligned} \tag{2.13}$$

The equivalent fractional-order capacitor is defined as

$$\frac{1}{C_{\alpha eq}} = \frac{1}{C_{\alpha 1}} + \frac{1}{C_{\alpha 2}} + \cdots + \frac{1}{C_{\alpha n}} \tag{2.14}$$

The series connection of fractional-order capacitors with different orders is presented in Fig. 2.5, where α_1, α_2, ..., and α_n are the orders of the fractional-order capacitors $C_{\alpha 1}$, $C_{\alpha 2}$, ..., $C_{\alpha n}$, respectively.

Since the currents flowing through each fractional-order capacitor are equal, $i_C = i_{C1} = i_{C2} = \cdots = i_{Cn}$, the total voltage by using the KVL can be derived as

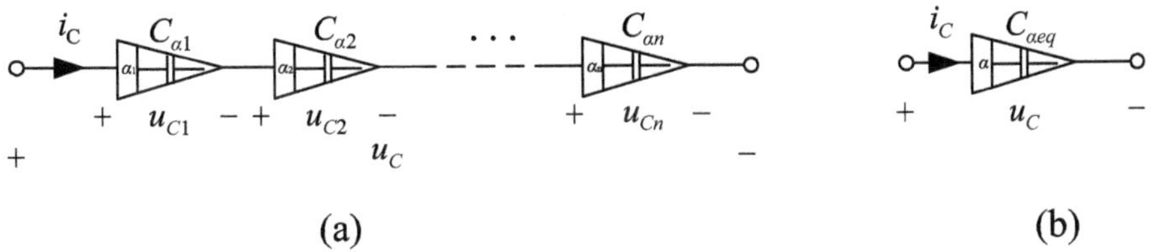

Fig. 2.5 Fractional-order capacitors in series: **a** series connection of fractional-order capacitors with different orders, **b** equivalent fractional-order capacitor

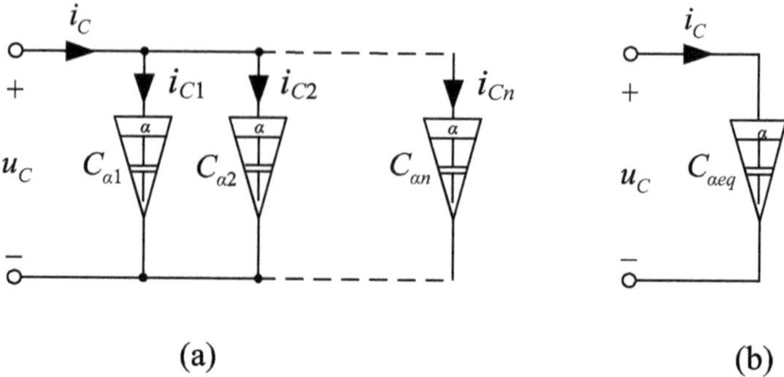

Fig. 2.6 Parallel connection of fractional-order capacitors: **a** parallel connection of fractional-order capacitors with the same order, **b** equivalent fractional-order capacitor

$$
\begin{aligned}
u_C &= u_{C1} + u_{C2} + \cdots + u_{Cn} = \frac{1}{C_{\alpha 1}}{}_0D_t^{-\alpha_1} i_{C1} + \frac{1}{C_{\alpha 2}}{}_0D_t^{-\alpha_2} i_{C2} + \cdots + \frac{1}{C_{\alpha n}}{}_0D_t^{-\alpha_n} i_{Cn} \\
&= \left(\frac{1}{C_{\alpha 1}}{}_0D_t^{-\alpha_1} + \frac{1}{C_{\alpha 2}}{}_0D_t^{-\alpha_2} + \cdots + \frac{1}{C_{\alpha n}}{}_0D_t^{-\alpha_n}\right) i_C
\end{aligned} \tag{2.15}
$$

Similarly, the parallel connection of fractional-order capacitors can be described from two aspects, including the same order and different orders.

Figure 2.6 shows the parallel connection of fractional-order capacitors with the same order, where $i_{C1}, i_{C2}, \ldots,$ and i_{Cn} are the currents flowing through the fractional-order capacitors $C_{\alpha 1}, C_{\alpha 2}, \ldots, C_{\alpha n}$, respectively, u_C is the voltage across the parallel branch, i_C is the total current flowing into all the parallel branches.

Since the voltages across each fractional-order capacitor are equal, and $u_C = u_{C1} = u_{C2} = \cdots = u_{Cn}$. Thus, based on the KCL, the total current is

$$
\begin{aligned}
i_C &= i_{C1} + i_{C2} + \cdots + i_{Cn} = C_{\alpha 1}\,{}_0D_t^{\alpha} u_{C1} + C_{\alpha 2}\,{}_0D_t^{\alpha} u_{C2} + \cdots + C_{\alpha n}\,{}_0D_t^{\alpha} u_{Cn} \\
&= (\alpha_{\alpha 1} + C_{\alpha 2} + \cdots + C_{\alpha n}){}_0D_t^{\alpha} u_C = C_{\alpha \mathrm{eq}}\,{}_0D_t^{\alpha} u_C
\end{aligned} \tag{2.16}
$$

Then the equivalent fractional-order capacitance can be derived as

$$
C_{\alpha eq} = C_{\alpha 1} + C_{\alpha 2} + \cdots + C_{\alpha n} \tag{2.17}
$$

The parallel connection of fractional-order capacitors with different orders is shown in Fig. 2.7. Then the total current is given by

$$
\begin{aligned}
i_C &= i_{C1} + i_{C2} + \cdots + i_{Cn} = C_{\alpha 1}\,{}_0D_t^{\alpha_1} u_{C1} + C_{\alpha 2}\,{}_0D_t^{\alpha_2} u_{C2} + \cdots + C_{\alpha n}\,{}_0D_t^{\alpha_n} u_{Cn} \\
&= \left(C_{\alpha 1}\,{}_0D_t^{\alpha_1} + C_{\alpha 2}\,{}_0D_t^{\alpha_2} + \cdots + C_{\alpha n}\,{}_0D_t^{\alpha_n}\right) u_C
\end{aligned} \tag{2.18}
$$

The series-parallel combination of fractional-order capacitors can be simplified by the above-mentioned principle of series connection and parallel connection of

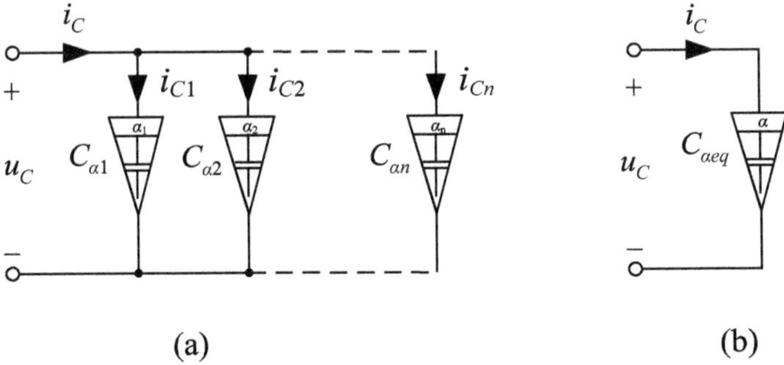

Fig. 2.7 Parallel connection of fractional-order capacitors: **a** parallel connection of fractional-order capacitors with different orders, **b** equivalent fractional-order capacitor

fractional-order capacitors [21]. Since the series-parallel combination of fractional-order capacitors with the same order is a particular case of different orders, without loss of generality, the series-parallel connection of three fractional-order capacitors with different orders is taken as an example for analysis, as shown in Fig. 2.8. $C_{\alpha 1}$ and $C_{\alpha 2}$ are connected in series, which can be equivalent to a fractional-order capacitor $C_{\alpha 12}$. Then, $C_{\alpha 12}$ and $C_{\alpha 3}$ are connected in parallel to form C_{α}. Here, u_{C1}, u_{C2} and u_{C3} are the voltages across the fractional-order capacitors $C_{\alpha 1}$, $C_{\alpha 2}$ and $C_{\alpha 3}$, u_{C12} and u_C are the voltages across the equivalent fractional-order capacitors $C_{\alpha 12}$ and C_{α}, respectively. i_{C1} is not only the current flowing through the series branch of $C_{\alpha 1}$ and $C_{\alpha 2}$ but also the current of $C_{\alpha 12}$. i_C is the total current flowing into the series-parallel circuit of $C_{\alpha 1}$, $C_{\alpha 2}$ and $C_{\alpha 3}$, which is also the current of C_{α}. i_{C3} is the current flowing through $C_{\alpha 3}$. α_1, α_2, α_2, α_{12} and α are the orders of $C_{\alpha 1}$, $C_{\alpha 2}$, $C_{\alpha 3}$, $C_{\alpha 12}$ and C_{α}, respectively.

Based on the series connection of $C_{\alpha 1}$ and $C_{\alpha 2}$, the currents flowing through the two fractional-order capacitors are equal, that is, $i_{C1} = i_{C2}$. According to KVL, the voltage u_C can be found as

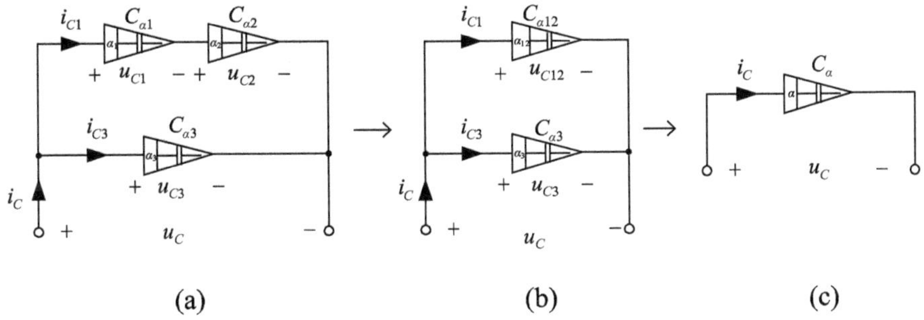

Fig. 2.8 Equivalent process of series-parallel combination of arbitrary-order fractional-order capacitors: **a** series-parallel connection of fractional-order capacitors, **b** parallel of fractional-order capacitors, **c** equivalent fractional-order capacitor

$$
\begin{aligned}
u_{C12} = u_{C1} + u_{C2} &= \frac{1}{C_{\alpha 1}} {}_0D_t^{-\alpha_1} i_{\alpha 1} + \frac{1}{C_{\alpha 2}} {}_0D_t^{-\alpha_2} i_{\alpha 2} \\
&= \left(\frac{1}{C_{\alpha 1}} {}_0D_t^{-\alpha_1} + \frac{1}{C_{\alpha 2}} {}_0D_t^{-\alpha_2} \right) i_{\alpha 1}
\end{aligned} \tag{2.19}
$$

Then, owing to the equal voltages across the equivalent fractional-order capacitor $C_{\alpha 12}$ and $C_{\alpha 3}$, the total current can be deduced by KCL as

$$
\begin{aligned}
i_C = i_{C1} + i_{C3} &= \frac{u_{C12}}{\left(\frac{1}{C_{\alpha 1}} {}_0D_t^{-\alpha_1} + \frac{1}{C_{\alpha 2}} {}_0D_t^{-\alpha_2} \right)} + C_{\alpha 3} {}_0D_t^{\alpha_3} u_{C3} \\
&= \left(\frac{C_{\alpha 1} C_{\alpha 2}}{C_{\alpha 2} {}_0D_t^{-\alpha_1} + C_{\alpha 1} {}_0D_t^{-\alpha_2}} + C_{\alpha 3} {}_0D_t^{\alpha_3} \right) u_C
\end{aligned} \tag{2.20}
$$

2.2.2 Circuits of Fractional-Order Inductors

The description of the series connection of fractional-order inductors can be divided into the following two cases: the same order and different orders.

The series connection of fractional-order inductors with the same order is shown in Fig. 2.9, where u_{L1}, u_{L2}, …, and u_{Ln} are the voltages across the fractional-order inductors $L_{\beta 1}$, $L_{\beta 2}$, …, and $L_{\beta n}$, respectively, u_L and i_L are the total voltage and the current of the series branch, $L_{\beta eq}$ is the equivalent fractional-order inductor. β is the order of all fractional-order inductors.

The currents flowing through each fractional-order inductor are equal, so $i_L = i_{L1} = i_{L2} = \cdots = i_{Ln}$, and the total voltage can be written by using KVL as

$$
\begin{aligned}
u_L = u_{L1} + u_{L2} + \cdots + u_{Ln} &= L_{\beta 1} {}_0D_t^{\beta} i_{L1} + L_{\beta 2} {}_0D_t^{\beta} i_{L2} + \cdots + L_{\beta n} {}_0D_t^{\beta} i_{Ln} \\
&= \left(L_{\beta 1} + L_{\beta 2} + \cdots + L_{\beta n} \right) {}_0D_t^{\beta} i_L = L_{\beta eq} {}_0D_t^{\beta} i_L
\end{aligned} \tag{2.21}
$$

Consequently, the equivalent fractional-order inductor as

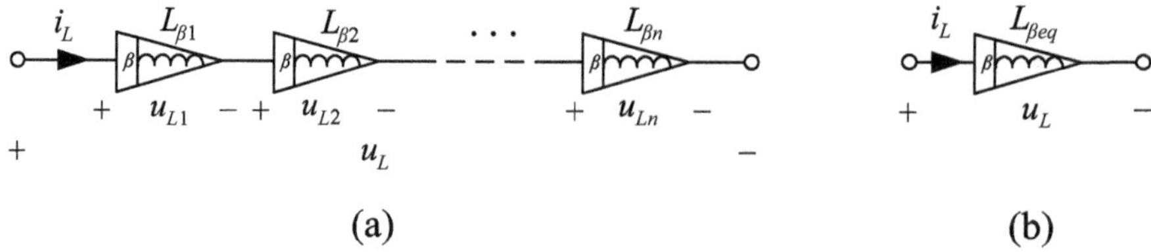

Fig. 2.9 Fractional-order inductors in series: **a** series connection of fractional-order inductors with the same order, **b** equivalent fractional-order inductor

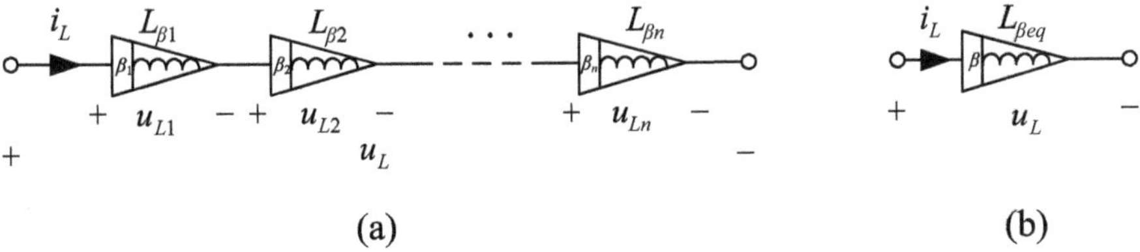

Fig. 2.10 Fractional-order inductors in series: **a** series connection of fractional-order inductors of different orders, **b** equivalent fractional-order inductor

$$L_{\beta eq} = L_{\beta 1} + L_{\beta 2} + \cdots + L_{\beta n} \tag{2.22}$$

The series connection of fractional-order inductors with different orders is depicted in Fig. 2.10, where β_1, β_2, ..., and β_n are the orders of $L_{\beta 1}$, $L_{\beta 2}$, ..., and $L_{\beta n}$, respectively.

The currents flowing through each fractional-order inductor are equal, so $i_L = i_{L1} = i_{L2} = \cdots = i_{Ln}$, and the total voltage can be obtained by using KVL as

$$\begin{aligned} u_L &= u_{L1} + u_{L2} + \cdots + u_{Ln} = L_{\beta 1}\,{}_0D_t^{\beta_1} i_{L1} + L_{\beta 2}\,{}_0D_t^{\beta_2} i_{L2} + \cdots + L_{\beta n}\,{}_0D_t^{\beta_n} i_{Ln} \\ &= \left(L_{\beta 1}\,{}_0D_t^{\beta_1} + L_{\beta 2}\,{}_0D_t^{\beta_2} + \cdots + L_{\beta n}\,{}_0D_t^{\beta_n}\right) i_L \end{aligned} \tag{2.23}$$

Likewise, the parallel connection of fractional-order inductors can also be described from two aspects, including the same order and different orders.

Figure 2.11 shows the parallel connection of fractional-order inductors of the same order. i_{L1}, i_{L2}, ..., and i_{Ln} are the currents flowing through the fractional-order inductors $L_{\beta 1}$, $L_{\beta 2}$, ..., and $L_{\beta n}$, respectively. i_L is the total current flowing into the parallel branch of $L_{\beta 1}$, $L_{\beta 2}$, ..., and $L_{\beta n}$. u_L is the voltage across the parallel branch.

The voltages across each fractional-order inductor are equal, so $u_L = u_{L1} = u_{L2} = \cdots = u_{Ln}$. Then the total current can be derived by using KCL as

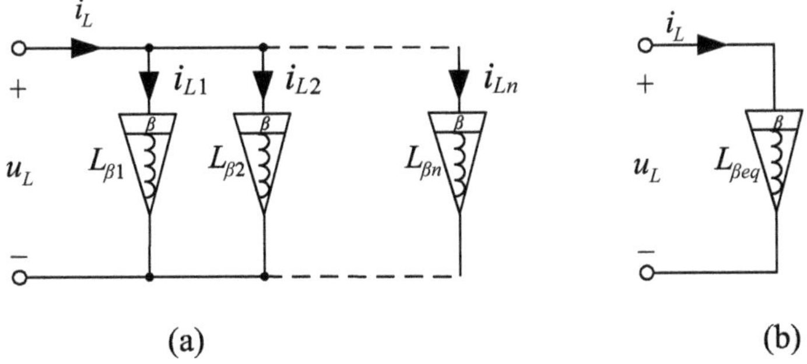

Fig. 2.11 Parallel connection of fractional-order inductors: **a** parallel connection of fractional-order inductors of the same order, **b** equivalent fractional-order inductor

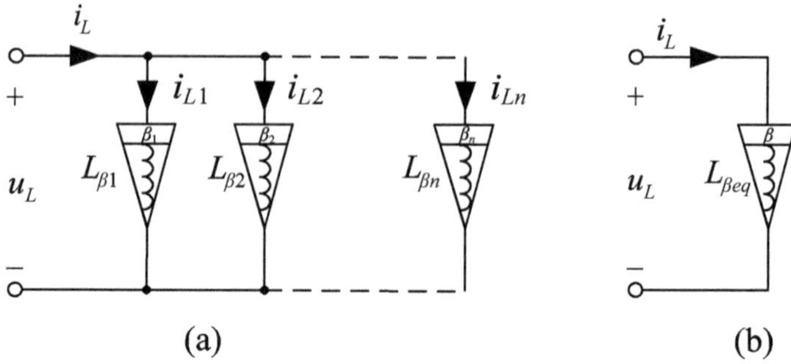

Fig. 2.12 Parallel connection of fractional-order inductors: **a** parallel connection of fractional-order inductors of different orders, **b** equivalent fractional-order inductor

$$
\begin{aligned}
i_L &= i_{L1} + i_{L2} + \cdots + i_{Ln} = \frac{1}{L_{\beta 1}} {}_0D_t^{-\beta} u_{L1} + \frac{1}{L_{\beta 2}} {}_0D_t^{-\beta} u_{L2} + \cdots + \frac{1}{L_{\beta n}} {}_0D_t^{-\beta} u_{Ln} \\
&= \left(\frac{1}{L_{\beta 1}} + \frac{1}{L_{\beta 2}} + \cdots + \frac{1}{L_{\beta n}} \right) {}_0D_t^{-\beta} u_L = \frac{1}{L_{\beta eq}} {}_0D_t^{-\beta} u_L
\end{aligned}
\tag{2.24}
$$

The equivalent fractional-order inductance can be written as

$$
\frac{1}{L_{\beta eq}} = \frac{1}{L_{\beta 1}} + \frac{1}{L_{\beta 2}} + \cdots + \frac{1}{L_{\beta n}}
\tag{2.25}
$$

Figure 2.12 presents the parallel connection of fractional-order inductors of different orders, where β_1, β_2, ..., and β_n are the orders of $L_{\beta 1}$, $L_{\beta 2}$, ..., and $L_{\beta n}$, respectively.

The voltages across each fractional inductor are equal, so $u_L = u_{L1} = u_{L2} = \cdots = u_{Ln}$. Then, according to KCL, the total current is given by

$$
\begin{aligned}
i_L &= i_{L1} + i_{L2} + \cdots + i_{Ln} = \frac{1}{L_{\beta 1}} {}_0D_t^{-\beta_1} u_{L1} + \frac{1}{L_{\beta 2}} {}_0D_t^{-\beta_2} u_{L2} + \cdots + \frac{1}{L_{\beta n}} {}_0D_t^{-\beta_n} u_{Ln} \\
&= \left(\frac{1}{L_{\beta 1}} {}_0D_t^{-\beta_1} + \frac{1}{L_{\beta 2}} {}_0D_t^{-\beta_2} + \cdots + \frac{1}{L_{\beta n}} {}_0D_t^{-\beta_n} \right) u_L
\end{aligned}
\tag{2.26}
$$

The series-parallel combination of fractional-order inductors can be simplified based on the above principle of the series and parallel combination of fractional-order inductors. Based on the fact that the series-parallel connection of fractional-order inductors of the same order is a particular case of different orders, without loss of generality. The series-parallel connection of three fractional-order inductors with different orders is taken as an example for analysis, as shown in Fig. 2.13. $L_{\beta 1}$, $L_{\beta 2}$ and $L_{\beta 3}$ are fractional-order inductors with different orders, and their orders are β_1, β_2 and β_3, respectively. $L_{\beta 1}$ and $L_{\beta 2}$ are connected in series with each other and in parallel with $L_{\beta 3}$, $L_{\beta 12}$ is the equivalent fractional-order inductor of the series branch

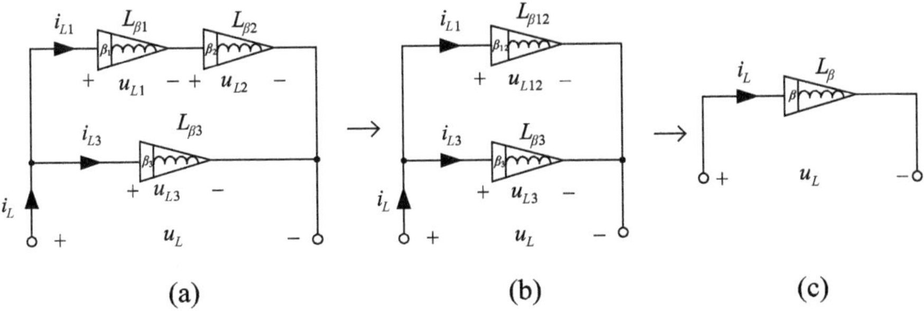

Fig. 2.13 Equivalent process of series-parallel combination of arbitrary-order fractional-order inductors: **a** series and parallel of fractional-order inductors, **b** parallel of fractional-order inductors, **c** equivalent fractional-order inductor

of $L_{\beta1}$ and $L_{\beta2}$, and its order is β_{12}. L_β is the equivalent fractional-order inductor of the entire series-parallel combination of $L_{\beta1}$, $L_{\beta2}$ and $L_{\beta3}$, and its order is β. u_{L1}, u_{L2} and u_{L3} are the voltages across $L_{\beta1}$, $L_{\beta2}$ and $L_{\beta3}$, u_{L12} and u_L are the voltages across $L_{\beta12}$ and L_β, respectively. i_{L1} is the current flowing through the series branch of $L_{\beta1}$ and $L_{\beta2}$, which is also the current of $L_{\beta12}$. i_L is the total current flowing into the series-parallel circuit of $L_{\beta1}$, $L_{\beta2}$ and $L_{\beta3}$, and it is also the current of L_β. i_{L3} is the current flowing through $L_{\beta3}$.

Based on the series connection of $L_{\beta1}$ and $L_{\beta2}$, the currents flowing through $L_{\beta1}$ and $L_{\beta2}$ are equal, that is, $i_{L1} = i_{L2}$. Then the total voltage across the series branch of $L_{\beta1}$ and $L_{\beta2}$ can be derived from KVL as

$$\begin{aligned} u_{L12} &= u_{L1} + u_{L2} = L_{\beta1}\,{}_0D_t^{\beta_1} i_{L1} + L_{\beta2}\,{}_0D_t^{\beta_2} i_{L2} \\ &= \left(L_{\beta1}\,{}_0D_t^{\beta_1} + L_{\beta2}\,{}_0D_t^{\beta_2}\right) i_{L1} \end{aligned} \tag{2.27}$$

The voltages across the equivalent fractional-order inductors $L_{\beta12}$ and $L_{\beta3}$ are equal, $u_{L12} = u_{L3}$. Applying KCL to the series-parallel circuit of Fig. 2.13, the total current can be calculated by

$$\begin{aligned} i_L = i_{L1} + i_{L3} &= \frac{u_{L12}}{L_{\beta1}\,{}_0D_t^{\beta1} + L_{\beta2}\,{}_0D_t^{\beta2}} + \frac{1}{L_{\beta3}}\,{}_0D_t^{-\beta_3} u_{L3} \\ &= \left(\frac{1}{L_{\beta1}\,{}_0D_t^{\beta1} + L_{\beta2}\,{}_0D_t^{\beta2}} + \frac{1}{L_{\beta3}}\,{}_0D_t^{-\beta_3}\right) u_L \end{aligned} \tag{2.28}$$

where $u_{L12} = u_L$.

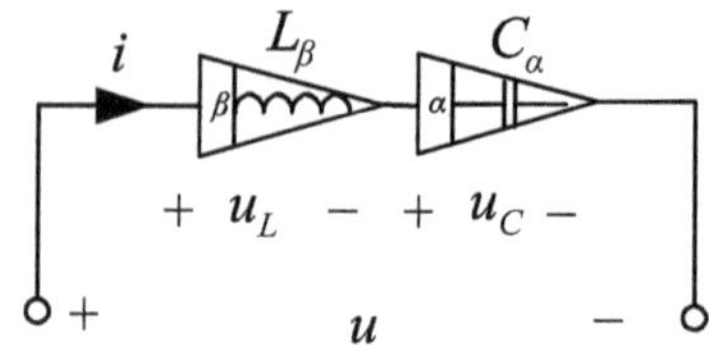

Fig. 2.14 Series connection of fractional-order inductor and capacitor

2.2.3 Circuits of Fractional-Order Inductors and Capacitors

Figure 2.14 describes the series circuit of a fractional-order inductor and a fractional-order capacitor. u_L and u_C are the voltages across the fractional-order inductor L_β and fractional-order capacitors C_α, β and α are the orders of L_β and C_α, respectively. u is the total voltage across the series circuit of L_β and C_α, i is the current flowing through the series circuit.

Owing to the series connection of L_β and C_α, the currents flowing through L_β and C_α, are equal, that is, $i = i_L = i_C$. Then the total voltage can be obtained by using KVL as

$$u = u_L + u_C = L_\beta \frac{\mathrm{d}^\beta i_L}{\mathrm{d}t^\beta} + \frac{1}{C_\alpha} {}_0D_t^{-\alpha} i_C = \left(L_\beta \, {}_0D_t^\beta + \frac{1}{C_\alpha} {}_0D_t^{-\alpha} \right) i \tag{2.29}$$

The derivation and analysis can be extended to the series combination with multiple fractional-order inductors and capacitors.

Figure 2.15 shows the parallel circuit of a fractional-order inductor and a fractional-order capacitor [22], where i_L and i_C are the currents flowing through L_β and C_α, respectively, i is the total current flowing into the parallel circuit of L_β and C_α, and u is the voltage across the parallel circuit.

The parallel connection of L_β and C_α, the voltage across L_β is equal to that of C_α, that is, $u = u_L = u_C$. Then, applying KVL to the parallel circuit of Fig. 2.15, the current i can be written as

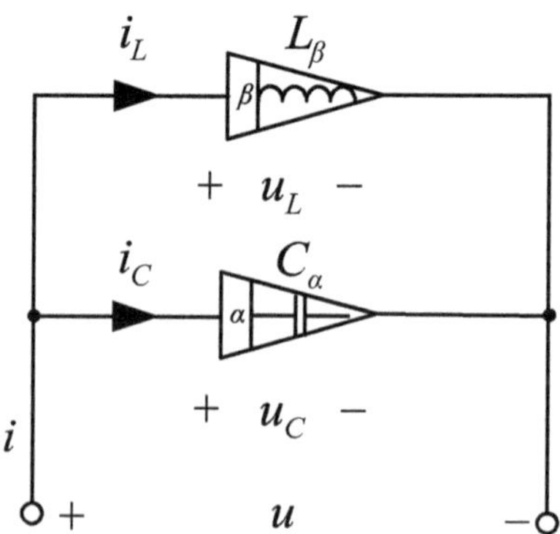

Fig. 2.15 Parallel connection of fractional-order inductor and capacitor

$$i = i_L + i_C = \frac{1}{L_\beta} {}_0D_t^{-\beta} u_L + C_\alpha \frac{\mathrm{d}^\alpha u_C}{\mathrm{d}t^\alpha} = \frac{1}{L_\beta} {}_0D_t^{-\beta} u + C_\alpha \, {}_0D_t^\alpha u \tag{2.30}$$

The above derivation and analysis can be generalized into the parallel combination with a large number of fractional-order inductors and fractional-order capacitors.

2.3 Summary

This chapter firstly introduces the definition and basic characteristics of the fractional-order inductor, fractional-order capacitor, and fractional-order mutual inductor are discussed. Then, the equivalent transformations of series, parallel, and series-parallel combinations of fractional-order inductors and capacitors are elaborated, which provides a theoretical basis for the study of the following chapters.

References

1. Freeborn TJ, Maundy B, Elwakil AS (2013) Measurement of supercapacitor fractional-order model parameters from voltage-excited step response. IEEE J Emerging Sel Top Circ Syst 3(3):367–376
2. Radwan AG, Salama KN (2012) Fractional-order RC and RL circuits. Circ Syst Signal Process 31(6):1901–1915
3. Rong C, Zhang B, Jiang Y (2020) Analysis of a fractional-order wireless power transfer system. IEEE Trans Circ Syst II Express Briefs 67(10):1755–1759
4. Jiang Y, Zhang B, Shu X (2020) Fractional-order autonomous circuits with order larger than one. J Adv Res 25:217–215
5. Jeffery D (2008) Capacitors, capacitance, and dielectrics, pp 2–6
6. Matsumori H, Urata K, Shimizu T et al (2018) Capacitor loss analysis method for power electronics converters. Microelectron Reliab 88–90:443–446
7. Westerlund S, Ekstam L (1994) Capacitor theory. IEEE Trans Dielectr Electr Insul 1(5):826–839
8. El-Khazali R, Nabeel T (2012) Realization of fractional-order capacitors and inductors. In: 5th-IFAC symposium on fractional differentiation and its applications, Nanjing, China
9. Radwan AG (2013) Resonance and quality factor of the $RL_\beta C_\alpha$ fractional circuit. IEEE J Emerging Sel Top Circ Syst 3(3):377–385
10. Jiang Y, Zhang B (2018) High-power fractional-order capacitor with $1 < \alpha < 2$ based on power converter. IEEE Trans Ind Electron 65(4):3157–3164
11. Monje C, Chen Y, Vinagre B et al (2010) Fractional order systems and control—Fundamentals and applications. Springer, London, pp 15–17
12. Radwan AG, Salama KN (2011) Passive and active elements using fractional $L_\beta C_\alpha$ circuit. IEEE Trans Circuits Syst I Regul Pap 58(10):2388–2397
13. Allagui A, Zhang D, Khakpour I, Elwakil A, Wang C (2019) Quantification of memory in fractional-order capacitors. J Phys D: Appl Phys 53(2)
14. Fouda ME, Elwakil AS, Radwan AG, Allagui A (2016) Power and energy analysis of fractional-order electrical energy storage devices. Energy 111:785–792
15. Álvarez Bel CM, Alcázar OM (2012) Electrical circuit theory. Universitat Politecnica de Valencia, Spain, pp 32–37

16. Tsirimokou G, Psychalinos C (2017) Design of CMOS analog integrated fractional-order circuits: applications in medicine and biology. Springer, pp 1–10
17. Petráš I, Chen Y, Coopmans C (2009) Fractional-order memristive systems. In: 2009 IEEE conference on emerging technologies & factory automation, Spain, pp 1–8
18. Soltan A, Radwan A, Soliman AM (2015) Fractional-order mutual inductance: analysis and design. Int J Circuit Theory Appl 44(1):85–97
19. Jakubowska-Ciszek A, Walczak J (2019) Modified method of identification of mutual fractional-order inductance. In: ITM web of conferences, vol 28, p 01025
20. Kartci A, Agambayev A, Herencsar N, Salama KN (2018) Series-, parallel-, and inter-connection of solid-state arbitrary fractional-order capacitors: theoretical study and experimental verification. IEEE Access 6:10933–10943
21. Tsirimokou G, Psychalinos C, Elwakil AS, Salama KN (2017) Experimental behavior evaluation of series and parallel connected constant phase elements. AEU-Int J Electron C 74:5–12
22. Chen P, He S (2013) Analysis of the fractional-order parallel tank circuit. J Circ Syst Comput 22(06)

Chapter 3
Theorems of Fractional-Order Circuits

In traditional integer-order circuit theory, there are many fundamental theorems, including Superposition theorem, Substitution theorem, Thevenin's theorem, Norton's theorem, Tellegen's theorem, Reciprocity theorem, Duality theorem, Compensation theorem, and Bisection theorem. According to Kirchhoff's laws and the basic characteristics of fractional-order components, these theorems can be extended to fractional-order circuits. This chapter analyzes fractional-order circuits and discusses the applicability of these theorems, which is helpful for exploring the basic characteristics of fractional-order circuits.

Learning theorems of fractional-order circuits (circuit theorems) can not only deepen the understanding of the internal rule of circuits but also help to solve circuit parameters and prove some conclusions. The theorem of fractional-order circuit is one of the main analysis methods of the circuit and it is the basic content of circuit theory. When the circuit theorems are used, it is not only necessary to clearly grasp the content of each theorem, but also to understand the connotation, usage conditions, and scope of the theorem.

Based on the fundamental theorems, the corresponding equations could be obtained to analyze the characteristics of the fractional-order circuits. The Superposition theorem, Substitution theorem, and Reciprocity theorem are often used together to analyze general fractional-order networks. The Bisection theorem is more convenient to study the symmetric two-port fractional-order networks [1–5].

3.1 Superposition Theorem

The most basic property of the fractional-order linear circuit is linearity, including additivity and homogeneity. The superposition theorem satisfies homogeneity and additivity, which is an important theorem for analyzing fractional-order linear circuits.

B. Zhang and X. Shu, *Fractional-Order Electrical Circuit Theory*, CPSS Power Electronics Series, https://doi.org/10.1007/978-981-16-2822-1_3

Superposition theorem states that, in any fractional-order linear circuit, which contains more than one power source, the voltage or current of any component is the algebraic sum of the voltages or currents of the component due to the independent operation of each independent power supply. However, if there is a controlled source in the circuit, it cannot be regarded as an excitation source to work alone, but can only be used as an element to describe the coupling relationship between branches, which should be retained in the electrical circuit [6].

As shown in Fig. 3.1a, there are two independent power supplies in the circuit. According to KCL and KVL, the circuit equation can be represented by i_2 and u_1 respectively as

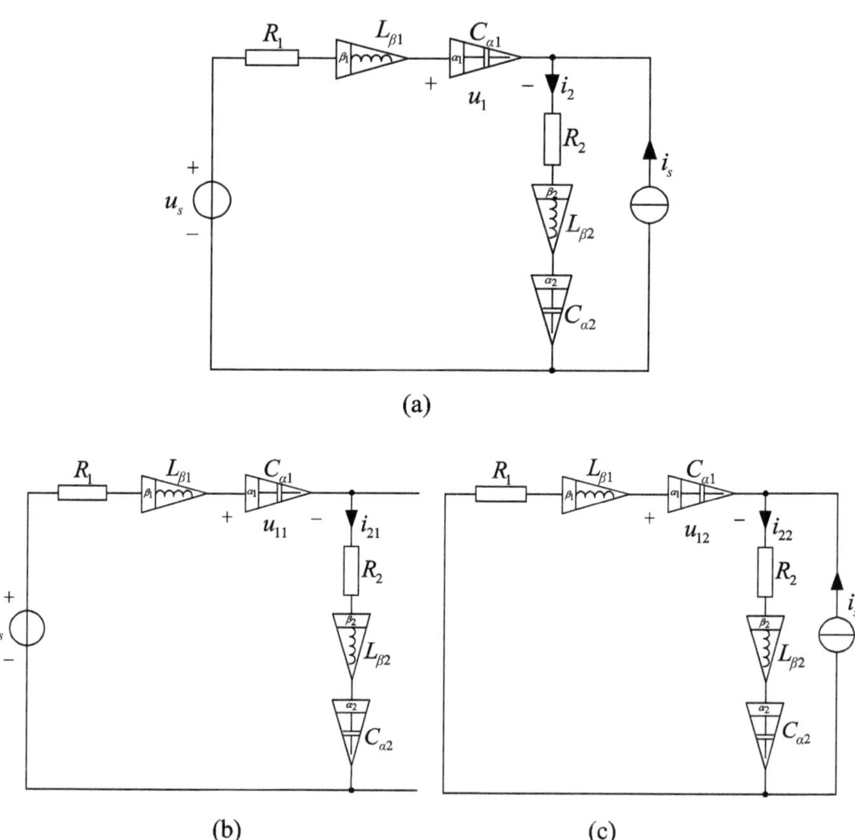

Fig. 3.1 Superposition theorem: **a** original circuit, **b** voltage source acting alone, **c** current source acting alone

$$u_s = R_1(i_2 - i_s) + L_{\beta 1}\, {}_0D_t^{\beta_1}(i_2 - i_s) + \frac{1}{C_{\alpha 1}}\, {}_0D_t^{-\alpha_1}(i_2 - i_s) + R_2 i_2 + L_{\beta 2}\, {}_0D_t^{\beta_2} i_2 + \frac{1}{C_{\alpha 2}}\, {}_0D_t^{-\alpha_2} i_2 \tag{3.1}$$

or

$$\begin{aligned} u_s &= R_1 C_{\alpha 1}\, {}_0D_t^{\alpha_1} u_1 + L_{\beta 1}\, {}_0D_t^{\beta_1}\left(C_{\alpha 1}\, {}_0D_t^{\alpha_1} u_1\right) + u_1 + R_2\left(C_{\alpha 1}\, {}_0D_t^{\alpha_1} u_1 + i_s\right) \\ &+ L_{\beta 2}\, {}_0D_t^{\beta_2}\left(C_{\alpha 1}\, {}_0D_t^{\alpha_1} u_1 + i_s\right) + \frac{1}{C_{\alpha 2}}\, {}_0D_t^{-\alpha_2}\left(C_{\alpha 1}\, {}_0D_t^{\alpha_1} u_1 + i_s\right) \\ &= R_1 C_{\alpha 1}\, {}_0D_t^{\alpha_1} u_1 + L_{\beta 1} C_{\alpha 1}\, {}_0D_t^{\alpha_1+\beta_1} u_1 + u_1 \\ &+ R_2 C_{\alpha 1}\, {}_0D_t^{\alpha_1} u_1 + R_2 i_s + L_{\beta 2} C_{\alpha 1}\, {}_0D_t^{\alpha_1+\beta_2} u_1 \\ &+ L_{\beta 2}\, {}_0D_t^{\beta_2} i_s + \frac{1}{C_{\alpha 2}}\, {}_0D_t^{-\alpha_2}\left(C_{\alpha 1}\, {}_0D_t^{\alpha_1} u_1 + i_s\right) \end{aligned} \tag{3.2}$$

where u_s and i_s are independent power sources, $L_{\beta 1}$ and $L_{\beta 2}$ are fractional-order inductors, $C_{\alpha 1}$ and $C_{\alpha 2}$ are fractional-order capacitors, R_1 and R_2 are resistors. u_1 is the voltage across $C_{\alpha 1}$, and i_2 is the current flowing through $L_{\beta 2}$, which are generated by all power sources operating simultaneously.

Then, considering the case that independent voltage sources work alone, as shown in Fig. 3.1b, the circuit equations can be written as

$$u_s = R_1 i_{21} + L_{\beta 1}\, {}_0D_t^{\beta_1} i_{21} + \frac{1}{C_{\alpha 1}}\, {}_0D_t^{-\alpha_1} i_{21} + R_2 i_{21} + L_{\beta 2}\, {}_0D_t^{\beta_2} i_{21} + \frac{1}{C_{\alpha 2}}\, {}_0D_t^{-\alpha_2} i_{21} \tag{3.3}$$

$$\begin{aligned} u_s &= R_1 C_{\alpha 1}\, {}_0D_t^{\alpha_1} u_{11} + L_{\beta 1} C_{\alpha 1}\, {}_0D_t^{\alpha_1+\beta_1} u_{11} + u_{11} \\ &+ R_2 C_{\alpha 1}\, {}_0D_t^{\alpha_1} u_{11} + L_{\beta 2} C_{\alpha 1}\, {}_0D_t^{\alpha_1+\beta_2} u_{11} \\ &+ \frac{1}{C_{\alpha 2}}\, {}_0D_t^{-\alpha_2}\left(C_{\alpha 1}\, {}_0D_t^{\alpha_1} u_1\right) \end{aligned} \tag{3.4}$$

where u_{11} is the voltage across $C_{\alpha 1}$, and i_{21} is the current flowing through $L_{\beta 2}$, which are generated by the independent voltage source u_s acting alone.

From Fig. 3.1c, the current source acts alone, and the following equations can be acquired as

$$\begin{aligned} 0 &= R_1(i_{22} - i_s) + L_{\beta 1}\, {}_0D_t^{\beta_1}(i_{22} - i_s) \\ &+ \frac{1}{C_{\alpha 1}}\, {}_0D_t^{-\alpha_1}(i_{22} - i_s) + R_2 i_{22} + L_{\beta 2}\, {}_0D_t^{\beta_2} i_{22} \\ &+ \frac{1}{C_{\alpha 2}}\, {}_0D_t^{-\alpha_2} i_{22} \end{aligned} \tag{3.5}$$

$$\begin{aligned}0 &= R_1 C_{\alpha 1}\, {}_0D_t^{\alpha_1} u_{12} + L_{\beta 1} C_{\alpha 1}\, {}_0D_t^{\alpha_1+\beta_1} u_{12} + u_{12} \\ &+ R_2 C_{\alpha 1}\, {}_0D_t^{\alpha_1} u_{12} + L_{\beta 2} C_{\alpha 1}\, {}_0D_t^{\alpha_1+\beta_2} u_{12} \\ &+ R_2 i_s + L_{\beta 2}\, {}_0D_t^{\beta_2} i_s + \frac{1}{C_{\alpha 2}}\, {}_0D_t^{-\alpha_2}\left(C_{\alpha 1}\, {}_0D_t^{\alpha_1} u_{12} + i_s\right)\end{aligned} \tag{3.6}$$

where u_{12} is the voltage across $C_{\alpha 1}$, and i_{22} is the current flowing through $L_{\beta 2}$, which are generated by the independent current source i_s acting alone.

Combining (3.3) with (3.5), (3.7) can be obtained as

$$\begin{aligned}u_s &= R_1(i_{21} + i_{22} - i_s) + L_{\beta 1}\, {}_0D_t^{\beta_1}(i_{21} + i_{22} - i_s) + \frac{1}{C_{\alpha 1}}\, {}_0D_t^{-\alpha_1}(i_{21} + i_{22} - i_s) \\ &+ R_2(i_{21} + i_{22}) + L_{\beta 2}\, {}_0D_t^{\beta_2}(i_{21} + i_{22}) + \frac{1}{C_{\alpha 2}}\, {}_0D_t^{-\alpha_2}(i_{21} + i_{22})\end{aligned} \tag{3.7}$$

According to (3.1) and (3.7), the branch current can be calculated by

$$i_2 = i_{21} + i_{22} \tag{3.8}$$

In the same way, from (3.4) and (3.6), there is

$$\begin{aligned}u_s &= R_1 C_{\alpha 1}\, {}_0D_t^{\alpha_1}(u_{11} + u_{12}) + L_{\beta 1} C_{\alpha 1}\, {}_0D_t^{\alpha_1+\beta_1}(u_{11} + u_{12}) + (u_{11} + u_{12}) \\ &+ R_2 C_{\alpha 1}\, {}_0D_t^{\alpha_1}(u_{11} + u_{12}) + R_2 i_s + L_{\beta 2} C_{\alpha 1}\, {}_0D_t^{\alpha_1+\beta_2}(u_{11} + u_{12}) \\ &+ L_{\beta 2}\, {}_0D_t^{\beta_2} i_s + \frac{1}{C_{\alpha 2}}\, {}_0D_t^{-\alpha_2}\left[C_{\alpha 1}\, {}_0D_t^{\alpha_1}(u_{11} + u_{12}) + i_s\right]\end{aligned} \tag{3.9}$$

Similarly, (3.10) can be deduced as

$$u_1 = u_{11} + u_{12} \tag{3.10}$$

From (3.8) and (3.10), it can be revealed that the branch voltage and current of the fractional-order linear circuit have a linear relationship with the input excitation, which demonstrates that the superposition theorem is valid. Therefore, the superposition theorem is applicable to fractional-order linear circuits. It should be noted that in each superimposed branch when an independent power supply acts alone, the other power supplies need to be set to zero. That is, the independent voltage source is replaced by the short circuit, and the independent current source is replaced by the open circuit, while the connection of other components remains unchanged, and the controlled sources remain in each branch.

Moreover, when applying the superposition theorem, it is necessary to pay attention to whether the reference direction of the branch voltage and current generated by all independent power supplies working independently is consistent with the reference direction generated by all the power supplies together acting simultaneously. If it is consistent, the meaning of the superposition is addition. Nonetheless, if it is opposite, then the meaning of the superposition is subtraction. It is also worth noting that the power does not satisfy the superposition theorem because the power is a quadratic function of voltage or current, which does not satisfy homogeneity and additivity. When solving the power, the superposition theorem must first be used to find the voltage and current of the branch, and then the power can be calculated based on the product of the branch voltage and current.

By the way, the homogeneity theorem is also an important theorem of the linear circuit. When all independent power supplies (i.e., independent voltage and current sources) increase or decrease K times simultaneously (where K is a real constant), the circuit response (i.e., voltage and current) will also increase or decrease K times, which can easily be derived from the superposition theorem. It should be pointed out that all power supplies must be increased or decreased by K times simultaneously. Otherwise, the homogeneity theorem is not applicable.

In conclusion, the superposition theorem is an important theorem for analyzing fractional-order linear circuits, which contains two basic properties of the linear circuit: homogeneity and additivity [7, 8].

3.2 Substitution Theorem

Substitution theorem states that, in a fractional-order circuit where all branch voltages and currents have unique solutions, if the voltage u_k or current i_k of a certain branch k is known, the branch can be replaced by a voltage source with a voltage of $u_s = u_k$ or a current source with a current of $i_s = i_k$, and the voltage and current values of other parts of the network remain unchanged after replacement.

As shown in Fig. 3.2, N_A and N_B are two one-port networks containing fractional-order components. Figure 3.2a shows the original circuit, Fig. 3.2b depicts the equivalent circuit in which N_B is replaced by a voltage source $u_s = u_k$, and Fig. 3.2c presents the equivalent circuit where N_B is replaced by a current source $i_s = i_k$. Besides, the voltage and current in Fig. 3.2a–c are the same.

Taking the voltage source instead of N_B as an example, Fig. 3.2d gives proof of the substitution theorem in the fractional-order circuit. Firstly, two voltage sources with opposite voltage directions and the same amplitude u_s are connected in series between terminals a and c of network N_B, which does not affect the voltages and currents in the fractional-order network N_A and N_B. Let $u_s = u_k$, then the voltage between b and d is $u_{bd} = 0$, which means that the two points b and d could be short-circuited. Therefore, the same circuit as shown in Fig. 3.2b can be obtained, in which the network N_B is replaced by a voltage source $u_s = u_k$. Similarly, Fig. 3.2c shows the proof of a current source replacing N_B. It can be proved by connecting two current sources with opposite current directions and the same amplitude i_s between terminals a and d.

Considering an application example of the substitution theorem shown in Fig. 3.3 with the given resistors R_1, R_2 and R_3, fractional-order inductors $L_{\beta 1}$, $L_{\beta 2}$ and $L_{\beta 3}$, fractional-order capacitors $C_{\alpha 1}$, $C_{\alpha 2}$ and $C_{\alpha 3}$, voltage sources u_{s1} and u_{s2}, the circuit

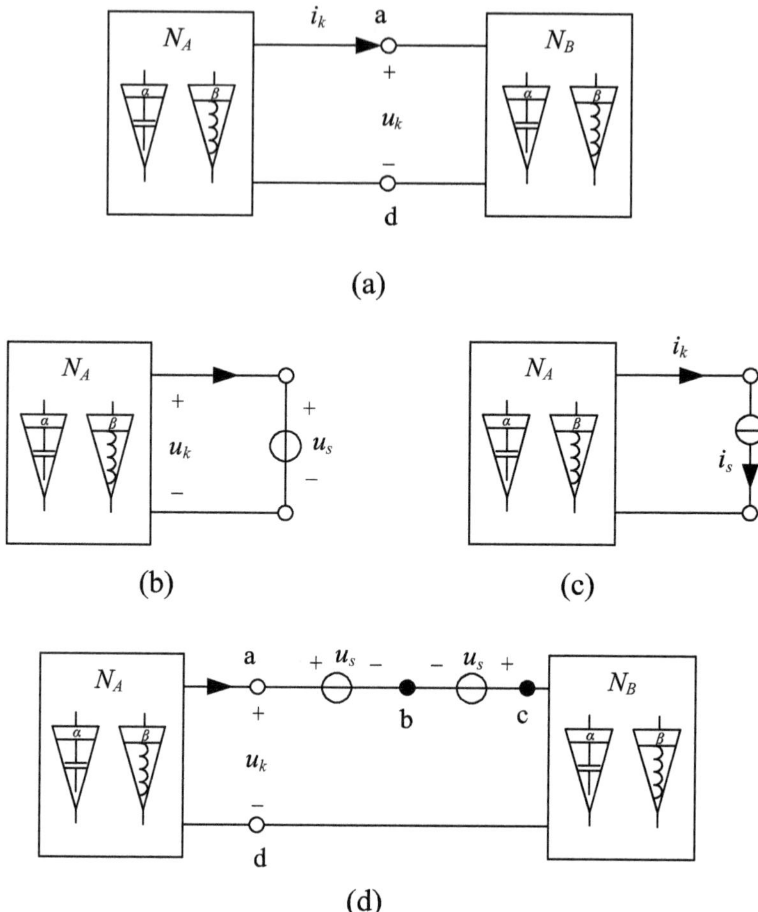

Fig. 3.2 Substitution theorem and its proof: **a** original circuit, **b** voltage source instead of N_B, **c** current source instead of N_B, **d** Proof circuit

equation can be written as

$$\begin{cases} u_{s1} = u_3 + R_1 i_1 + L_{\beta 1}\, {}_0D_t^{\beta_1} i_1 + \dfrac{1}{C_{\alpha 1}}\, {}_0D_t^{-\alpha_1} i_1 \\ u_3 = u_{s2} + R_3 i_3 + L_{\beta 3}\, {}_0D_t^{\beta_3} i_3 + \dfrac{1}{C_{\alpha 3}}\, {}_0D_t^{-\alpha_3} i_3 \\ u_3 = R_2 i_2 + L_{\beta 2}\, {}_0D_t^{\beta_2} i_2 + \dfrac{1}{C_{\alpha 2}}\, {}_0D_t^{-\alpha_2} i_2 \end{cases} \tag{3.11}$$

In Fig. 3.3b, if the voltage of the branch 3 is known to be u_3, the whole branch 3 can be replaced by a voltage source $u_s = u_3$, and its external circuit still satisfies the

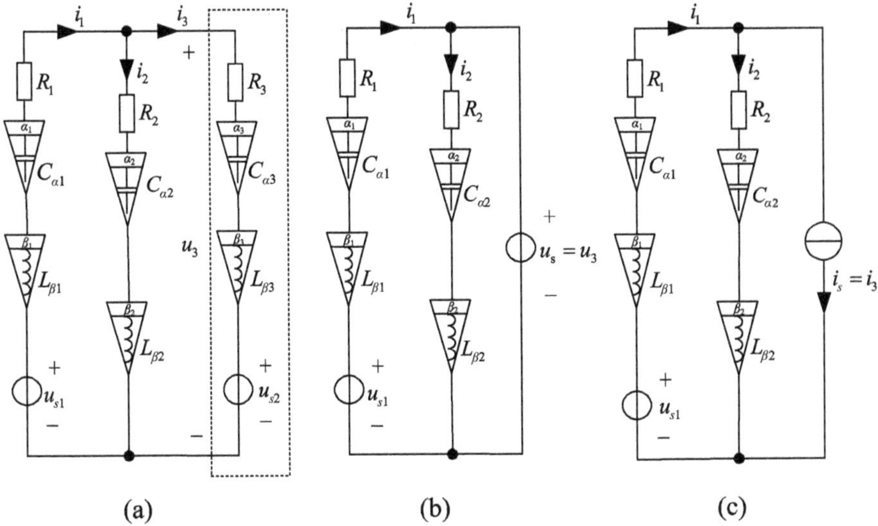

Fig. 3.3 An application example of substitution theorem: **a** original circuit, **b** voltage source alternative circuit, **c** current source alternative circuit

relationship of (3.11). Therefore, the currents of branches 1 and 2 can be represented by the voltage u_3 of branch 3.

For Fig. 3.3c, there is

$$\begin{cases} u_s = R_1 i_1 + L_{\beta 1}\, {}_0D_t^{\beta_1} i_1 + \dfrac{1}{C_{\alpha 1}}\, {}_0D_t^{-\alpha_1} i_1 + R_2 i_2 + L_{\beta 2}\, {}_0D_t^{\beta_2} i_2 + \dfrac{1}{C_{\alpha 2}}\, {}_0D_t^{-\alpha_2} i_2 \\ i_1 = i_2 + i_3 \end{cases} \tag{3.12}$$

In Fig. 3.3c, if the current of the branch 3 is known to be i_3, the whole branch 3 can be replaced by a current source $i_s = i_3$, and its external circuit still satisfies the relationship of (3.12). Therefore, the currents of branches 1 and 2 can be represented by the current i_3 of branch 3.

When using the substitution theorem, the following requirements should be noted. In general, to determine whether a branch is replaced by a voltage source or a current source, it is necessary to combine other network theorems adopted in the next step. After the replacement, whether the network is linear or nonlinear, the network must have a unique solution as before the replacement. Besides, there is no coupling relationship between the replaced branch and other branches. It is not allowed to erase the original coupling relationship after replacing the branch with a voltage source $u_s = u_k$ or a current source $i_s = i_k$. For instance, if there is a controlled source in N_A, it is controlled by the parameters in N_B, then this control relationship cannot be expressed after replacing N_B, so N_B cannot be replaced in this case.

The substitution theorem is a widely used theorem, which can be applied to fractional-order linear and nonlinear circuits. It is often used to simplify fractional-order circuits so that the circuit can be easily analyzed or calculated.

3.3 Thevenin's and Norton's Theorems

Thevenin's theorem is one of the most important theorems in fractional-order circuits. Its importance is to help us analyze the problems of particular branches in the circuit and give us a general understanding of linear one-port active networks (i.e., a linear one-port network with independent power supplies). It states that any fractional-order linear one-port network with the independent power sources and unique solutions shown in Fig. 3.4a can be represented by an equivalent circuit consisting of a voltage source in series with an impedance, as shown in Fig. 3.4b. In the equivalent circuit, the voltage source is equal to the voltage across the open terminals of the fractional-order network, and the equivalent impedance is equal to the input impedance of the fractional-order one-port network in which the independent power sources are set to be zero.

Norton's theorem is an extension of Thevenin's theorem. There is a strong duality between these two theorems. Norton's theorem states that any fractional-order linear one-port network with the independent power sources and unique solutions can be

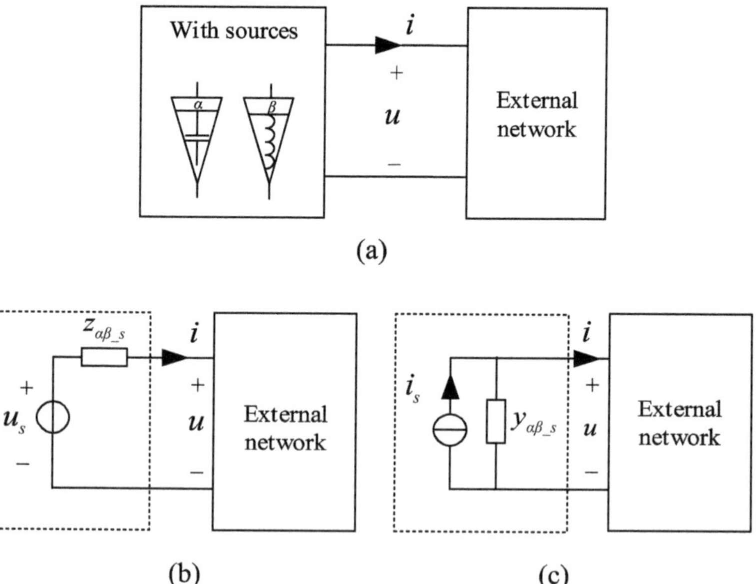

Fig. 3.4 Thevenin's and Norton's theorems of a fractional-order one-port active network: **a** original circuit, **b** Thevenin's circuit, **c** Norton's circuit

replaced by an equivalent circuit, which consists of a current source in parallel with an admittance, as shown in Fig. 3.4c. In the equivalent circuit, the current source is equal to the short-circuit current of the fractional-order one-port network. The admittance is equal to the input admittance of the fractional-order one-port network where the independent power sources are set to be zero.

When Thevenin's or Norton's theorems are used, it is necessary to note that the fractional-order network should be linear and has unique solutions. As for its external network, it can be linear and nonlinear. Besides, there is no coupling relationship between the internal and external branches of the fractional-order one-port active network. If a coupling relationship exists, the fractional-order linear one-port active network can be simplified as a Thevenin's branch or Norton's branch, only after converting the coupling variables into a representation of port voltages or currents. However, since this transformation generally requires the relationship between the internal and external branches, the applicability of the equivalent Thevenin's or Norton's branch after such transformation has certain limitations. It is only applicable to fractional-order networks whose external network structure and parameters remain unchanged after transformation.

To verify the applicability of Thevenin's and Norton's theorems in fractional-order circuits, Fig. 3.5 is an example. Here, Fig. 3.5b is Thevenin's equivalent circuit and Fig. 3.5c is Norton's equivalent circuit.

As can be seen from Fig. 3.6a, the circuit equations can be written as

$$\begin{cases} u_s = R_1 i_3 + L_\beta\, {}_0D_t^\beta i_3 + \dfrac{1}{C_\alpha}\, {}_0D_t^{-\alpha} i_3 + R_2 i_3 \\ u_{oc} = \dfrac{1}{C_\alpha}\, {}_0D_t^{-\alpha} i_3 + R_2 i_3 \end{cases} \tag{3.13}$$

Then, the open-circuit voltage u_{oc} can be derived as

$$u_{oc} = \frac{{}_0D_t^{-\alpha} + R_2 C_\alpha}{R_1 C_\alpha + C_\alpha L_\beta\, {}_0D_t^\beta + {}_0D_t^{-\alpha} + R_2 C_\alpha} u_s \tag{3.14}$$

Based on KVL and KCL, the equivalent impedance $z_{\alpha\beta\_eq}$ can be calculated by

$$\begin{cases} u_0 = R_1 i_L + L_\beta\, {}_0D_t^\beta i_L \\ u_0 = R_2 i_C + \dfrac{1}{C_\alpha}\, {}_0D_t^{-\alpha} i_C \\ i_0 = i_L + i_C \end{cases} \tag{3.15}$$

Thus, the equivalent impedance $z_{\alpha\beta\_eq}$ can be expressed as

$$z_{\alpha\beta\_eq} = \frac{u_0}{i_0} = \frac{\left(R_1 + L_\beta\, {}_0D_t^\beta\right)\left(R_2 C_\alpha + {}_0D_t^{-\alpha}\right)}{R_2 C_\alpha + {}_0D_t^{-\alpha} + C_\alpha\left(R_1 + L_\beta\, {}_0D_t^\beta\right)} \tag{3.16}$$

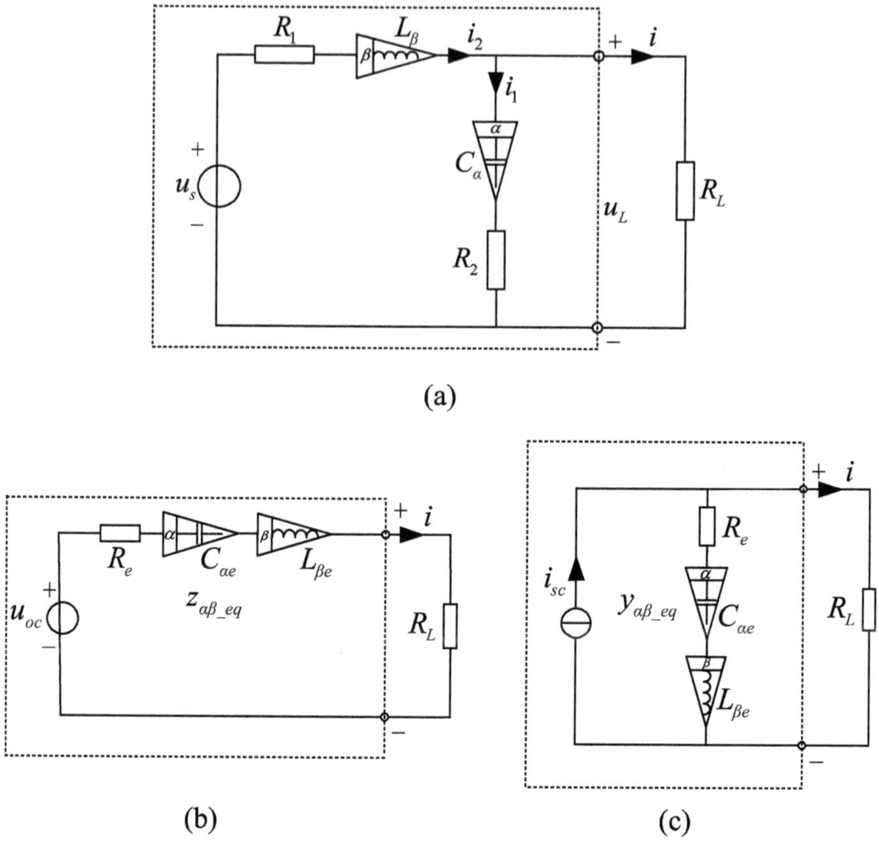

(a)

(b)

(c)

Fig. 3.5 An application example of the Thevenin's and Norton's theorems: **a** original circuit, **b** Thevenin's theorem circuit, **c** Norton's theorem circuit

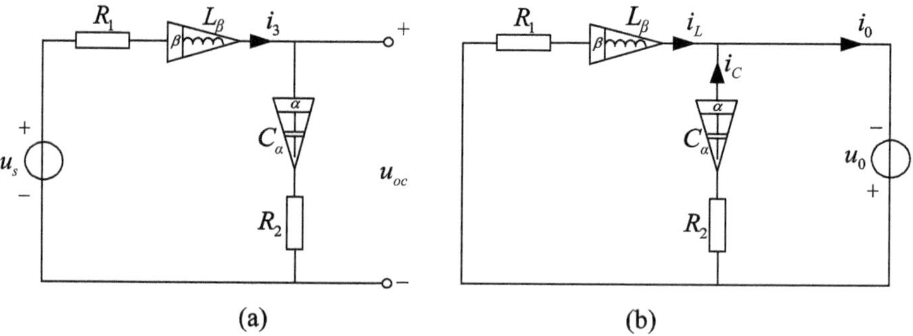

(a) (b)

Fig. 3.6 Proof process of Thevenin's and Norton's theorems. **a** open-circuit voltage diagram. **b** equivalent impedance diagram

According to Thevenin's equivalent circuit shown in Fig. 3.5b, the current i flowing through R_{L} can be calculated as

$$
\begin{aligned}
i &= \frac{u_{oc}}{z_{\alpha\beta\_eq} + R_L} = \frac{\frac{{}_0D_t^{-\alpha} + R_2C_\alpha}{R_1C_\alpha + C_\alpha L_\beta\, {}_0D_t^\beta + {}_0D_t^{-\alpha} + R_2C_\alpha} u_s}{\frac{\left(R_1 + L_\beta\, {}_0D_t^\beta\right)\left(R_2C_\alpha + {}_0D_t^{-\alpha}\right)}{R_2C_\alpha + {}_0D_t^{-\alpha} + C_\alpha\left(R_1 + L_\beta\, {}_0D_t^\beta\right)} + R_L} \\
&= \frac{\left({}_0D_t^{-\alpha} + R_2C_\alpha\right)u_s}{\left(R_1 + L_\beta\, {}_0D_t^\beta\right)\left(R_2C_\alpha + {}_0D_t^{-\alpha}\right) + R_L\left(R_1C_\alpha + C_\alpha L_\beta\, {}_0D_t^\beta + {}_0D_t^{-\alpha} + R_2C_\alpha\right)}
\end{aligned}
\tag{3.17}
$$

Then, the current obtained by the equivalent circuit can be brought back to the original circuit to verify its correctness, that is

$$
\begin{aligned}
u_L &= R_L i \\
&= \frac{\left({}_0D_t^{-\alpha} + R_2C_\alpha\right)R_L u_s}{\left(R_1 + L_\beta\, {}_0D_t^\beta\right)\left(R_2C_\alpha + {}_0D_t^{-\alpha}\right) + R_L\left(R_1C_\alpha + C_\alpha L_\beta\, {}_0D_t^\beta + {}_0D_t^{-\alpha} + R_2C_\alpha\right)}
\end{aligned}
\tag{3.18}
$$

Besides, the load voltage u_L can be written by KVL as

$$
u_L = R_2 i_1 + \frac{1}{C_\alpha}\, {}_0D_t^{-\alpha} i_1 \tag{3.19}
$$

The current i_1 can be expressed as

$$
\begin{aligned}
i_1 &= \frac{u_L}{R_2 + \frac{1}{C_\alpha}\, {}_0D_t^{-\alpha}} \\
&= \frac{C_\alpha R_L u_s}{\left(R_1 + L_\beta\, {}_0D_t^\beta\right)\left(R_2C_\alpha + {}_0D_t^{-\alpha}\right) + R_L\left(R_1C_\alpha + C_\alpha L_\beta\, {}_0D_t^\beta + {}_0D_t^{-\alpha} + R_2C_\alpha\right)}
\end{aligned}
\tag{3.20}
$$

Hence, by combining (3.17) and (3.20), i_1 can be given by

$$
\begin{aligned}
i_2 &= i + i_1 \\
&= \frac{\left(C_\alpha R_L + R_2C_\alpha + {}_0D_t^{-\alpha}\right)u_s}{\left(R_1 + L_\beta\, {}_0D_t^\beta\right)\left(R_2C_\alpha + {}_0D_t^{-\alpha}\right) + R_L\left(R_1C_\alpha + C_\alpha L_\beta\, {}_0D_t^\beta + {}_0D_t^{-\alpha} + R_2C_\alpha\right)}
\end{aligned}
\tag{3.21}
$$

and u_s can be derived as

$$
\begin{aligned}
u_s &= R_1 i_2 + L_\beta\, {}_0D_t^\beta i_2 + u_L \\
&= \frac{\left(R_1 + L_\beta\, {}_0D_t^\beta\right)\left(C_\alpha R_L + R_2 C_\alpha + {}_0D_t^{-\alpha}\right)u_s}{\left(R_1 + L_\beta\, {}_0D_t^\beta\right)\left(R_2 C_\alpha + {}_0D_t^{-\alpha}\right) + R_L\left(R_1 C_\alpha + C_\alpha L_\beta\, {}_0D_t^\beta + {}_0D_t^{-\alpha} + R_2 C_\alpha\right)} \\
&\quad + \frac{\left({}_0D_t^{-\alpha} + R_2 C_\alpha\right) R_L u_s}{\left(R_1 + L_\beta\, {}_0D_t^\beta\right)\left(R_2 C_\alpha + {}_0D_t^{-\alpha}\right) + R_L\left(R_1 C_\alpha + C_\alpha L_\beta\, {}_0D_t^\beta + {}_0D_t^{-\alpha} + R_2 C_\alpha\right)}
\end{aligned}
\tag{3.22}
$$

Therefore, Thevenin's theorem is valid in fractional-order circuits. Similarly, the applicability of Norton's theorem in fractional-order circuits can also be verified.

Due to the existence of the controlled source and fractional-order components, the input impedance may be zero or infinite, which is caused by the negative resistance effect of the controlled source and the fractional-order components. To solve this problem, there are three general methods, which are.

1. Network equivalent transformation method. The main core of this method is to obtain the input impedance $z_{\alpha\beta\_in}$ of the port by the series–parallel connection of resistors and fractional-order components, the Y/Δ transformation, and Thevenin's-Norton's equivalent transformation and simplification of the branch containing the controlled source.
2. External power source method. A current source i_s is used to excite the port of the one-port fractional-order relaxation network, and the input impedance is derived by calculating the ratio of the port voltage u to the port current i_s, as shown in Fig. 3.7. Here, the relaxation network refers to the network without independent power sources.

$$
z_{\alpha\beta\_in} = \frac{u}{i_s} \tag{3.23}
$$

3. Using the inference of the Thevenin's or Norton's theorems to find the input impedance. The input impedance $z_{\alpha\beta\_in}$ is calculated by the ratio of the port open-circuit voltage u_{oc} to its port short-circuit current i_k. For a one-port fractional-order network with only one independent power supply and a controlled source, it is better to use this method to calculate the input impedance.

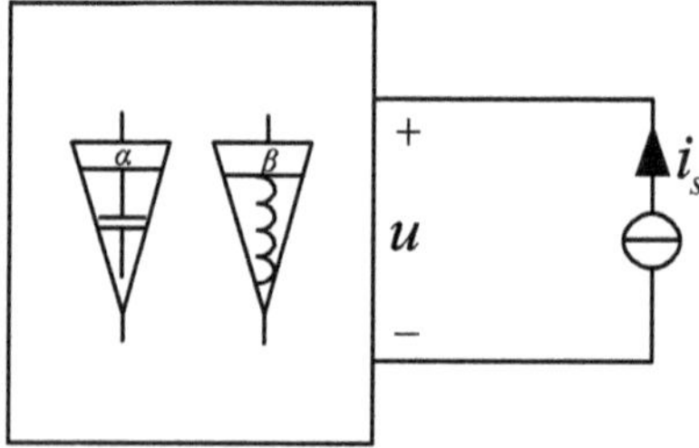

Fig. 3.7 One-port fractional-order relaxation network

The open-circuit voltage u_{oc} and the short-circuit current i_k of the port are solved by disconnecting the external network connected to the port and keeping the port open or short-circuited.

The equivalence of the Thevenin's and Norton's theorems is local, which are equivalent only for the voltage and current of the port. This equivalent simplification can only keep the relationship between the voltage and current of the port unchanged. If the voltage and current of each branch in the one-port fractional-order network with independent power sources are required, the port's voltage and current can only be solved by taking them as known conditions and substituting them into the original equations of the unaltered one-port network.

Finally, it should be pointed out that the concepts of the Thevenin's and Norton's theorems and substitution theorem are essentially different. The substitution theorem is to replace a branch with a certain voltage or current source, while the Thevenin's and Norton's theorems are to simplify the fractional-order linear one-port active network into a Thevenin's or Norton's branch. This simplified version of the Thevenin's or Norton's branch is applicable to any external network where the control variable of the controlled source is not converted through the external branch of the network during the simplification process.

3.4 Tellegen's Theorem

Tellegen's theorem is a fundamental theorem that is universally applicable to fractional-order circuits. It states that for any fractional-order circuit network with n nodes and b branches, the algebraic sum of the product of the branch voltage u_k and the branch current i_k of the whole network is zero at any time, that is

$$\sum_{k=1}^{\mathrm{b}} u_k i_k = 0 \tag{3.24}$$

where $(u_1, u_2, \ldots, u_b)$, $(i_1, i_2, \ldots, i_b)$ are the voltages and currents of the b branches, respectively. The reference directions of the voltage u_k and current i_k are taken as the associated reference direction.

The specific proof is as follows: assuming that the column matrix of the branch voltages of the fractional-order network is $\boldsymbol{u}_b$, the column matrix of the node potential is $\boldsymbol{\varphi}_n$, the column matrix of the branch current is $\boldsymbol{i}_b$, and the node association matrix of the network is $\boldsymbol{A}$, there are

$$\sum_{k=1}^{b} u_k i_k = \boldsymbol{u}_b^T \boldsymbol{i}_b \tag{3.25}$$

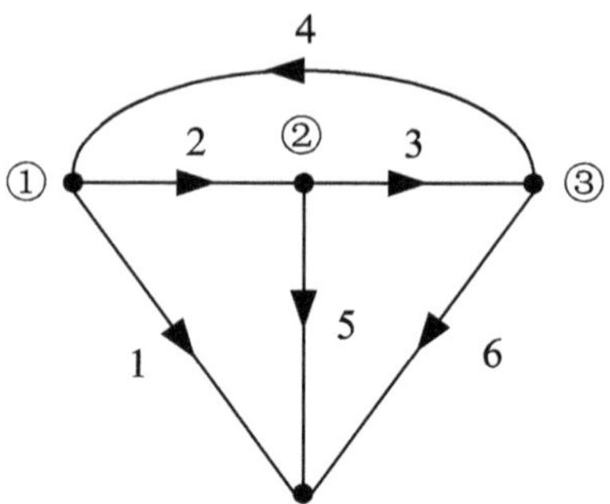

Fig. 3.8 Proof of Tellegen's theorem

According to the KVL and KCL, it can be found that

$$\begin{cases} \boldsymbol{u}_b = \boldsymbol{A}^{\mathrm{T}} \boldsymbol{\varphi}_n \\ \boldsymbol{u}_b^{\mathrm{T}} = \boldsymbol{\varphi}_n \boldsymbol{A} \\ \boldsymbol{A} \cdot \boldsymbol{i}_b = 0 \end{cases} \tag{3.26}$$

Then, substituting (3.26) into (3.25), (3.25) can be described as

$$\sum_{k=1}^{b} u_k i_k = \boldsymbol{u}_b^{\mathrm{T}} \boldsymbol{i}_b = \boldsymbol{\varphi}_n \boldsymbol{A} \boldsymbol{i}_b = 0 \tag{3.27}$$

Further, taking the line graph shown in Fig. 3.8 as an example, let u_{n1}, u_{n2}, and u_{n3} be the node voltages of nodes ①, ②, and ③, respectively. The relationship between the branch voltages $u_1, u_2, \ldots, u_6$ and the node voltages u_{n1}, u_{n2}, u_{n3} can be expressed by

$$\begin{cases} u_1 = u_{n1} \\ u_2 = u_{n1} - u_{n2} \\ u_3 = u_{n2} - u_{n3} \\ u_4 = -u_{n1} + u_{n3} \\ u_5 = u_{n2} \\ u_6 = u_{n3} \end{cases} \tag{3.28}$$

Applying KCL to nodes ①, ②, ③, there is

$$\begin{cases} i_1 + i_2 - i_4 = 0 \\ -i_2 + i_3 + i_5 = 0 \\ -i_3 + i_4 + i_6 = 0 \end{cases} \tag{3.29}$$

The algebraic sum of the product of the branch voltage u_k and the branch current i_k of the whole network is

$$\sum_{k=1}^{6} u_k i_k = u_1 i_1 + u_2 i_2 + u_3 i_3 + u_4 i_4 + u_5 i_5 + u_6 i_6 \tag{3.30}$$

Substituting (3.28) into (3.30), (3.30) can be written as

$$\sum_{k=1}^{6} u_k i_k = u_{n1} i_1 + (u_{n1} - u_{n2}) i_2 + (u_{n2} - u_{n3}) i_3 \\ + (-u_{n1} + u_{n3}) i_4 + u_{n2} i_5 + u_{n3} i_6 \tag{3.31}$$

Then, by sorting out (3.31), (3.31) can be rewritten as

$$\sum_{k=1}^{6} u_k i_k = u_{n1}(i_1 + i_2 - i_4) + u_{n2}(-i_2 + i_3 + i_5) + u_{n3}(-i_3 + i_4 + i_6) \tag{3.32}$$

Last, by substituting (3.29) into (3.32), there is

$$\sum_{k=1}^{6} u_k i_k = 0 \tag{3.33}$$

This proof can be generalized to any fractional-order circuit with n nodes and b branches. In the above process of proving Tellegen's theorem, only KCL and KVL are applied, which are only based on the topological properties of the circuit and do not involve the specific composition of the branches. Therefore, Tellegen's theorem can be applied to any circuit with linear, nonlinear, time-invariant, or time-varying components. This theorem is essentially a mathematical expression of power conservation, which indicates that the total energy of any fractional-order circuit can neither increase nor decrease, and the sum of power absorbed by all branches is equal to zero. Furthermore, Tellegen's theorem also demonstrates that $\boldsymbol{u}_b^{\mathrm{T}} \boldsymbol{i}_b = 0$ or $\boldsymbol{i}_b^{\mathrm{T}} \boldsymbol{u}_b = 0$, which means that the column vector $\boldsymbol{u}_b$ of the branch voltage and the column vector $\boldsymbol{i}_b$ of the branch current is a set of orthogonal vectors.

There are also several other forms of Tellegen's theorem. The content and certification are as follows:

1. Tellegen's theorem in the power-like form

There are two fractional-order networks, as shown in Fig. 3.9, the two networks have the same topological structure and the same number of branches, each branch voltage and circuit take the associated reference direction, the corresponding numbers of the branch and the node are the same, and the node incidence matrix is equal, that is, ($\boldsymbol{A} = \hat{\boldsymbol{A}}$). Here, $(u_1, u_2, \ldots, u_b)$, $(i_1, i_2, \ldots, i_b)$ and $(\hat{u}_s, \hat{u}_2, \ldots, \hat{u}_{\mathrm{b}})$, $(\hat{i}_1, \hat{i}_2, \ldots, \hat{i}_{\mathrm{b}})$ represent the voltages and currents of the b branches in the two fractional-order circuits N and $\hat{N}$, respectively. Then, according to Tellegen's theorem, there is

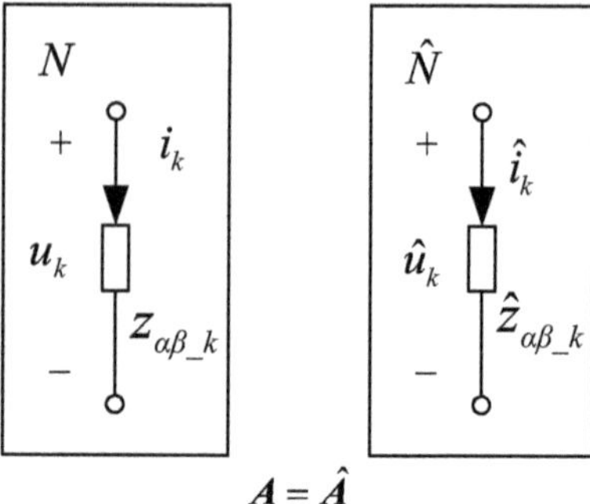

Fig. 3.9 Proof of Tellegen's theorem in the power-like form

$$\begin{cases} \sum_{k=1}^{b} u_k \hat{i}_k = 0 \\ \sum_{k=1}^{b} \hat{u}_k i_k = 0 \end{cases} \tag{3.34}$$

It can be proved by the following process: assuming that the column matrix of the branch voltages, branch currents, and node potential of the two fractional-order circuit networks are $\boldsymbol{u}_b$, $\hat{\boldsymbol{u}}_b$, $\boldsymbol{i}_b$, $\hat{\boldsymbol{i}}_b$, $\boldsymbol{\varphi}_n$ and $\hat{\boldsymbol{\varphi}}_n$, respectively. Then, by using KCL and KVL, (3.35) can be derived.

$$\begin{cases} \sum_{k=1}^{b} u_k \hat{i}_k = \boldsymbol{u}_b^{\mathrm{T}} \hat{\boldsymbol{i}}_b = \boldsymbol{\varphi}_n^{\mathrm{T}} \boldsymbol{A} \hat{\boldsymbol{i}}_b = \boldsymbol{\varphi}_n^{\mathrm{T}} \hat{\boldsymbol{A}} \hat{\boldsymbol{i}}_b = 0 \\ \sum_{k=1}^{b} \hat{u}_k i_k = \hat{\boldsymbol{u}}_b^{\mathrm{T}} \boldsymbol{i}_b = \hat{\boldsymbol{\varphi}}_n^{\mathrm{T}} \hat{\boldsymbol{A}} \boldsymbol{i}_b = \hat{\boldsymbol{\varphi}}_n^{\mathrm{T}} \boldsymbol{A} \boldsymbol{i}_b = 0 \end{cases} \tag{3.35}$$

Similarly, taking the line graph shown in Fig. 3.8 as an example, the circuit equations of the fractional-order circuit $\hat{N}$ can be expressed as

$$\begin{cases} \hat{i}_1 + \hat{i}_2 - \hat{i}_4 = 0 \\ -\hat{i}_2 + \hat{i}_3 + \hat{i}_5 = 0 \\ -\hat{i}_3 + \hat{i}_4 + \hat{i}_6 = 0 \end{cases} \tag{3.36}$$

By combining (3.28) and (3.35), (3.37) can be deduced.

$$\begin{aligned} \sum_{k=1}^{6} u_k \hat{i}_k = u_{n1}\left(\hat{i}_1 + \hat{i}_2 - \hat{i}_4\right) + u_{n2}\left(-\hat{i}_2 + \hat{i}_3 + \hat{i}_5\right) \\ + u_{n3}\left(-\hat{i}_3 + \hat{i}_4 + \hat{i}_6\right) \end{aligned} \tag{3.37}$$

By substituting (3.36) into (3.37), (3.37) can be rewritten as

$$\sum_{k=1}^{6} u_k \hat{i}_k = 0 \tag{3.38}$$

Extending the above proof to any two fractional-order circuits with n nodes and b branches, (3.34) can be obtained.

It is worth noting that the power-like form of Tellegen's theorem cannot be explained by power conservation. It is only used to describe the mathematical relationship that the branch voltages of one circuit and the branch currents of the other circuit or the corresponding branch voltages and branch currents of a fractional-order circuit at different times must be followed, in which the two fractional-order circuits have the same topological structure. Since it has the form of the sum of power, it is called Tellegen's theorem in power-like form. In addition, the theorem has no restriction on the composition of the branch.

2. Tellegen's theorem in the port form

If the branches of a fractional-order network are divided into two categories: one is the internal branch of the network, whose branch voltages and branch currents are u_{mi} and i_{mi} $(i = 1, 2, \ldots, m)$, respectively. The branch voltages and branch currents take the associated reference direction. The other is the port branch of the network, whose branch voltages and branch currents are u_{nk} and i_{nk} $(k = 1, 2, \ldots, n)$, respectively. The voltages and currents of the port branches adopt the inverse associated reference direction, as shown in Fig. 3.10.

According to Tellegen's theorem shown in (3.24), (3.39) can be obtained.

$$\sum_{i=1}^{m} u_{mi} i_{mi} + \sum_{k=1}^{n} (-u_{nk}) i_{nk} = 0 \tag{3.39}$$

and (3.39) can be rewritten as

$$\sum_{k=1}^{n} u_{nk} i_{nk} = \sum_{i=1}^{m} u_{mi} i_{mi} \tag{3.40}$$

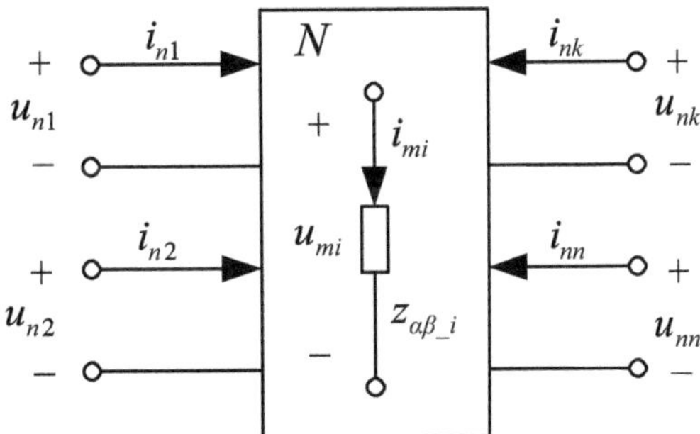

Fig. 3.10 Proof of Tellegen's theorem in port form

(3.40) is the expression of Tellegen's theorem in port form, which shows that the output power is equal to the total power consumption for any fractional-order network.

In addition, Tellegen's theorem in power-like form can also be written in the port form. Supposing that the branches of the fractional-order networks N and $\hat{N}$ can also be divided into two types: the internal branches of the network and the port branches of the network. In these cases, the voltages and currents of the internal branches take associated reference direction, and the voltages and currents of the port branches take the reverse associated reference direction. According to (3.34), the Tellegen's theorem in the power-like form of the port can be derived as

$$\begin{cases} \sum_{k=1}^{n} u_{nk}\hat{i}_{nk} = \sum_{i=1}^{m} u_{mi}\hat{i}_{mi} \\ \sum_{k=1}^{n} \hat{u}_{nk}i_{nk} = \sum_{i=1}^{m} \hat{u}_{mi}i_{mi} \end{cases} \tag{3.41}$$

3. Tellegen's theorem in the port-difference form

The port-difference form is a particular Tellegen's theorem, also known as the weak form of the Tellegen's theorem. Only when the fractional-order network satisfies certain conditions, the port-difference form of Tellegen's theorem can be applied. Otherwise, the port-difference form of Tellegen's theorem cannot be established. The applicable condition of the theorem is that if two fractional-order networks have the same topological structure $\boldsymbol{A} = \hat{\boldsymbol{A}}$, and the branches of these two fractional-order networks are divided into two parts, one is called the internal branches of the network, the other is the port branches of the network. When the internal branches of the two networks do not contain independent power supplies, the relationship between the branch voltages and currents can be written as the following general matrix equation.

$$\begin{bmatrix} \boldsymbol{u}_p \\ \boldsymbol{i}_q \end{bmatrix} = \begin{bmatrix} \boldsymbol{z}_{pp} & \boldsymbol{h}_{pq} \\ \boldsymbol{h}_{qp} & \boldsymbol{y}_{qq} \end{bmatrix} \begin{bmatrix} \boldsymbol{i}_p \\ \boldsymbol{u}_q \end{bmatrix} \tag{3.42}$$

$$\begin{bmatrix} \hat{\boldsymbol{u}}_p \\ \hat{\boldsymbol{i}}_q \end{bmatrix} = \begin{bmatrix} \hat{\boldsymbol{z}}_{\alpha\beta\_pp} & \hat{\boldsymbol{h}}_{pq} \\ \hat{\boldsymbol{h}}_{qp} & \hat{\boldsymbol{y}}_{qq} \end{bmatrix} \begin{bmatrix} \hat{\boldsymbol{i}}_p \\ \hat{\boldsymbol{u}}_q \end{bmatrix} \tag{3.43}$$

where $p + q = m$.

If the parameter matrixes of (3.42) and (3.43) satisfy the following relationship

$$\begin{cases} \boldsymbol{z}_{pp} = \hat{\boldsymbol{z}}_{pp}^{\mathrm{T}} \\ \boldsymbol{y}_{pp} = \hat{\boldsymbol{y}}_{pp}^{\mathrm{T}} \\ \boldsymbol{h}_{pq} = -\hat{\boldsymbol{h}}_{qp}^{\mathrm{T}} \\ \boldsymbol{h}_{qp} = -\hat{\boldsymbol{h}}_{pq}^{\mathrm{T}} \end{cases} \tag{3.44}$$

Then, the two fractional-order networks will satisfy Tellegen's theorem in the port-difference form. Obviously, the internal branches of two fractional-order networks satisfying the Tellegen's theorem in the port-difference form do not contain independent power supplies, and the coefficient matrix of their VCR matrix equations satisfies (3.44).

Assuming that two fractional-order circuit networks N and $\hat{N}$ satisfy the above conditions, and their port voltages and port currents are defined as u_{nk}, i_{nk}, $\hat{u}_{nk}$ and $\hat{i}_{nk}$ ($k = 1, 2, \ldots, n$), respectively. Then, according to (3.41), there are

$$\begin{aligned}\sum_{i=1}^{m} u_{mi}\hat{i}_{mi} &= \hat{\boldsymbol{i}}_p^{\mathrm{T}}\boldsymbol{u}_p + \hat{\boldsymbol{i}}_q^{\mathrm{T}}\boldsymbol{u}_q \\ &= \hat{\boldsymbol{i}}_p^{\mathrm{T}}\left[\boldsymbol{z}_{pp}\boldsymbol{i}_p + \boldsymbol{h}_{pq}\boldsymbol{u}_q\right] + \left[\hat{\boldsymbol{h}}_{qp}\hat{\boldsymbol{i}}_p + \hat{\boldsymbol{y}}_{qq}\hat{\boldsymbol{u}}_q\right]^{\mathrm{T}}\boldsymbol{u}_q \\ &= \hat{\boldsymbol{i}}_p^{\mathrm{T}}\boldsymbol{z}_{pp}\boldsymbol{i}_p + \hat{\boldsymbol{i}}_p^{\mathrm{T}}\boldsymbol{h}_{pq}\boldsymbol{u}_q + \hat{\boldsymbol{i}}_p^{\mathrm{T}}\hat{\boldsymbol{h}}_{qp}^{\mathrm{T}}\boldsymbol{u}_q + \hat{\boldsymbol{u}}_q^{\mathrm{T}}\hat{\boldsymbol{y}}_{qq}^{\mathrm{T}}\boldsymbol{u}_q\end{aligned} \tag{3.45}$$

$$\begin{aligned}\sum_{i=1}^{m} \hat{u}_{mi}i_{mi} &= \hat{\boldsymbol{u}}_p^{\mathrm{T}}\boldsymbol{i}_p + \hat{\boldsymbol{u}}_q^{\mathrm{T}}\boldsymbol{i}_q \\ &= \left[\hat{\boldsymbol{z}}_{pp}\hat{\boldsymbol{i}}_p + \hat{\boldsymbol{h}}_{pq}\hat{\boldsymbol{u}}_q\right]^{\mathrm{T}}\boldsymbol{I}_p + \hat{\boldsymbol{u}}_q^{\mathrm{T}}\left[\boldsymbol{h}_{qp}\boldsymbol{i}_p + \boldsymbol{y}_{qq}\boldsymbol{u}_q\right] \\ &= \hat{\boldsymbol{i}}_p^{\mathrm{T}}\hat{\boldsymbol{z}}_{pp}^{\mathrm{T}}\boldsymbol{i}_p + \hat{\boldsymbol{u}}_q^{\mathrm{T}}\hat{\boldsymbol{h}}_{pq}^{\mathrm{T}}\boldsymbol{i}_p + \hat{\boldsymbol{u}}_q^{\mathrm{T}}\boldsymbol{h}_{qp}\boldsymbol{i}_p + \hat{\boldsymbol{u}}_q^{\mathrm{T}}\boldsymbol{y}_{qq}\boldsymbol{u}_q\end{aligned} \tag{3.46}$$

From $\boldsymbol{h}_{pq} = -\hat{\boldsymbol{h}}_{qp}^{\mathrm{T}}$, $\boldsymbol{h}_{qp} = -\hat{\boldsymbol{h}}_{pq}^{\mathrm{T}}$, (3.45) and (3.46) can be expressed as

$$\begin{cases}\displaystyle\sum_{\mathrm{i}=1}^{m} u_{mi}\hat{i}_{mi} = \hat{\boldsymbol{i}}_p^{\mathrm{T}}\boldsymbol{z}_{pp}\boldsymbol{i}_p + \hat{\boldsymbol{u}}_q^{\mathrm{T}}\hat{\boldsymbol{y}}_{qq}^{\mathrm{T}}\boldsymbol{u}_q \\ \displaystyle\sum_{i=1}^{m} \hat{u}_{mi}i_{mi} = \hat{\boldsymbol{i}}_p^{\mathrm{T}}\hat{\boldsymbol{z}}_{pp}^{\mathrm{T}}\boldsymbol{i}_p + \hat{\boldsymbol{u}}_q^{\mathrm{T}}\boldsymbol{y}_{qq}\boldsymbol{u}_q\end{cases} \tag{3.47}$$

By substituting $\boldsymbol{z}_{pp} = \hat{\boldsymbol{z}}_{pp}^{\mathrm{T}}$ and $\boldsymbol{y}_{pp} = \hat{\boldsymbol{y}}_{pp}^{\mathrm{T}}$ into (3.47), there is

$$\hat{\boldsymbol{i}}_p^{\mathrm{T}}\boldsymbol{z}_{pp}\boldsymbol{i}_p + \hat{\boldsymbol{u}}_q^{\mathrm{T}}\hat{\boldsymbol{y}}_{qq}^{\mathrm{T}}\boldsymbol{u}_q = \hat{\boldsymbol{i}}_p^{\mathrm{T}}\hat{\boldsymbol{z}}_{pp}^{\mathrm{T}}\boldsymbol{i}_p + \hat{\boldsymbol{u}}_q^{\mathrm{T}}\boldsymbol{y}_{qq}\boldsymbol{u}_q \tag{3.48}$$

Therefore, $\sum_{k=1}^{n} u_{nk}\hat{i}_{nk} = \sum_{k=1}^{n} \hat{u}_{nk}i_{nk}$ can be obtained, the theorem is proved.

If the internal networks of the fractional-order networks N and $\hat{N}$ are passive linear impedance networks that do not contain independent power supplies and controlled sources, (3.49) can be obtained.

$$\begin{cases}\boldsymbol{h}_{pq} = \hat{\boldsymbol{h}}_{pq} = 0 \\ \boldsymbol{h}_{qp} = \hat{\boldsymbol{h}}_{qp} = 0\end{cases} \tag{3.49}$$

It can also be written as

$$\begin{cases} \boldsymbol{h}_{pq} = -\boldsymbol{h}_{qp}^{\mathrm{T}} \\ \boldsymbol{h}_{qp} = -\boldsymbol{h}_{pq}^{\mathrm{T}} \end{cases} \tag{3.50}$$

Moreover, there are

$$\begin{cases} \boldsymbol{z}_{pp} = \hat{\boldsymbol{z}}_{pp} = \boldsymbol{z}_{pp}^{\mathrm{T}} \\ \boldsymbol{y}_{pp} = \hat{\boldsymbol{y}}_{ppp} = \boldsymbol{y}_{pp}^{\mathrm{T}} \end{cases} \tag{3.51}$$

Then, (3.44) is satisfied, and these two fractional-order networks N and $\hat{N}$ satisfy the port-difference form of Tellegen's theorem. It should be noted that the prerequisite for the establishment of all the above forms of Tellegen's theorem is that the values of voltage and current should form a pair of orthogonal vectors, and they should satisfy the KVL and KCL of the same topology respectively [9, 10].

3.5 Reciprocity Theorem

In a two-port network, reciprocity theorem states that if a voltage is applied to one terminal of a fractional-order linear passive network and a current is produced at the other terminal, then the same voltage applied to the second terminal will produce the same current at the first terminal. In other words, the ratio of the response to the excitation remains unchanged after the excitation and response exchanging position. The reciprocity theorem is essentially the reification of Tellegen's theorem in the port-difference form in a two-port reciprocal network.

Before introducing the reciprocity theorem, it is necessary to explain the concept of reciprocal components. For any multi-port component that does not contain the independent power supplies, the performance equation between its voltage and current can be expressed by

$$\begin{bmatrix} \boldsymbol{u}_a \\ \boldsymbol{i}_b \end{bmatrix} = \begin{bmatrix} \boldsymbol{z}_{aa} & \boldsymbol{h}_{ab} \\ \boldsymbol{h}_{ba} & \boldsymbol{y}_{bb} \end{bmatrix} \begin{bmatrix} \boldsymbol{i}_a \\ \boldsymbol{u}_b \end{bmatrix} \tag{3.52}$$

If (3.52) satisfies the following conditions,

$$\begin{cases} \boldsymbol{z}_{aa} = \boldsymbol{z}_{aa}^{\mathrm{T}} \\ \boldsymbol{y}_{bb} = \boldsymbol{y}_{bb}^{\mathrm{T}} \\ \boldsymbol{h}_{ab} = -\boldsymbol{h}_{ba}^{\mathrm{T}} \end{cases} \tag{3.53}$$

such a multi-port component is called a reciprocal component.

For a fractional-order linear time-invariant $RL_\beta C_\alpha$ component, its performance equation is a scalar equation, and its characteristic matrix is also a scalar parameter matrix, so that (3.53) is satisfied. Such components are reciprocal components. For linear time-invariant coupled inductors and ideal transformers, they are two-port components and also reciprocal components. However, as for the gyrator and the four types of controlled sources, they are not reciprocal components.

The network composed of reciprocal components must satisfy the reciprocity theorem. Its performance equation has the form of (3.52) and meets the requirements of (3.53). In addition, for a fractional-order network containing a plurality of controlled power sources, if the matrix form of the performance equation of the branch voltages and currents is the same as (3.52) and satisfies (3.53), the network also satisfies the reciprocity theorem. In fact, the network can be viewed as a multi-port reciprocal component. For the reciprocal network, when using the reciprocity theorem, it is also required that the parameter matrix in the matrix equation of the voltages and currents in the network satisfies (3.53) under different excitations.

There are three forms of reciprocity theorem. In order to use the reciprocity theorem to analyze network problems flexibly, the basic contents of these three forms of reciprocity theorem must be grasped firmly and skillfully.

1. The first form of reciprocity theorem

In the reciprocal network shown in Fig. 3.11, the short-circuit current i_2 generated by the independent voltage source of branch 1 on the short-circuited branch 2 is equal to the short-circuited current $\hat{i}_1$ generated when the independent voltage source of branch 1 is connected in series to the short-circuit branch 2 and the port after removing the independent voltage source in branch 1 is shorted. In Fig. 3.11b, the network and variables after exchanging positions are marked with "^".

According to the reciprocity theorem, there is

$$i_2 = \hat{i}_1 \tag{3.54}$$

The first form of reciprocity theorem can be intuitively described as: when the positions of the independent voltage source and ammeter are interchanged, the reading of the ammeter remains the same.

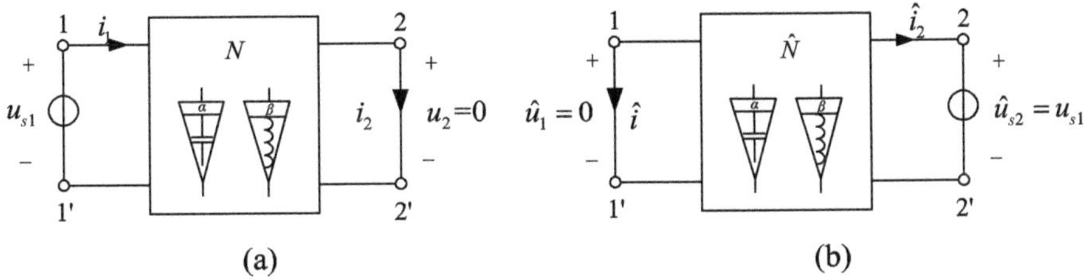

Fig. 3.11 The first form of reciprocity theorem: **a** original circuit, **b** voltage source moving circuit

The specific proof is as follows: since the two fractional-order networks satisfy Tellegen's theorem in the port-difference form, the port voltages and currents shown in Fig. 3.11 can be expressed as

$$\begin{cases} u_1 = u_{s1} \\ u_2 = 0 \\ \hat{u}_1 = 0 \\ \hat{u}_2 = \hat{u}_{s2} = u_{s1} \end{cases} \tag{3.55}$$

Then, there is

$$-u_1\hat{i}_1 - u_2\hat{i}_2 = \hat{u}_1 i_1 - \hat{u}_2 i_2 \tag{3.56}$$

which can be represented by

$$-u_{s1}\hat{i}_1 - 0 = 0 - u_{s1} i_2 \tag{3.57}$$

Thus, $i_2 = \hat{i}_1$ is obtained, the theorem is proved.

When applying the first form of the reciprocity theorem, the associated reference direction rules of the branch voltages and currents should be noted. When $\hat{i}_1$ and u_1 take the associated reference direction (or vice versa), i_2 and $\hat{u}_2$ also take the associated reference direction (or vice versa). In addition, there is $\hat{u}_2 = \hat{u}_{s2} \neq u_{s1}$, the first form of reciprocity theorem can be derived as

$$\frac{i_2}{u_{s1}} = \frac{\hat{i}_1}{\hat{u}_{s2}} \tag{3.58}$$

which means that $Y_{21} = Y_{12}$. Then, from the reciprocity theorem and the homogeneity of the superposition theorem, it can be known that (3.58) states that the transfer admittances of the two networks are equal.

2. The second form of the reciprocity theorem

As for the reciprocal network shown in Fig. 3.12, the open-circuit voltage u_2 generated by the independent current source of branch 1 on the open branch 2 is equal to the open-circuit voltage $\hat{u}_1$ generated when the independent current source of branch 1 is connected in parallel to the open-circuit branch 2 and the port after removing the independent current source in branch 1 is maintained open, that is

$$u_2 = \hat{u}_1 \tag{3.59}$$

The second form of the reciprocity theorem declares that when the positions of the independent current source and the voltmeter are interchanged, the reading of the voltmeter remains unchanged. Here, the current source and voltmeter are ideal, whose internal resistors are infinite.

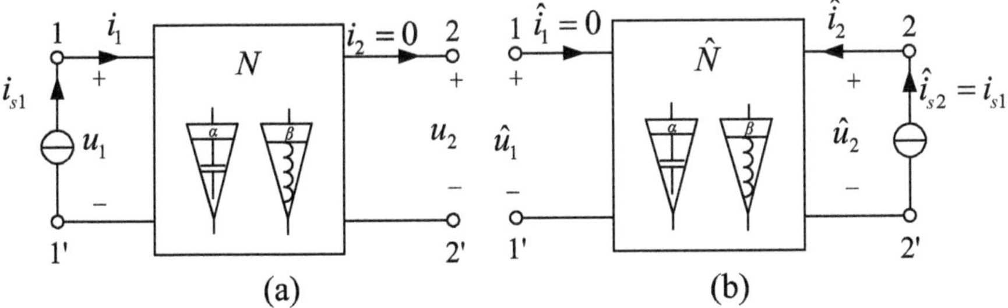

Fig. 3.12 The second form of reciprocity theorem: **a** original circuit, **b** source moving circuit

Since these two fractional-order networks satisfy Tellergen's theorem in the port-difference form, the port voltages and currents shown in Fig. 3.12 can be described by

$$\begin{cases} i_1 = i_{s1} \\ i_2 = 0 \\ \hat{i}_1 = 0 \\ \hat{i}_2 = \hat{i}_{s2} = i_{s1} \end{cases} \tag{3.60}$$

Then, (3.61) can be derived.

$$u_1\hat{i}_1 + u_2\hat{i}_2 = \hat{u}_1 i_1 - \hat{u}_2 i_2 \tag{3.61}$$

by substituting (3.60) into (3.61), (3.61) can be written as

$$u_2\hat{i}_{s1} = u_1 i_{s1} \tag{3.62}$$

Therefore, there is $u_2 = \hat{u}_1$, and the theorem is proved.

When applying the second form of the reciprocity theorem, it is worth noting that the associated reference directions of the branch voltages and currents. When $\hat{u}_1$ and i_1 take the associated reference direction (or vice versa), $\hat{u}_2$ and i_2 also take the associated reference direction (or vice versa), and there is $\hat{i}_2 = \hat{i}_{s2} \neq \hat{i}_{s1}$, the second form of reciprocity theorem can be defined by

$$\frac{u_1}{i_{s1}} = \frac{\hat{u}_1}{\hat{i}_{s2}} \tag{3.63}$$

which indicates that $z_{21} = z_{12}$. Then, considering the reciprocity theorem and the homogeneity of superposition theorem, (3.63) also demonstrates that the transfer impedances of the two networks are equal.

3. The third form of the reciprocity theorem

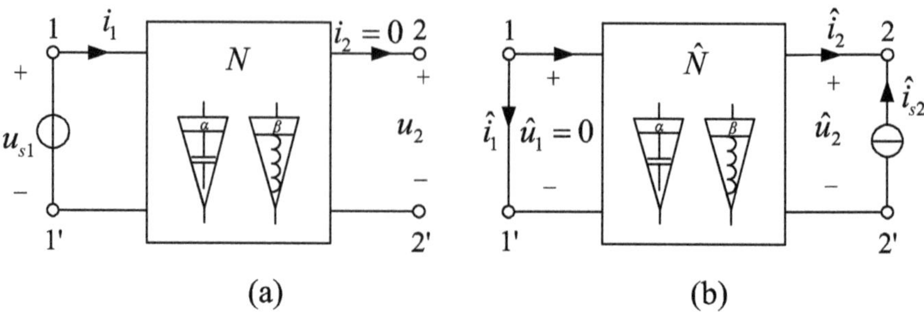

Fig. 3.13 The third form of reciprocity theorem: **a** original circuit, **b** current source alternative circuit

In the reciprocal network, the open-circuit voltage u_2 generated by the independent voltage source u_{s1} of branch 1 on the open branch 2 is proportional to the short-circuit current i_1 obtained by connecting the independent current source $\hat{i}_{s2}$ in parallel to the open branch 2 and setting the independent voltage source of the branch 1 to zero (i.e., short-circuited), as shown in Fig. 3.13, there is

$$\frac{u_2}{u_{s1}} = \frac{\hat{i}_2}{\hat{i}_{s2}} \tag{3.64}$$

which is equivalent to $K_u = \hat{K}_i$. K_u represents the voltage gain of the network N, $\hat{K}_i$ denotes the current gain of the network $\hat{N}$.

Since the above two fractional-order networks satisfy Tellegen's theorem in the port-difference form, the port voltages and currents can be written as

$$\begin{cases} u_1 = u_{s1} \\ i_2 = 0 \\ \hat{u}_1 = 0 \\ \hat{i}_2 = \hat{i}_{s2} \end{cases} \tag{3.65}$$

Then, (3.66) can be derived.

$$-u_1\hat{i}_1 + u_2\hat{i}_2 = \hat{u}_1 i_1 - \hat{u}_2 i_2 \tag{3.66}$$

which can also be rewritten as

$$-u_{s1}\hat{i}_1 + u_2\hat{i}_{s2} = 0 \tag{3.67}$$

From (3.67), there is $\frac{u_2}{u_{s1}} = \frac{\hat{i}_1}{\hat{i}_{s2}}$, and the theorem is proved.

When using the third form of the reciprocity theorem, the associated reference direction of the branch voltages and currents should be noted, which are different

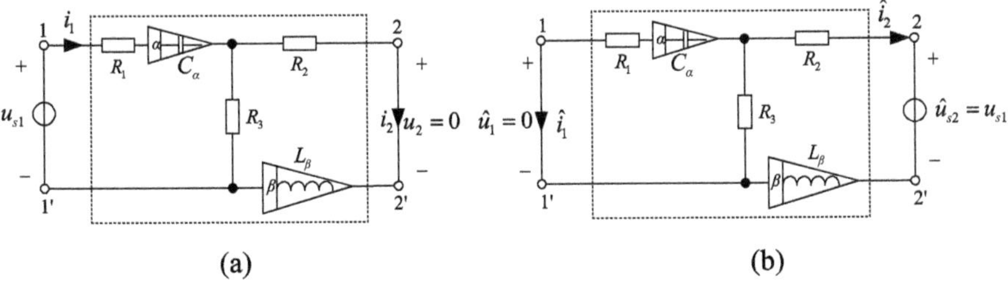

Fig. 3.14 An application example of the first form of reciprocity theorem: **a** original circuit, **b** equivalent reciprocal circuit

from the first and second forms of the reciprocity theorem. Here, u_1 and $\hat{i}_1$ take the associated reference direction (or opposite), while u_2 and $\hat{i}_2$ take the inverse associated reference direction (or the same). The third form of reciprocity theorem has another statement: the forward voltage transfer ratio is equal to the reverse current transfer ratio, or the reverse voltage transfer ratio is equal to the forward current transfer ratio.

Figure 3.14 is an application example of the first form of the reciprocity theorem, which shows a fractional-order linear circuit consisting of resistors R_1, R_2, R_3, fractional-order inductor L_β, and fractional-order capacitor C_α.

Based on KVL and KCL, the circuit equation of the fractional-order circuit shown in Fig. 3.14a can be denoted as

$$\begin{cases} u_{s1} = \left(R_1 + \frac{1}{C_\alpha}{}_0D_t^{-\alpha}\right)i_1 + \left(R_2 + L_\beta\,{}_0D_t^{\beta}\right)i_2 \\ 0 = -R_3 i_1 + \left(R_2 + R_3 + L_\beta\,{}_0D_t^{\beta}\right)i_2 \end{cases} \tag{3.68}$$

Then, i_2 can be expressed as

$$i_2 = \frac{R_3 u_{s1}}{R_3\left(R_2 + L_\beta\,{}_0D_t^{\beta}\right) + \left(R_2 + R_3 + L_\beta\,{}_0D_t^{\beta}\right)\left(R_1 + \frac{1}{C_\alpha}{}_0D_t^{-\alpha}\right)} \tag{3.69}$$

Similarly, the circuit equation of the fractional-order circuit in Fig. 3.14b can be represented as

$$\begin{cases} \hat{u}_{s2} = \left(R_1 + \frac{1}{C_\alpha}{}_0D_t^{-\alpha}\right)\hat{i}_1 - \left(R_2 + L_\beta\,{}_0D_t^{\beta}\right)\hat{i}_2 \\ 0 = R_3\hat{i}_2 + \left(R_1 + R_3 + \frac{1}{C_\alpha}{}_0D_t^{-\alpha}\right)\hat{i}_1 \end{cases} \tag{3.70}$$

Then, (3.71) can be derived.

$$\begin{aligned}\hat{i}_1 &= \frac{R_3\hat{u}_{s2}}{R_3\left(R_1+\frac{1}{C_\alpha}{}_0D_t^{-\alpha}\right)+\left(R_2+L_\beta\,{}_0D_t^{\beta}\right)\left(R_1+R_3+\frac{1}{C_\alpha}{}_0D_t^{-\alpha}\right)}\\ &= \frac{R_3\hat{u}_{s2}}{R_3\left(R_2+L_\beta\,{}_0D_t^{\beta}\right)+\left(R_1+\frac{1}{C_\alpha}{}_0D_t^{-\alpha}\right)\left(R_2+R_3+L_\beta\,{}_0D_t^{\beta}\right)}\end{aligned} \tag{3.71}$$

From (3.69) and (3.71), it can be seen that $i_2 = \hat{i}_1$, which is consistent with the first form of reciprocity theorem (3.53).

3.6 Duality Theorem

Duality theorem is a generalization and summary of a large number of similarities in the analysis of fractional-order circuits. When the corresponding elements in the two circuit equations are interchanged, the equations can be converted to each other, the exchanged elements are called dual elements, and the fractional-order circuits corresponding to these two equations are also dual to each other.

A series circuit of n impedance elements is shown in Fig. 3.15, and a parallel circuit of n admittance elements is depicted in Fig. 3.15b, which are denoted by N and $\overline{N}$, respectively. There are the following similarities between the circuit equations of N and $\overline{N}$:

In the fractional-order circuit N and $\overline{N}$, there is

$$\begin{cases} z = \sum\limits_{k=1}^{n} z_k \\ i = \frac{u}{z} \\ u_k = \frac{z_k}{z} u \end{cases} \tag{3.72}$$

and

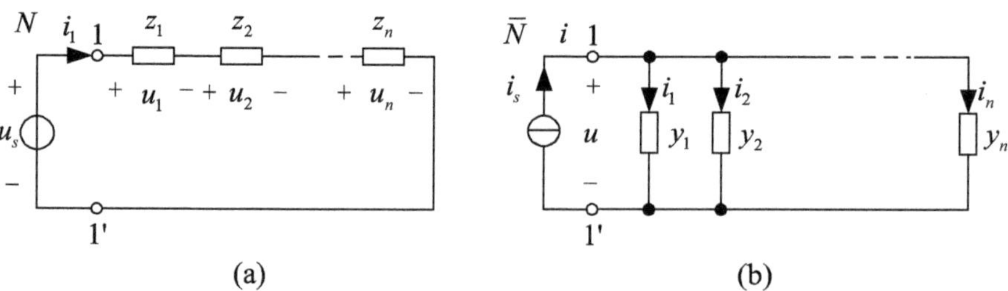

Fig. 3.15 The duality of series and parallel: **a** series circuit, **b** parallel circuit

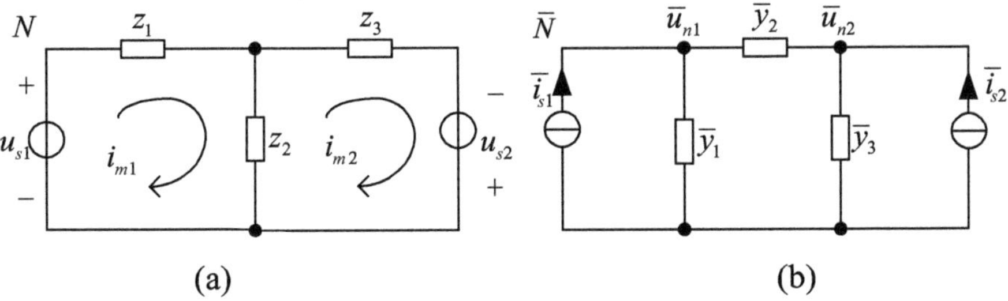

Fig. 3.16 The dual circuits: **a** original circuit, **b** equivalent dual circuit

$$\begin{cases} y = \sum\limits_{k=1}^{n} y_k \\ u = \frac{i}{y} \\ i_k = \frac{y_k}{y} i \end{cases} \tag{3.73}$$

As can be observed from (3.72) and (3.73), if the voltage u and the current i are interchanged, and the impedance z is interchanged with the admittance y, then the circuit equation of N shown in (3.72) is converted to the circuit equation of $\overline{N}$ shown in (3.73), and vice versa. This correspondence is called duality, and these interchangeable elements are dual elements. Generally, series and parallel, voltage and current, impedance and admittance satisfy duality, and the above circuits N and $\overline{N}$ are called dual circuits.

If the reference directions of the mesh currents and node voltages are given, as shown in Fig. 3.16a, b, the mesh current equation of $\overline{N}$ can be described as

$$\begin{cases} (z_1 + z_2)i_{m1} - z_2 i_{m2} = u_{s1} \\ -z_2 i_{m1} + (z_2 + z_3)i_{m2} = u_{s2} \end{cases} \tag{3.74}$$

and the node voltage equation of N can be derived as

$$\begin{cases} (\bar{y}_1 + \bar{y}_2)\bar{u}_{n1} - \bar{y}_2\bar{u}_{n2} = \bar{i}_{s1} \\ -\bar{y}_2\bar{u}_{n1} + (\bar{y}_2 + \bar{y}_3)\bar{u}_{n2} = \bar{i}_{s2} \end{cases} \tag{3.75}$$

When the corresponding elements, such as z_i and $\bar{y}_i$ $(i = 1, 2, 3)$, u_{sj} and $\bar{i}_{sj}$, i_{mj} and $\bar{u}_{nj}$ $(j = 1, 2)$, are interchanged, (3.74) and (3.75) can also be converted to each other, so the mesh currents and node voltages are also dual elements and the two fractional-order circuits N and $\overline{N}$ are dual circuits.

In addition, the voltage-current relationships of the fractional-order capacitor and inductor are

$$i_C = C_\alpha \frac{\mathrm{d}^\alpha u_C}{\mathrm{d}t^\alpha} \tag{3.76}$$

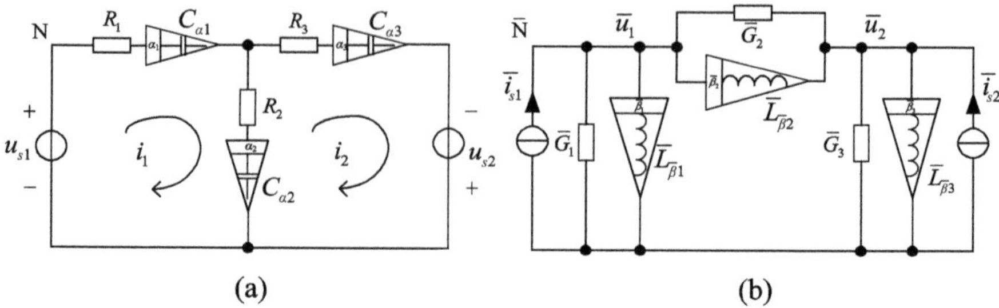

Fig. 3.17 An application example of duality theorem: **a** original circuit, **b** equivalent dual circuit

$$u_L = L_\beta \frac{\mathrm{d}^\beta i_L}{\mathrm{d}t^\beta} \tag{3.77}$$

As can be seen from (3.76) and (3.77), the fractional-order capacitor and inductor are also dual elements.

Figure 3.17 shows an application example of the duality theorem. The fractional-order circuit N shown in Fig. 3.17a is composed of voltage sources u_{s1} and u_{s2}, resistors R_1, R_2 and R_3, and fractional-order capacitors $C_{\alpha 1}$, $C_{\alpha 2}$ and $C_{\alpha 3}$, which can be analyzed by the mesh current equation. The fractional-order circuit $\overline{N}$ shown in Fig. 3.17b consists of current sources $\bar{i}_{s1}$ and $\bar{i}_{s2}$, conductance $\overline{G}_1$, $\overline{G}_2$ and $\overline{G}_3$, and fractional-order inductors $\overline{L}_{\beta 1}$, $\overline{L}_{\beta 2}$ and $\overline{L}_{\beta 3}$, which can be analyzed by the node voltage equation.

The mesh current equation of the fractional-order circuit shown in Fig. 3.17a is

$$\begin{cases} u_{s1} = i_1\left(R_1 + R_2 + \dfrac{1}{C_{\alpha 1}}{}_0D_t^{-\alpha_1} + \dfrac{1}{C_{\alpha 2}}{}_0D_t^{-\alpha_2}\right) - i_2\left(R_2 + \dfrac{1}{C_{\alpha 2}}{}_0D_t^{-\alpha_2}\right) \\ u_{s2} = i_1\left(R_2 + \dfrac{1}{C_{\alpha 2}}{}_0D_t^{-\alpha_2}\right) - i_2\left(R_2 + R_3 + \dfrac{1}{C_{\alpha 2}}{}_0D_t^{-\alpha_2} + \dfrac{1}{C_{\alpha 3}}{}_0D_t^{-\alpha_3}\right) \end{cases} \tag{3.78}$$

The node voltage equation of the fractional-order circuit shown in Fig. 3.17b is

$$\begin{cases} \bar{i}_{s1} = \bar{u}_{n1}\left(\overline{G}_1 + \overline{G}_2 + \dfrac{1}{\overline{L}_{\bar{\beta}1}}{}_0D_t^{\bar{\beta}_1} + \dfrac{1}{\overline{L}_{\bar{\beta}2}}{}_0D_t^{\bar{\beta}_2}\right) - \bar{u}_{n2}\left(\overline{G}_2 + \dfrac{1}{\overline{L}_{\bar{\beta}2}}{}_0D_t^{\bar{\beta}_2}\right) \\ \bar{i}_{s2} = \bar{u}_{n1}\left(\overline{G}_2 + \dfrac{1}{\overline{L}_{\bar{\beta}2}}{}_0D_t^{\bar{\beta}_2}\right) - \bar{u}_{n2}\left(\overline{G}_2 + \overline{G}_3 + \dfrac{1}{\overline{L}_{\bar{\beta}2}}{}_0D_t^{\bar{\beta}_2} + \dfrac{1}{\overline{L}_{\bar{\beta}3}}{}_0D_t^{\bar{\beta}_3}\right) \end{cases} \tag{3.79}$$

From the above analysis, it can be seen that the resistor R_m, the fractional-order capacitor $C_{\alpha m}$, the fractional order α_m, the mesh current i_n, the voltage source u_{sn} in (3.78) and the conductance $\overline{G}_m$, fractional-order inductor $\overline{L}_{\bar{\beta}m}$, fractional order

$\bar{\beta}_m$, node voltage $\bar{u}_n$, and current source $\bar{i}_{sn}$ in (3.79) are dual elements (where $m = 1, 2, 3$, $n = 1, 2$.). If these dual elements are interchanged, (3.78) and (3.79) would be transformed into each other, which indicates that the two fractional-order circuits are dual circuits.

The duality theorem points out that in dual circuits, the relationship (or equation) between certain elements can be converted to each other by exchanging dual elements. According to the duality theorem, if the expressions and conclusions are derived in a fractional-order circuit, it is equivalent to solving the relational expressions and conclusions in another fractional-order circuit that is dual with it. The content of the duality also includes the topological structure, variables, components, and equations of the fractional-order circuit, and even theorems. For example, Thevenin's theorem and Norton's theorem are dual to each other, KVL and KCL are also dual. Besides, it should be noted that duality and equivalence are two different concepts.

3.7 Compensation Theorem

Compensation theorem is very convenient for the analysis of the fractional-order network with variable parameters. It can attribute the variations of branch voltages and currents in a fractional-order network, which is caused by the changes in the parameters of a branch, to the relaxation of the network. Network relaxation is to remove the independent power sources in the network and connect an incremental voltage source in series to the branch or connect an incremental current source in parallel on the branch after the parameter changes. In other words, the compensation theorem is to transform the influence of the variation of branch parameters on the branch voltages and currents to the effect of an incremental voltage (or current) source that connected in series (or parallel) on the branch after the parameter changes on the branch voltages and currents [11].

For a fractional-order linear active network, if the impedance parameter of a certain branch changes from z_k to $z_k' = z_k + \Delta z_k$, then the voltage and current of each branch in the network must change from u_m, i_m to $u_m' = u_m + \Delta u_m$, $i_m' = i_m + \Delta i_m$. If a compensated voltage source is added and connected in series with the branch whose impedance changes, the voltage and current of all branches in the network can be restored to their original values, as shown in Fig. 3.18.

In this case, the direction of the compensated voltage source is opposite to the direction of the branch current, and their values satisfy

$$\Delta u_s = i_k \left(R' + \frac{1}{C_\alpha'} {}_0D_t^{-\alpha} + L_\beta' \, {}_0D_t^{\beta} \right) = i_k \Delta z_k \tag{3.80}$$

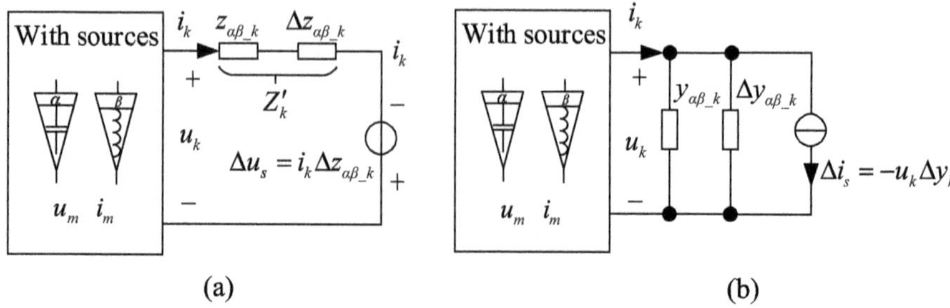

Fig. 3.18 Compensation theorem: **a** Thevenin's circuit, **b** Norton's circuit

The fractional-order circuit shown in Fig. 3.18a is the compensation theorem of Thevenin form of the circuit. If the Thevenin's branch is equivalently transformed into a Norton's branch, the compensation theorem in Norton form can be obtained, as shown in Fig. 3.18b. The direction of Δi_s is consistent with the direction of the branch current i_k, and there is

$$y_k' = \frac{1}{z_k'} = \frac{1}{z_k + \Delta z_k} = y_k + \Delta y_k \tag{3.81}$$

$$\Delta i_s = -\frac{u_k}{R' + \frac{1}{C_\alpha'} {_0D_t^{-\alpha}} + L_\beta' {_0D_t^\beta}} = -u_k \Delta y_k \tag{3.82}$$

Obviously, the current Δi_s of compensation current source compensates the current $u_k \Delta y_k$ of the incremental admittance Δy_k.

According to Thevenin's theorem, the fractional-order one-port active network shown in Fig. 3.18a can be simplified to a Thevenin's branch. Let its open-circuit voltage be u_{ko} and the equivalent input impedance be z_{in}, as shown in Fig. 3.19. Then, the voltage u_{ko} can be derived as

$$u_{ko} = i_k \left(z_{in} + R + \frac{1}{C_\alpha} {_0D_t^{-\alpha}} + L_\beta {_0D_t^\beta} \right) = i_k (z_k + z_{in}) \tag{3.83}$$

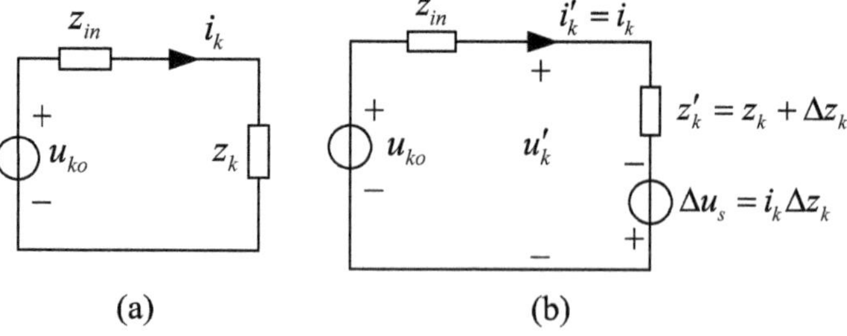

Fig. 3.19 Proof of compensation theorem: **a** original circuit, **b** compensation circuit

From Fig. 3.19b, there is

$$\begin{aligned} u_{ko} + \Delta u_s &= i'_k\left(z_{in} + R + \frac{1}{C_\alpha}{}_0D_t^{-\alpha} + L_\beta\,{}_0D_t^\beta + R' + \frac{1}{C'_\alpha}{}_0D_t^{-\alpha} + L'_\beta\,{}_0D_t^\beta\right) \\ &= i'_k(z_k + z_{in} + \Delta z_k) \end{aligned} \tag{3.84}$$

and the branch current i'_k is

$$i'_k = \frac{u_{ko} + \Delta u_s}{z_{in} + z_k + \Delta z_k} = \frac{i_k(z_k + z_{in}) + i_k\Delta z_k}{z_{in} + z_k + \Delta z_k} = i_k \tag{3.85}$$

The above equations show that the current of the branch is restored to its original value, and its voltage also returns to its original value. Thus, the theorem is proved.

By using the compensation theorem, it is easy to calculate the variation of the branch voltages and currents in the fractional-order network that the circuit parameters have been changed.

Figure 3.20 is an application example of compensation theorem. If the impedance of the fractional-order branch k changes from z_k to $z'_k = z_k + \Delta z_k$, that is

$$\begin{cases} z_{\alpha\beta\_k} = R_k + s^\beta L_{\beta k} \\ \Delta z_{\alpha\beta\_k} = \Delta R_k + s^{\Delta\beta}(\Delta L_{\beta k}) \\ z'_{\alpha\beta\_k} = z_{\alpha\beta\_k} + \Delta z_{\alpha\beta\_k} = R_k + s^\beta L_{\beta k} + \Delta R_k + s(\Delta L_{\beta k}) \end{cases} \tag{3.86}$$

Then, the voltage and current of each branch in the network would change from u_m, i_m, u_k, i_k to

$$\begin{cases} u'_m = u_m + \Delta u_m \\ i'_m = i_m + \Delta i_m \\ u'_k = u_k + \Delta u_k \\ i'_k = i_k + \Delta i_k \end{cases} \tag{3.87}$$

Connecting two voltage sources with equal magnitude and opposite directions in series on the branch k of the fractional-order circuit shown in Fig. 3.20b, as shown in Fig. 3.20c, and applying the superposition theorem to the fractional-order circuit of Fig. 3.20c, as shown in Fig. 3.20d, e, u' and i'_k can be represented as

$$\begin{cases} u'_k = u_k^{(1)} + u_k^{(2)} = u_k + \Delta u_k \\ i'_k = i_k^{(1)} + i_k^{(2)} = i_k + \Delta i_k \end{cases} \tag{3.88}$$

As mentioned before, based on the compensation theorem, there is

$$u_k^{(1)} = u_k, i_k^{(1)} = i_k \tag{3.89}$$

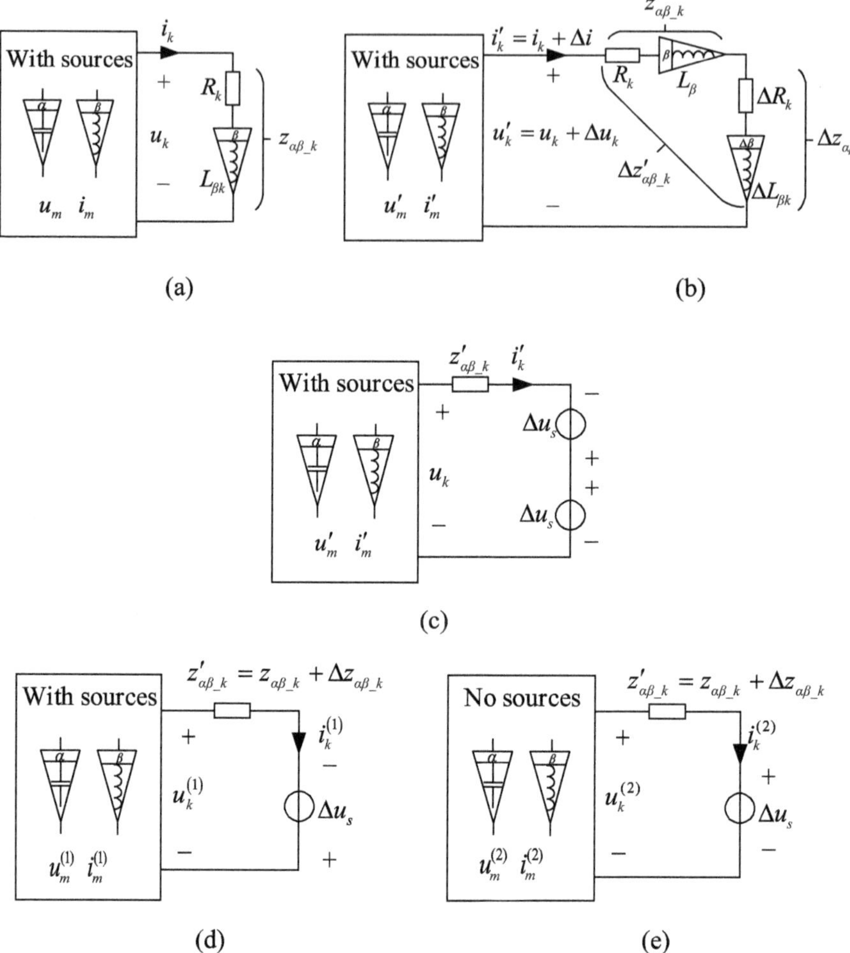

Fig. 3.20 An application example of compensation theorem: **a** original circuit, **b** compensation impedance circuit, **c** compensation voltage sources, **d** reverse compensation voltage source working circuit, **e** positive compensation voltage source working circuit

Then, (3.90) can be obtained.

$$\Delta u_k = u_k^{(2)},\ \Delta i_k = i_k^{(2)} \tag{3.90}$$

From the above analysis, it can be seen that the fractional-order circuit shown in Fig. 3.20e can be used to calculate the variations of the branch voltages and currents, including Δu_m, Δi_m, Δu_k, Δi_k, which are caused by the variation in impedance from z_k to $z'_k = z_k + \Delta z_k$. The voltage direction of the incremental voltage source Δu_s should be consistent with the direction of the original current of the branch.

Furthermore, the Thevenin's branch shown in Fig. 3.20e can also be equivalent to the Norton's branch, in which the equivalent current source is $\Delta i_s = \Delta u_s / \left[R' + (1/C'_\alpha)_0 D_t^{-\alpha} + L'_\beta{}_0 D_t^\beta \right]$, the direction of the current source is opposite to the direction of the branch voltage u_k, and its parallel admittance is $y'_k = 1/z'_k$.

3.8 Bisection Theorem

Bisection theorem is applicable to the transfer symmetric two-port fractional-order network. If the two-port fractional-order network exists a plane that can split it into two completely symmetrical parts, as shown in Fig. 3.21, such a two-port fractional-order network is called the transfer symmetric two-port fractional-order network. The plane oo' splitting this two-port network is called the mid-plane or the transfer symmetric plane.

The bisection theorem on the transfer symmetric two-port fractional-order network can be illustrated from two aspects, including symmetrical and anti-symmetrical excitation, which are described as follows:

1. Transfer symmetric two-port network under symmetrical excitation

Symmetrical excitation refers to the two power sources with equal magnitude and the same direction added to the two ports of the transfer symmetrical two-port network, as shown in Fig. 3.22.

In the transfer symmetrical two-port fractional-order network with symmetrical excitation, the docking branch current at plane oo' is equal to zero, so it can be disconnected, which means

$$\begin{cases} i_1 = 0 \\ i_k = 0 \end{cases} \tag{3.91}$$

and the port voltage formed by the cross-connection at plane oo' is also equal to zero, so it can be short-circuited, which can be represented by

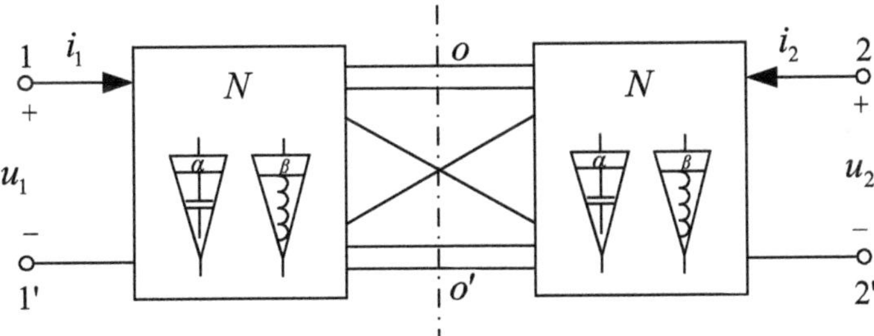

Fig. 3.21 Transfer symmetric two-port fractional-order network

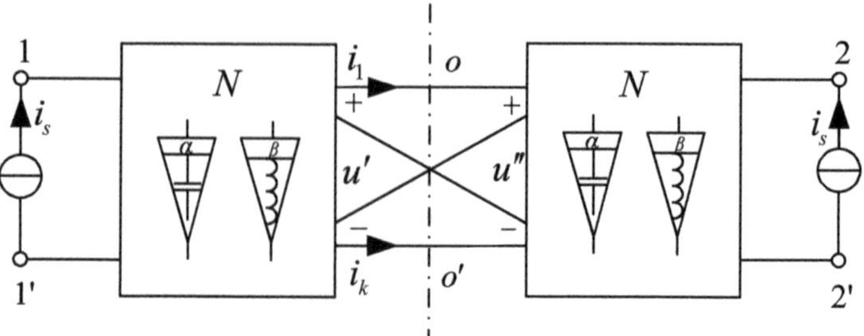

Fig. 3.22 Symmetrical excitation

$$u' = u'' = 0 \tag{3.92}$$

2. Transfer symmetric two-port fractional-order network under anti-symmetrical excitation

Anti-symmetrical excitation is to add two power sources with equal magnitude and opposite directions on the two ports of the symmetrical network, as shown in Fig. 3.23.

In the transfer symmetrical two-port fractional-order network with anti-symmetrical excitation, the intersection between the mid-plane and the docking branch is the equipotential point, which can be short-circuited. As shown in Fig. 3.23, points o and o' are equipotential points and can be short-circuited.

If the transfer symmetrical two-port fractional-order network is operated under the single-port power supply, that is, only one port is powered by the power source, as shown in Fig. 3.24. The current flowing out of the cross-connected ports will be equal to the incoming current, that is, $i_1 = i_2$. Then, under anti-symmetrical excitation, the cross-connected ports can be disconnected. In this case, the port currents satisfy

$$i' = i'' = 0 \tag{3.93}$$

It should be emphasized that only when the transfer symmetrical two-port fractional-order network is excited by a single-port power supply, the current flowing

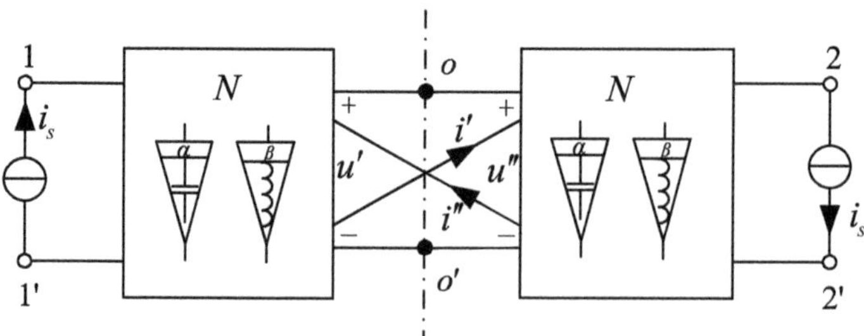

Fig. 3.23 Anti-symmetrical excitation

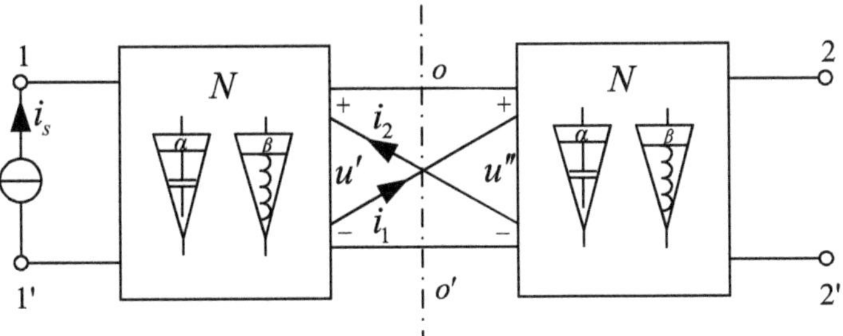

Fig. 3.24 Single-port excitation

out of the cross-connected port is equal to the incoming current, and the current flowing into and out of the cross-connected port under the anti-symmetric excitation will be equal to zero. Otherwise, the inference is not true.

In the fractional-order circuit shown in Fig. 3.25b, under the single-port excitation, the current flowing out of the cross-connected port is not equal to the incoming current, that is, $i_1 \neq i_2$.

In order to facilitate analysis, the circuit of Fig. 3.25b under a single-port excitation can be transformed to the one shown in Fig. 3.25c, there is

$$\begin{cases} 2i_2\Big(R_3 + L_{\beta 3\,0}D_t^{\beta_3}\Big) + i_2\Big(R_2 + L_{\beta 2\,0}D_t^{\beta_2}\Big) = (i_s - 2i_2)\Big(R_3 + L_{\beta 3\,0}D_t^{\beta_3}\Big) \\ i_s = 2i_2 + i_3 \\ i_1 = i_2 + i_3 \end{cases} \tag{3.94}$$

Then, (3.95) can be obtained.

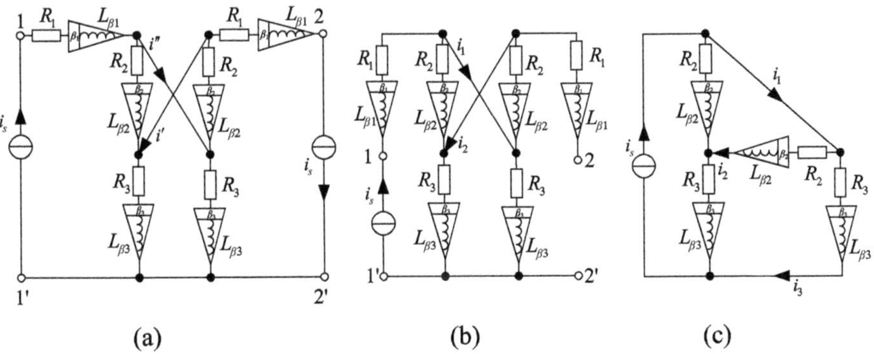

Fig. 3.25 An application example of non-zero cross-connected port current: **a** original circuit, **b** single-port excitation circuit, **c** modified circuit

$$\begin{cases} i_2 = \frac{1}{2} i_s \frac{R_3 + L_{\beta 3\, 0}D_t^{\beta_3}}{0.5\left(R_2 + L_{\beta 2\, 0}D_t^{\beta_2}\right) + 2\left(R_3 + L_{\beta 3\, 0}D_t^{\beta_3}\right)} = i_s \frac{z_3}{z_2 + 4z_3} \\ i_1 = i_2 + i_s \frac{0.5\left(R_2 + L_{\beta 2\, 0}D_t^{\beta_2}\right) + \left(R_3 + L_{\beta 3\, 0}D_t^{\beta_3}\right)}{0.5\left(R_2 + L_{\beta 2\, 0}D_t^{\beta_2}\right) + 2\left(R_3 + L_{\beta 3\, 0}D_t^{\beta_3}\right)} = i_s \frac{z_2 + 3z_3}{z_2 + 4z_3} \end{cases} \quad (3.95)$$

where $z_m = R_m + L_{\beta m\, 0}D_t^{\beta_m}$ $(m = 1, 2, 3)$.

Obviously, there is $i_1 \neq i_2$. Therefore, for the fractional-order circuit shown in Fig. 3.25a, under anti-symmetrical excitation, its cross-connected port branch cannot be disconnected, which means that i' and i'' are not equal to zero.

The conclusions of the above-mentioned bisection theorem under symmetrical and anti-symmetrical excitation can be proved by the superposition theorem. If the two power sources i_{s1}, i_{s2}, or u_{s1}, u_{s2} of the two ports do not constitute the symmetrical or anti-symmetrical excitation, then the excitation of the port can be decomposed into the superposition of a set of symmetrical excitation (i_s' or u_s') and a set of anti-symmetrical excitation (i_s'' or u_s''). It can be described by

$$\begin{cases} i_{s1} = i_s' + i_s'' \\ i_{s2} = i_s' - i_s'' \end{cases} \text{or} \quad \begin{cases} u_{s1} = u_s' + u_s'' \\ u_{s2} = u_s' - u_s'' \end{cases} \quad (3.96)$$

Then, the symmetrical excitation can be written as

$$i_s' = \frac{i_{s1} + i_{s2}}{2} \quad \text{or} \quad u_s' = \frac{u_{s1} + u_{s2}}{2} \quad (3.97)$$

and the anti-symmetrical excitation can be derived as

$$i_s'' = \frac{i_{s1} - i_{s2}}{2} \quad \text{or} \quad u_s'' = \frac{u_{s1} - u_{s2}}{2} \quad (3.98)$$

Furthermore, when using the bisection theorem to analyze fractional-order networks, it is necessary to note that the voltage or current excitation must be added to the two ports simultaneously. It is not allowed to add a voltage source to one port and add a current source to the other port [12, 13].

3.9 Summary

Circuit theorems are the embodiment of the basic properties of the fractional-order network. Learning these circuit theorems can deepen the understanding to the inherent regularity of the fractional-order circuit and directly apply these theorems to solve the fractional-order circuit. The circuit theorems are the basic contents of fractional-order circuit theory, which mainly includes the superposition theorem, substitution theorem, Thevenin's and Norton's theorems, Tellegen's theorem, reciprocity theorem, duality theorem, compensation theorem, and bisection theorem. Through the brief introduction of these circuit theorems of the fractional-order circuit, beginners can have a preliminary understanding of the fractional-order circuit theory and use it as a theoretical basis for in-depth research.

References

1. Johnston JS (1978) Electric circuits. U.S. Patent 4080821
2. Bird J (2003) Electrical circuit theory and technology. Newnes, US
3. Otsuka S (2009) Electric circuit and method for designing electric circuit. U.S. Patent 0108893
4. Gosling G (2006) Electronics texts for engineers and scientists. Cambridge University Press, Cambridge, New York
5. Hayt WH, Durbin SM (2012) Engineering circuit analysis. McGraw-Hill
6. Bohannan GW, Hurst SK, Spangler L (2006) Electrical component with fractional order impedance. U.S. Patent 0267595
7. Cua LO (1987) Linear and nonlinear circuits. McGraw-Hill
8. Bell DA (1978) Fundamentals of electric circuits. Reston Publishing, US
9. Smon I, Verbic G, Gubina F (2006) Local voltage-stability index using Tellegen's theorem. IEEE Trans Power Syst 21(3):1267–1275
10. Gledhill CS (1971) Book review: Tellegen's theorem and electrical networks. Int J Electr Eng Educ 9(5)
11. Monteath GD (2010) Application of the compensation theorem to certain radiation and propagation problems. In: Proceedings of the IEEE—Part IV: institution monographs, vol 98, no 1, pp 23–30
12. Zadeh LA (1962) From circuit theory to system theory. Proc IRE 50(5):856–865
13. Boylestad RL (1987) Introductory circuit analysis. Merrill Publishing Company, London

Chapter 4
Time-Domain Analysis of Fractional-Order Circuits

Fractional-order capacitors and inductors can be regarded as energy storage components since they can store energy like integer-order components while the order influences their impedance. When the switching state of switches or the excitation sources of the circuit containing one or more energy storage components changes, a transient state will occur in the circuit until the new steady-state condition is reached. This chapter will analyze the dynamic behaviors of fractional-order circuits containing energy storage components in the time domain. Fractional-order circuits have some unique characteristics that are quite different from integer-order circuits because of fractional-order components, such as fractional-order capacitors and fractional-order inductors. The time-domain behaviors of fractional-order circuits can be described by fractional differential equations, which are inseparable from the constraints of classical circuit theories such as Kirchhoff's laws. Laplace transform techniques are introduced and then applied to solve fractional differential equations, including analytical and numerical methods. The time-domain analysis will expand from simple circuit analysis with one fractional-order component to complex circuit analysis containing two fractional-order components. The fundamental analysis procedure of fractional-order circuits includes the following four aspects:

(1) Establish fractional differential equations in the time domain according to the basic circuit theory and the property of components.
(2) Derive the fractional differential equation with regard to one variable of the circuit.
(3) Obtain the initial conditions of fractional differential equations according to the initial states of the circuit components.
(4) Solve the fractional differential equations by Laplace transforms and analyze the dynamic behaviors of fractional-order circuits.

B. Zhang and X. Shu, *Fractional-Order Electrical Circuit Theory*, CPSS Power Electronics Series, https://doi.org/10.1007/978-981-16-2822-1_4

4.1 Characteristics of Differential Equations for Fractional-Order Circuits

As shown in Fig. 4.1, there is an unknown potential node in the circuit, and the external excitation is $u(t)$. $C_{\alpha 1}$, $C_{\alpha 2}$, …, $C_{\alpha n}$ are fractional-order capacitors, their orders are $\alpha_1, \alpha_2, \ldots, \alpha_n$, and the initial voltages are $U_{\alpha 1}, U_{\alpha 2}, \ldots, U_{\alpha n}$, respectively. Besides, $L_{\beta 1}, L_{\beta 2}, \ldots, L_{\beta n}$ are fractional-order inductors, their orders are $\beta_1, \beta_2, \ldots, \beta_n$, and the initial currents are $I_{\beta 1}, I_{\beta 2}, \ldots, I_{\beta n}$, respectively. The resistor is R, and all the orders of the fractional-order components satisfy $0 < \alpha_n, \beta_n < 2$ ($n = 1, 2, \ldots, N$). When the switch S is closed at $t = 0$, the fractional differential equation, which describes the dynamic response of the circuit in the time domain, can be obtained by KCL and KVL. The specific process is as follows:

Assuming that the voltage of the node N is $\varphi(t)$, the sum of all the branch currents must satisfy

$$i_s(t) = i_{\alpha 1}(t) + i_{\beta 1}(t) + \ldots + i_{\alpha n}(t) + i_{\beta n}(t) \tag{4.1}$$

where $n = 1, 2, \ldots, N$, and

$$i_s(t) = \frac{u(t) - \varphi(t)}{R} \tag{4.2}$$

The current $i_{\alpha k}$ of each capacitor branch can be calculated as

$$i_{\alpha k}(t) = C_{\alpha k\,0}D_t^{\alpha_k} u_{Ck}(t) = C_{\alpha k}\frac{\mathrm{d}^{\alpha_k} u_{Ck}}{\mathrm{d}t^{\alpha_k}} \tag{4.3}$$

where $k \in N$ and $\alpha_k \in (0, 2)$.

The current $i_{\beta k}$ of each inductor branch can be calculated as

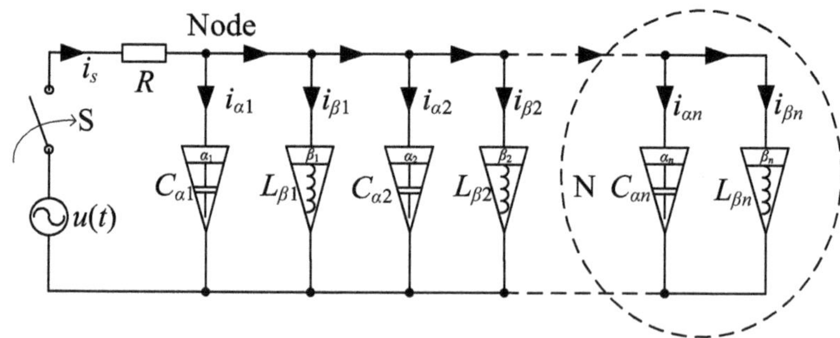

Fig. 4.1 General fractional-order circuits

$$i_{\beta k}(t) = \frac{1}{L_{\beta k}} {}_0D_t^{-\beta_k} u_{Lk}(t) = \frac{1}{L_{\beta k}} \int_0^t \frac{1}{\Gamma(\beta)} (t-\tau)^{\beta_k - 1} u_{Lk}(\tau) d\tau \tag{4.4}$$

where $k \in N$ and $\beta_k \in (0, 2)$.

While the node voltage satisfies $\varphi(t) = u_{Ck}(t) = u_{Lk}(t)$, the fractional differential equation with regard to node potential φ can be stated as

$$\begin{cases} \sum_{k=1}^{N} \left[\frac{R}{L_{\beta k}} {}_0D_t^{-\beta_k} \varphi(t) + RC_{\alpha k} {}_0D_t^{\alpha_k} \varphi(t) \right] + \varphi(t) = u(t) \\ u_{\alpha k}(0) = U_{\alpha k}, i_{\beta k}(0) = I_{\beta k} \end{cases} \tag{4.5}$$

Equation (4.5) contains fractional integral and differential operators, and its analytical solution can be solved by applying Laplace transform in the complex frequency domain [1].

In order to verify the correctness of the analytical solution, this chapter can also obtain numerical solutions and curves of fractional differential equations by applying inverse Laplace transform [2]. All analytical solutions are expressed as combinations of Mittag–Leffler function (MLF), while the associated curves are drawn by using MATLAB MLF function written by Igor Podlubny [3], and FOTF Toolbox (V1.3) written by Professor Dingyu Xue [4].

Based on the analysis of general fractional differential equations in Chap. 1, it can be found that the analytical solutions of time-domain response for fractional-order circuits are related to two-parameter Mittag–Leffler function $E_{\alpha,\beta}(\lambda t)$ [5, 6]. Hence, we can define this particular type of function family as eigenfunctions, which can illustrate response characteristics of the fractional-order circuit in the time domain. The eigenfunctions under the R-L and Caputo definitions are different, and they will be applied to analyze the behaviors of fractional-order circuits, respectively.

The eigenfunction of the R-L fractional-order operator can be defined as

$$f_{\mathrm{RL}}(t, \alpha, \lambda) = t^{\alpha-1} E_{\alpha,\alpha}(\lambda t^{\alpha}) = \sum_{k=0}^{\infty} \frac{\lambda^k t^{\alpha k + \alpha - 1}}{\Gamma(\alpha k + \alpha)} \tag{4.6}$$

and the eigenfunction of Caputo fractional-order operator can be defined as

$$f_{\mathrm{Cap}}(t, \alpha, \lambda) = E_{\alpha,1}(\lambda t^{\alpha}) = \sum_{k=0}^{\infty} \frac{\lambda^k t^{\alpha k}}{\Gamma(\alpha k + 1)} \tag{4.7}$$

where λ is the eigenvalue of the circuit, and the time constant τ is

$$\tau = \frac{1}{\lambda} \tag{4.8}$$

Eigenvalue λ and time constant τ are determined by the topology of the circuit and the values of circuit components, which determine the transient response speed [7, 8].

The eigenvalues of fractional-order circuits are always negative, and the property of eigenfunction f is determined by the fractional-order components and circuit topologies, which can be described as follows:

(1) As $\alpha \to 0$, $f_{\mathrm{RL}}(t, \alpha, \lambda)$ is closed to impulse function, and it is not significant in the real circuit. $f_{\mathrm{Cap}}(t, \alpha, \lambda)$ is approached to step function.
(2) When $\alpha = 1$, there is $f_{\mathrm{RL}}(t, \alpha, \lambda) = f_{\mathrm{Cap}}(t, \alpha, \lambda) = \mathrm{e}^{-\lambda t}$, which is the same as the case of integer-order circuits.
(3) As $1 < \alpha < 2$, both $f_{\mathrm{RL}}(t, \alpha, \lambda)$ and $f_{\mathrm{Cap}}(t, \alpha, \lambda)$ begin to oscillate and eventually decay to zero, which is the stable case.
(4) When $\alpha = 2$, the eigenfunction f will oscillate forever like the sinusoidal function, which is the conditionally stable case.

As for the general circuits containing fractional-order components, although the input is bounded, the stability of the output should be discussed case by case. The eigenfunction curves of different orders under R-L and Caputo definitions are shown in Fig. 4.2, respectively.

4.2 Zero-Input Response of Fractional-Order RC_α and RL_β Circuits

The time-domain response of the simple circuit containing one fractional-order element will be discussed in this section, including RC_α and RL_β circuits. Zero-input response refers to the response of the circuit containing non-zero state energy storage components without an external excitation source, which means that the initial energy of the inductor or capacitor is not equal to zero, so the transition is only generated by the initial internal energy. According to the definitions of fractional differential operators, the initial conditions required to solve the fractional differential equations based on R-L and Caputo definitions are different. Hence, there is a significant difference between the dynamic behaviors of the circuit with non-zero initial states, which are described by the two fractional differential definitions. The analytical solutions of the zero-input response of RC_α and RL_β circuits will be derived from R-L and Caputo definitions, respectively. And the results under different parameters are compared and analyzed.

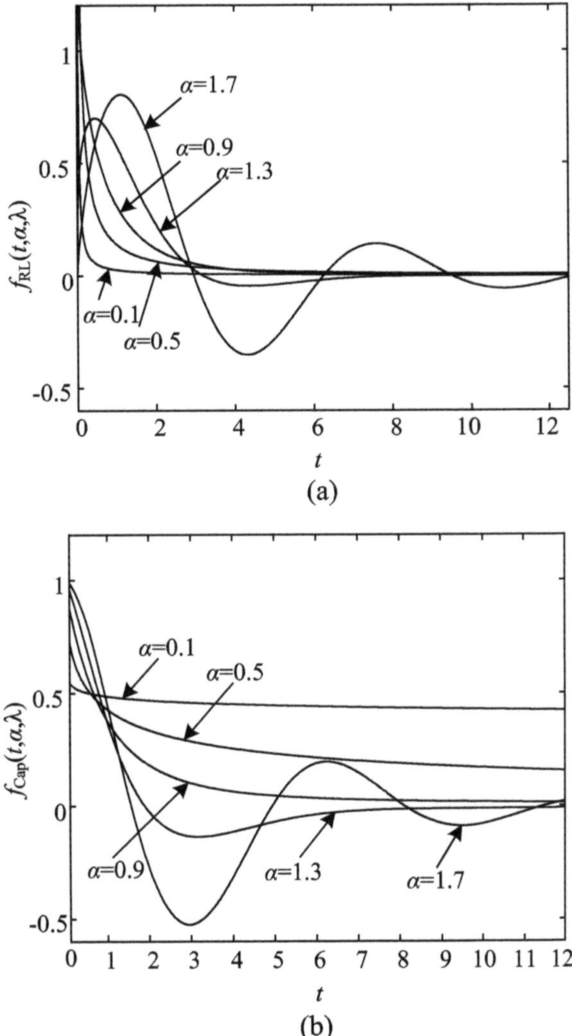

Fig. 4.2 Eigenfunctions of fractional-order operators: **a** Eigenfunction of R-L fractional operator. **b** Eigenfunction of Caputo fractional operator

4.2.1 Fractional-Order RC_α Circuit

As shown in Fig. 4.3, the original position of switch S in the RC_α circuit was terminal "1", and it does not change from "1" to "2" until the voltage across the fractional-order capacitor C_α is charged to U_0. At $t = 0_+$, let the initial charge stored in the capacitor be Q_0, then we will analyze the zero-input response of the RC_α circuit as $t > 0$. The zero-input response of the RC_α circuit represents the process of the capacitor discharging.

Based on KVL and the property of components, fractional differential equation with regard to the voltage $u_{C\alpha}$ across the capacitor satisfies

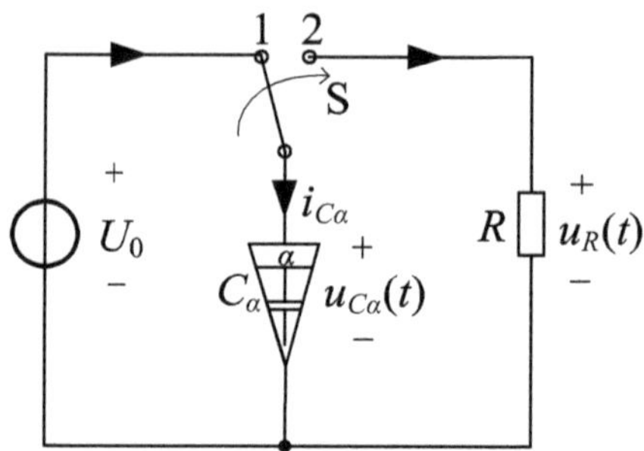

Fig. 4.3 Fractional-order RC_α circuit with zero input

$$\begin{cases} C_\alpha \frac{\mathrm{d}^\alpha u_{C\alpha}}{\mathrm{d}t^\alpha} + \frac{1}{R} u_{C\alpha} = 0 \\ u_{C\alpha}(0_+) = u_{C\alpha}(0_-) = U_0 \\ Q_{C\alpha}(0_+) = Q_{C\alpha}(0_-) = Q_0 \end{cases} \tag{4.9}$$

where $t > 0$.

The initial charge Q_0 of fractional-order capacitor is the integral of the charging current as $t < 0$, which is determined by the charging voltage

$$Q_0 = \int_{-\infty}^{0} i_{C\alpha}(\tau)\mathrm{d}\tau = \int_{-\infty}^{0} C_\alpha \frac{\mathrm{d}^\alpha u_{C\alpha}(\tau)}{\mathrm{d}t^\alpha}\mathrm{d}\tau \tag{4.10}$$

1. Solutions under R-L Fractional Definition

According to the R-L definition, using the Laplace transform to (4.9), it can be stated as

$$s^\alpha U_{C\alpha}(s) - \left[{}_0D_t^{\alpha-1} u_{C\alpha}(t)\right]_{t=0} + \frac{1}{RC_\alpha} U_{C\alpha}(s) = 0 \tag{4.11}$$

where $\alpha \in (0, 1)$.

Then, the capacitor voltage can be obtained by

$$U_{C\alpha}(s) = \frac{\left[{}_0D_t^{\alpha-1} u_{C\alpha}(t)\right]_{t=0}}{s^\alpha + 1/(RC_\alpha)} \tag{4.12}$$

where $\left[{}_0D_t^{\alpha-1} u_{C\alpha}(t)\right]_{t=0}$ is the initial condition required by Laplace transform, and the value is related to the initial energy stored in the capacitor. Since the current of the fractional-order capacitor is defined by $i_C = \frac{\mathrm{d}q_C}{\mathrm{d}t} = C_\alpha \frac{\mathrm{d}^\alpha u_C}{\mathrm{d}t^\alpha} = C_\alpha\, {}_0D_t^\alpha u_C$, the fractional-order operator ${}_0D_t^{\alpha-1}$can be considered as the integral of the charging current of the capacitor, and its result is the charge Q_0 stored in the capacitor [9]. Hence, the initial condition for solving the fractional differential equation is

$$\left[{}_0D_t^{\alpha-1}u_{C\alpha}(t)\right]_{t=0}=\frac{Q_0}{C_\alpha} \tag{4.13}$$

Applying the inverse Laplace transform to (4.12), the analytical solution of the capacitor voltage $u_{C\alpha}$ in the time domain can be obtained as

$$\begin{aligned}u_{C\alpha}(t)&=\mathcal{L}^{-1}[u_{C\alpha}(s)]=U_0t^{\alpha-1}E_{\alpha,\alpha}\left(-\frac{1}{RC_\alpha}t^\alpha\right)\\&=\frac{Q_0}{C_\alpha}\sum_{k=0}^{\infty}\left(-\frac{1}{RC_\alpha}\right)^k\frac{t^{\alpha k+\alpha-1}}{\Gamma(\alpha k+\alpha)}\end{aligned} \tag{4.14}$$

where $t > 0$.

Due to characteristics of the series circuit and the property of components, other circuit variables can be easily derived as follows:

The voltage u_R of the resistor is

$$u_R(t)=\frac{Q_0}{C_\alpha}t^{\alpha-1}E_{\alpha,\alpha}\left(-\frac{1}{RC_\alpha}t^\alpha\right) \tag{4.15}$$

The current i_R of the resistor is

$$i_R(t)=\frac{u_R}{R}=\frac{Q_0}{RC_\alpha}t^{\alpha-1}E_{\alpha,\alpha}\left(-\frac{1}{RC_\alpha}t^\alpha\right) \tag{4.16}$$

The current $i_{C\alpha}$ of the capacitor is

$$i_{C\alpha}(t)=-i_R(t)=-\frac{Q_0}{RC_\alpha}t^{\alpha-1}E_{\alpha,\alpha}\left(-\frac{1}{RC_\alpha}t^\alpha\right) \tag{4.17}$$

where $t > 0$.

Since the voltages and currents in RC_α circuits are not continuous at $t = 0$ (when $t < 0$, $u_{C\alpha}(t) = U_0$, $i_R(t) = 0$, $i_{C\alpha}(t) = 0$), the unit step function $\varepsilon(t)$ can be applied to the above equations, which can represent the instant switching action. Hence, the final expressions of circuit variables for all time can be written as follows:

The voltage $u_{C\alpha}$ of the capacitor is

$$u_{C\alpha}(t)=\frac{Q_0}{C_\alpha}t^{\alpha-1}E_{\alpha,\alpha}\left(-\frac{1}{RC_\alpha}t^\alpha\right)\varepsilon(t)+U_0\varepsilon(-t) \tag{4.18}$$

The voltage u_R of the resistor is

$$u_R(t)=\frac{Q_0}{C_\alpha}t^{\alpha-1}E_{\alpha,\alpha}\left(-\frac{1}{RC_\alpha}t^\alpha\right)\varepsilon(t) \tag{4.19}$$

The current i_R of the resistor is

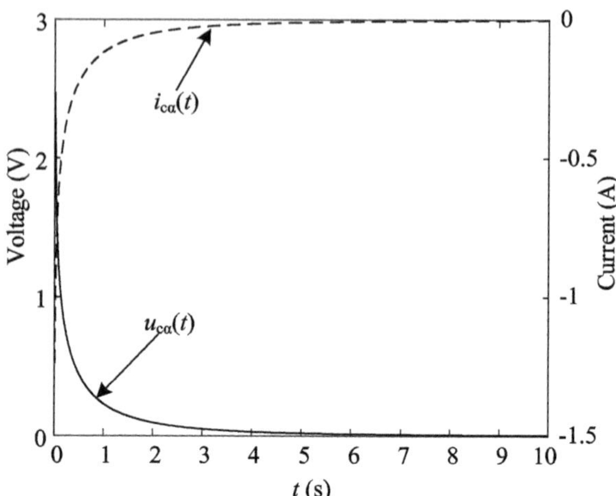

Fig. 4.4 RC_α circuit discharges under R-L fractional operator for $R = 2\ \Omega$, $C_\alpha = 0.5\ \text{F/s}^{1-\alpha}$, $\alpha = 0.75$, $U_0 = 1\ \text{V}$

$$i_R(t) = \frac{Q_0}{RC_\alpha} t^{\alpha-1} E_{\alpha,\alpha}\left(-\frac{1}{RC_\alpha} t^\alpha\right)\varepsilon(t) \tag{4.20}$$

The current $i_{C\alpha}$ of the capacitor is

$$i_{C\alpha}(t) = -\frac{Q_0}{RC_\alpha} t^{\alpha-1} E_{\alpha,\alpha}\left(-\frac{1}{RC_\alpha} t^\alpha\right)\varepsilon(t) \tag{4.21}$$

Figure 4.4 shows the transition state of RC_α circuit discharging under R-L definition, which illustrates that the current and voltage of capacitor eventually decay to zero.

2. Solutions under Caputo Fractional Definition

According to Caputo definition, applying the Laplace transform to (4.9), it can be stated as

$$s^\alpha U_{C\alpha}(s) - s^{\alpha-1} u_{C\alpha}(0) + \frac{1}{RC_\alpha} U_{C\alpha}(s) = 0 \tag{4.22}$$

where $\alpha \in (0, 1)$.

The capacitor voltage in the complex frequency domain is

$$U_{C\alpha}(s) = \frac{s^{\alpha-1} u_{C\alpha}(0)}{s^\alpha + 1/(RC_\alpha)} \tag{4.23}$$

where $u_{C\alpha}(0) = U_0$ is the initial condition required by Laplace transform, and its value is related to the initial voltage across the fractional-order capacitor.

The analytical solution of the capacitor voltage in the time domain can be derived by the inverse Laplace transform of (4.23) as

$$u_{C\alpha}(t) = \mathcal{L}^{-1}[u_{C\alpha}(s)] = U_0 E_{\alpha,1}\left(-\frac{1}{RC_\alpha}t^\alpha\right)$$
$$= U_0 \sum_{k=0}^{\infty} \frac{1}{\Gamma(\alpha k+1)}\left(-\frac{t^\alpha}{RC_\alpha}\right)^k \quad (4.24)$$

where $t > 0$.

Then, applying the unit step function $\varepsilon(t)$ to the above-mentioned equations, the final expressions of currents and voltages for all time can be written as follows:

The capacitor voltage $u_{C\alpha}$ is

$$u_{C\alpha}(t) = U_0 E_{\alpha,1}\left(-\frac{1}{RC_\alpha}t^\alpha\right)\varepsilon(t) + U_0\varepsilon(-t) \quad (4.25)$$

The resistor voltage u_R is

$$u_R(t) = U_0 E_{\alpha,1}\left(-\frac{1}{RC_\alpha}t^\alpha\right)\varepsilon(t) \quad (4.26)$$

The resistor current i_R is

$$i_R(t) = \frac{U_0}{R} E_{\alpha,1}\left(-\frac{1}{RC_\alpha}t^\alpha\right)\varepsilon(t) \quad (4.27)$$

The capacitor current $i_{C\alpha}$ is

$$i_{C\alpha}(t) = -\frac{U_0}{R} E_{\alpha,1}\left(-\frac{1}{RC_\alpha}t^\alpha\right)\varepsilon(t) \quad (4.28)$$

Figure 4.5 shows the transition state of RC_α circuit discharging under Caputo definition, which illustrates that the current and voltage of capacitor finally decay to zero.

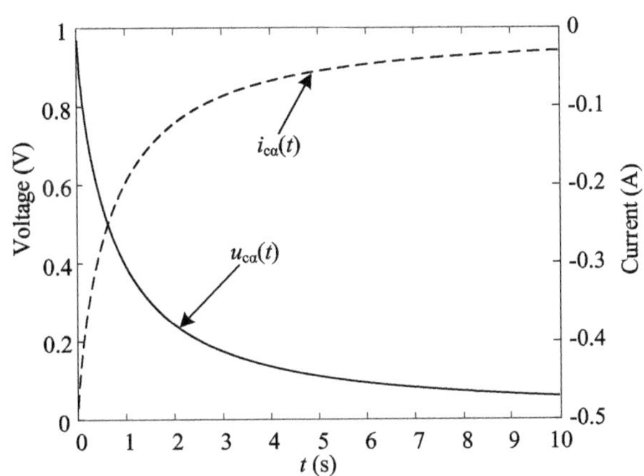

Fig. 4.5 RC_α circuit discharges under Caputo fractional operator for $R = 2\ \Omega$, $C_\alpha = 0.5$ F/s$^{1-\alpha}$, $\alpha = 0.75$, $U_0 = 1$ V

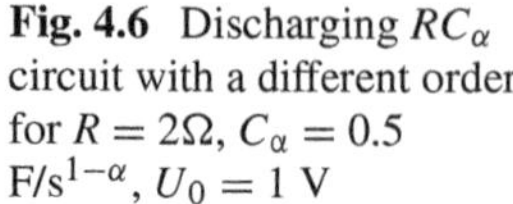

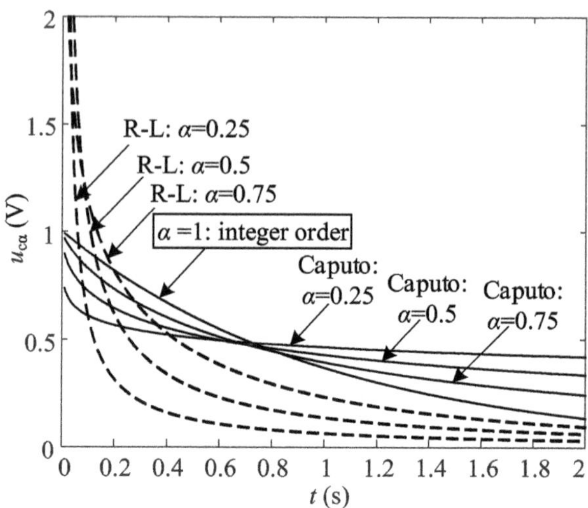

Fig. 4.6 Discharging RC_α circuit with a different order for $R = 2\Omega$, $C_\alpha = 0.5$ F/s$^{1-\alpha}$, $U_0 = 1$ V

3. Comparison between Solutions of R-L and Caputo Fractional Definitions

As shown in Fig. 4.6, there is an obvious voltage impulse under R-L definition as $t \to 0$, which is much higher than the initial voltage across the fractional-order capacitor, and it decays to zero rapidly. However, the maximum voltage of RC_α circuit discharging under Caputo definition is the same as the initial voltage across the fractional-order capacitor as $t \to 0$, and it decays to zero much slower than R-L definition. If the order of fractional-order capacitor is $\alpha = 1$, the analytical solutions based on R-L and Caputo definitions are both equal to the exponential function, which leads to the same results of classical integer-order circuits.

In order to analyze the discharging speed of RC_α circuit, we define the settling time t_{ss} as the time at which the response reaches 98% of its final value. As for the RC_α circuit, settling time t_{ss} is calculated by

$$\frac{u_{C\alpha}(0) - u_{C\alpha}(t)}{u_{C\alpha}(0)} = 0.98 \tag{4.29}$$

As shown in Fig. 4.7, the settling time of the RC_α circuit discharging under the R-L or Caputo definition is related to the order of the capacitor. Obviously, for the Caputo definition, the described settling time t_{ss} increases quickly with the decrease of order α, and $t_{ss} \to \infty$ as $\alpha \to 0$. For the R-L definition, although the settling time t_{ss} varies with the capacitor order α, it always takes a finite time, which is much shorter than the result of the Caputo definition.

The surfaces of the RC_α circuit discharged under R-L and Caputo definitions are shown in Fig. 4.8 and Fig. 4.9, respectively. With the same initial discharging voltage, the total discharging time is affected by the order and the time constant, the higher

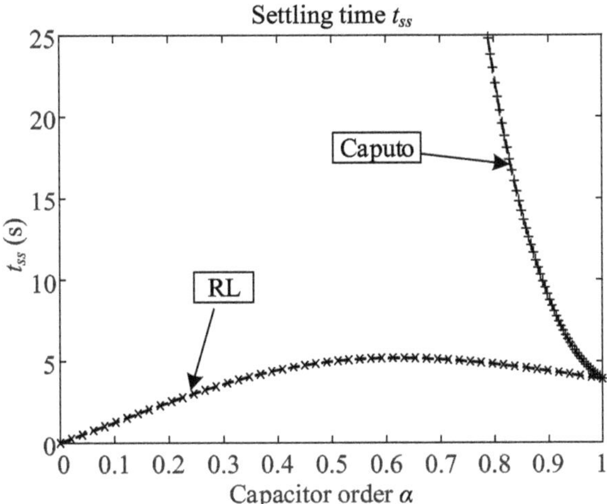

Fig. 4.7 Settling time t_{ss} of RC_α circuit with different orders

the order α is, the longer the discharge time, and the smaller the order α is, the shorter the discharge time [10]. The time constant τ of the RC_α circuit is defined as

$$\tau = RC_\alpha \tag{4.30}$$

According to the analytical solutions of RC_α circuit discharging, it can be concluded that the discharging time will increase with the increase of time constant τ, which demonstrates that a larger time constant will reduce the discharging speed.

4.2.2 Fractional-Order RL_β Circuit

As shown in Fig. 4.10, the switch S in the RL_β circuit was originally located at position "1" until the initial current flowing through the inductor L_β is I_0, and then it switches from "1" to "2" at $t = 0$. Assuming that the initial magnetic flux is ψ_0 at $t = 0_+$, we will analyze the zero-input response of the RL_β circuit as $t > 0$. The zero-input response of the RL_β circuit represents the process of the inductor discharging. According to KVL, the fractional differential equation with regard to inductor current $i_{L\beta}$ can be written as

$$\begin{cases} L_\beta \frac{\mathrm{d}^\beta i_{L\beta}}{\mathrm{d}t^\beta} + Ri_{L\beta} = 0 \\ i_{L\beta}(0_+) = i_{L\beta}(0_-) = I_0 \\ \psi_{L\beta}(0_+) = \psi_{L\beta}(0_-) = \psi_0 \end{cases} \tag{4.31}$$

where $t > 0$.

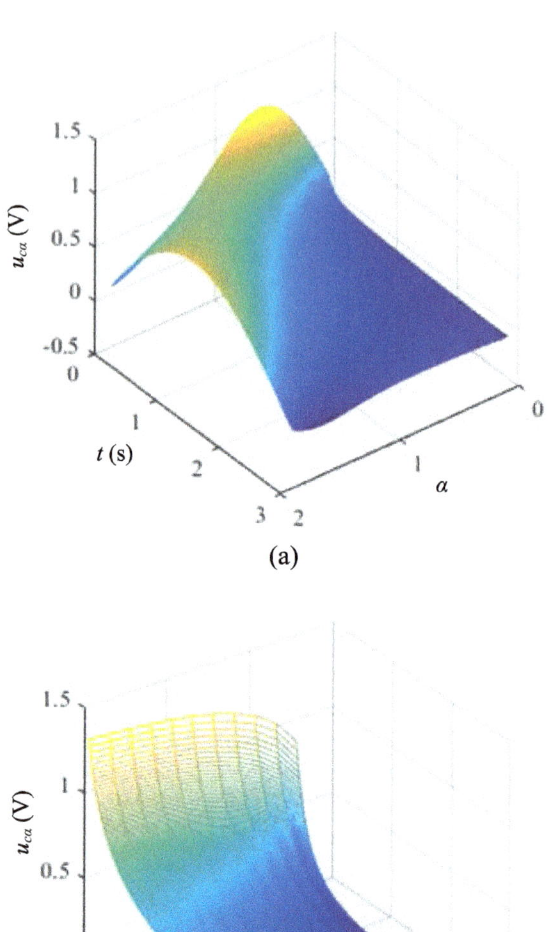

Fig. 4.8 The surfaces of the RC_α circuit discharging under R-L definition for initial voltage $U_0 = 1$ V: **a** Relationship between discharging time and order of the capacitor. **b** Relationship between discharging time and time constant

The initial magnetic flux ψ_0 of the fractional-order inductor is the integral of the charging voltage, which is determined by the charging current of the whole process of $t < 0$, that is

$$\psi_0 = \int_{-\infty}^{0} u_{L\beta}(\tau)\mathrm{d}\tau = \int_{-\infty}^{0} L_\beta \frac{\mathrm{d}^\beta i_{L\beta}(\tau)}{\mathrm{d}t^\beta}\mathrm{d}\tau \tag{4.32}$$

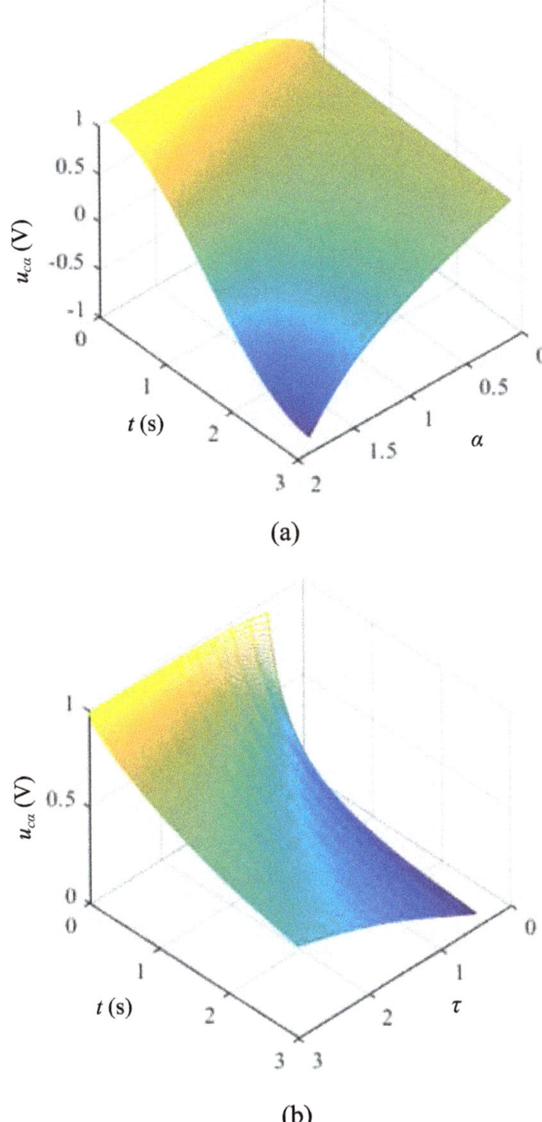

Fig. 4.9 The surfaces of the RC_α circuit discharging under Caputo definition for $U_0 = 1$ V: **a** Relationship between discharging time and capacitor order. **b** Relationship between discharging time and time constant

1. Solutions under R-L Fractional Definition

Based on R-L definition, taking Laplace transform of (4.31), it can be stated as

$$s^\beta I_{L\beta}(s) - \left[{}_0D_t^{\beta-1} i_{L\beta}(t)\right]_{t=0} + \frac{R}{L_\beta} I_{L\beta}(s) = 0 \tag{4.33}$$

where $0 < \beta \le 1$.

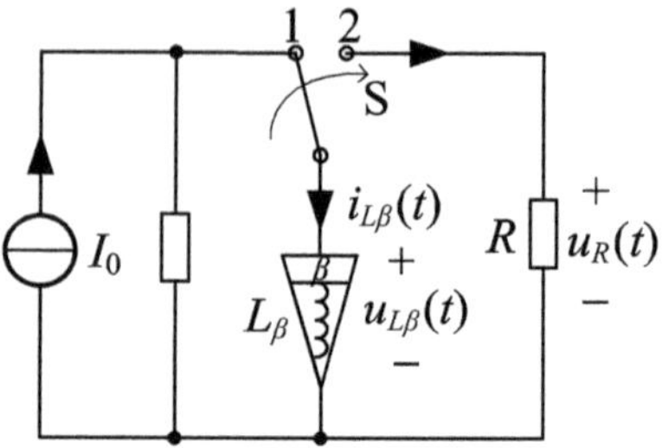

Fig. 4.10 Fractional-order RL_β circuit with zero input

The inductor current in the complex frequency domain is

$$I_{L\beta}(s) = \frac{\left[{}_0D_t^{\beta-1} i_{L\beta}(t)\right]_{t=0}}{s^\beta + R/L_\beta} \tag{4.34}$$

where $\left[{}_0D_t^{\beta-1} i_{L\beta}(t)\right]_{t=0}$ is the initial condition required by Laplace transform, and its value is related to the initial energy stored in the inductor. Since the voltage of the fractional-order inductor is defined by $u_L = \frac{\mathrm{d}\Psi_L}{\mathrm{d}t} = L_\beta \frac{\mathrm{d}^\beta i_L}{\mathrm{d}t^\beta} = L_\beta\, {}_0D_t^\beta i_L$, the fractional-order operator ${}_0D_t^{\beta-1}$ can be considered as the integral of charging voltage, and its result represents the charge stored in inductor ψ_0. Hence, the initial condition for solving the differential equation is

$$\left[{}_0D_t^{\beta-1} i_{L\beta}(t)\right]_{t=0} = \frac{\Psi_0}{L_\beta} \tag{4.35}$$

Applying the inverse Laplace transform to (4.34), the analytical solution of the inductor current $i_{L\beta}$ in the time domain can be obtained as

$$i_{L\beta}(t) = \mathcal{L}^{-1}\left[I_{L\beta}(s)\right] = I_0 t^{\beta-1} E_{\beta,\beta}\left(-\frac{R}{L_\beta} t^\beta\right) = I_0 \sum_{k=0}^{\infty} \left(-\frac{R}{L_\beta}\right)^k \frac{t^{\beta k+\beta-1}}{\Gamma(\beta k+\beta)} \tag{4.36}$$

where $t > 0$.

According to the characteristics of the parallel circuit, the currents and voltages of the circuit can be easily derived as follows:

The current $i_{L\beta}$ of the inductor is

$$i_{L\beta}(t) = \frac{\Psi_0}{L_\beta} t^{\beta-1} E_{\beta,\beta}\left(-\frac{R}{L_\beta} t^\beta\right) \varepsilon(t) + I_0 \varepsilon(-t) \tag{4.37}$$

The current i_R of the resistor is

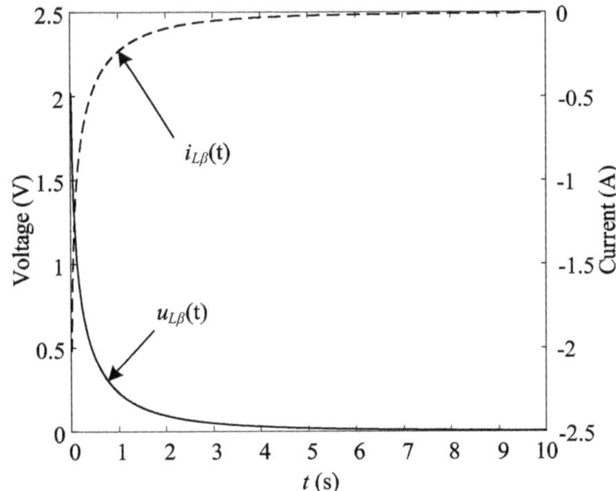

Fig. 4.11 RL_β circuit discharges under R-L fractional operator for $R = 1$ Ω, $L_\beta = 1$ H/s$^{1-\beta}$, $\beta = 0.75$, $I_0 = 1$ A

$$i_R(t) = -\frac{\Psi_0}{L_\beta} t^{\beta-1} E_{\beta,\beta}\left(-\frac{R}{L_\beta} t^\beta\right)\varepsilon(t) \tag{4.38}$$

The voltage u_R of the resistor is

$$u_R(t) = -\frac{\Psi_0}{L_\beta} R t^{\beta-1} E_{\beta,\beta}\left(-\frac{R}{L_\beta} t^\beta\right)\varepsilon(t) \tag{4.39}$$

The voltage $u_{L\beta}$ of the inductor is

$$u_{L\beta}(t) = -\frac{\Psi_0}{L_\beta} R t^{\beta-1} E_{\beta,\beta}\left(-\frac{R}{L_\beta} t^\beta\right)\varepsilon(t) \tag{4.40}$$

Figure 4.11 shows the transition state of RL_β circuit discharging under R-L definition, which illustrates that the current and voltage of the inductor finally decay to zero.

2. Solutions under Caputo Fractional Definition

Based on Caputo definition, applying Laplace transform to (4.31), it can be derived as

$$s^\beta I_{L\beta}(s) - s^{\beta-1} i_{L\beta}(0) + \frac{R}{L_\beta} I_{L\beta}(s) = 0 \tag{4.41}$$

The inductor current in the complex frequency domain is

$$I_{L\beta}(s) = \frac{s^{\beta-1} i_{L\beta}(0)}{s^\beta + R/L_\beta} \tag{4.42}$$

where $i_{L\beta}(0) = I_0$is the initial condition required by Laplace transform, and its value is related to the initial current flowing through the inductor.

The analytical solution of the current flowing through the inductor in the time domain can be obtained by the inverse Laplace transform of (4.42) as

$$i_{L\beta}(t) = \mathcal{L}^{-1}\big[I_{L\beta}(s)\big] = I_0 E_{\beta,1}\left(-\frac{R}{L_\beta}t^\beta\right) = I_0 \sum_{k=0}^{\infty} \frac{1}{\Gamma(\beta k + 1)}\left(-\frac{R}{L_\beta}t^\beta\right)^k \tag{4.43}$$

where $t > 0$.

According to the characteristics of the parallel circuit, the currents and voltages of the fractional-order circuit can be easily derived as follows:

The current $i_{L\beta}$ of the inductor is

$$i_{L\beta}(t) = I_0 E_{\beta,1}\left(-\frac{R}{L_\beta}t^\beta\right)\varepsilon(t) + I_0\varepsilon(-t) \tag{4.44}$$

The current i_R of the resistor is

$$i_R(t) = -I_0 E_{\beta,1}\left(-\frac{R}{L_\beta}t^\beta\right)\varepsilon(t) \tag{4.45}$$

The voltage u_R across the resistor is

$$u_R(t) = -I_0 R E_{\beta,1}\left(-\frac{R}{L_\beta}t^\beta\right)\varepsilon(t) \tag{4.46}$$

The voltage $u_{L\beta}$ across the inductor is

$$u_{L\beta}(t) = -I_0 R E_{\beta,1}\left(-\frac{R}{L_\beta}t^\beta\right)\varepsilon(t) \tag{4.47}$$

Figure 4.12 shows the transition state of $R_{L\beta}$ circuit discharging under Caputo definition. However, the voltage and current of the fractional-order inductor described under Caputo definition doesn't decay to zero even after a long time.

3. Comparison between Solutions of R-L and Caputo Fractional Definitions

As shown in Fig. 4.13, as $t \to 0$, there is an obvious current impulse during the discharge of the RL_β circuit under R-L definition, which is much higher than the initial current flowing through the fractional-order inductor, and it decays to zero quickly. However, as $t \to 0$, the maximum current of RL_β circuit discharging under Caputo definition is the same as the initial current of the fractional-order inductor, and it decays to zero much slower than that of the R-L definition. When the order of inductor satisfies $\alpha = 1$, the analytical solutions obtained by R-L and Caputo

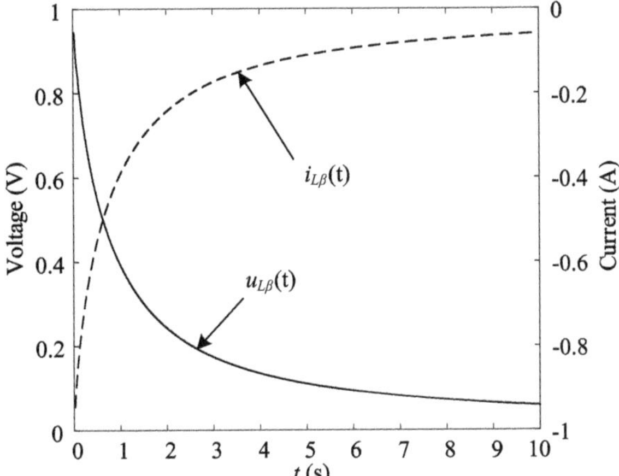

Fig. 4.12 RL_β circuit discharging under Caputo fractional operator for $R = 1$ Ω, $L_\beta = 1$ H/s$^{1-\beta}$, $\beta = 0.75$, $I_0 = 1$ A

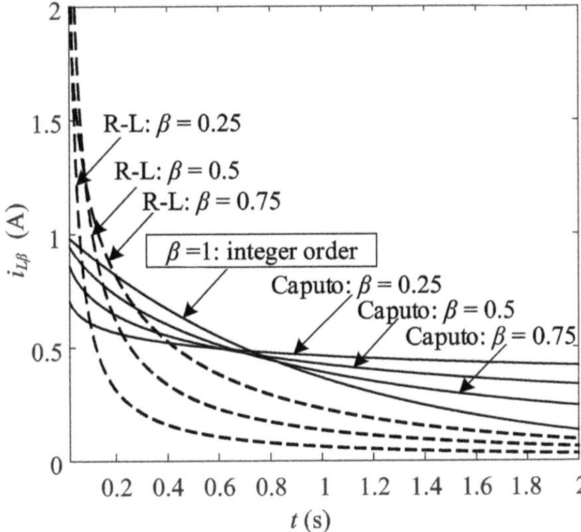

Fig. 4.13 Discharging RL_β circuit with different orders for $R = 1$ Ω, $L_\beta = 1$ H/s$^{1-\beta}$, $I_0 = 1$ A

definitions are equal to the exponential function, which are in agreement with the same results of the classical integer-order circuit.

The surfaces of the RL_β circuit discharging under R-L and Caputo are shown in Fig. 4.14 and Fig. 4.15, respectively. With the same initial discharging voltage, the total discharging time is affected by the order of the capacitor and the time constant [10]. While the higher the order β, the longer the discharge time. Namely, the smaller the order β, the shorter the discharge time. The time constant τ of the RL_β circuit is defined by

$$\tau = \frac{L_\beta}{R} \tag{4.48}$$

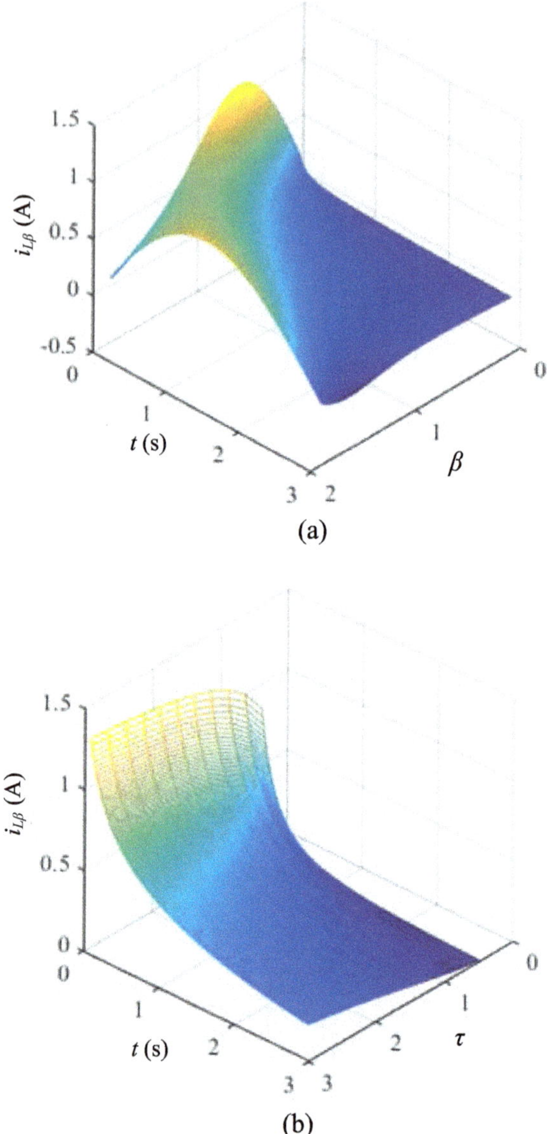

Fig. 4.14 The surfaces of the RL_β circuit discharging under R-L definition for initial current $I_0 = 1$ A: **a** Relationship between discharging time and the order of inductor. **b** Relationship between discharging time and time constant

According to the analytical solutions of RL_β circuit discharging, we can conclude that the discharging time of the RL_β circuit will increase with the increase of time constant τ, which means that a larger time constant will slow down the discharging speed.

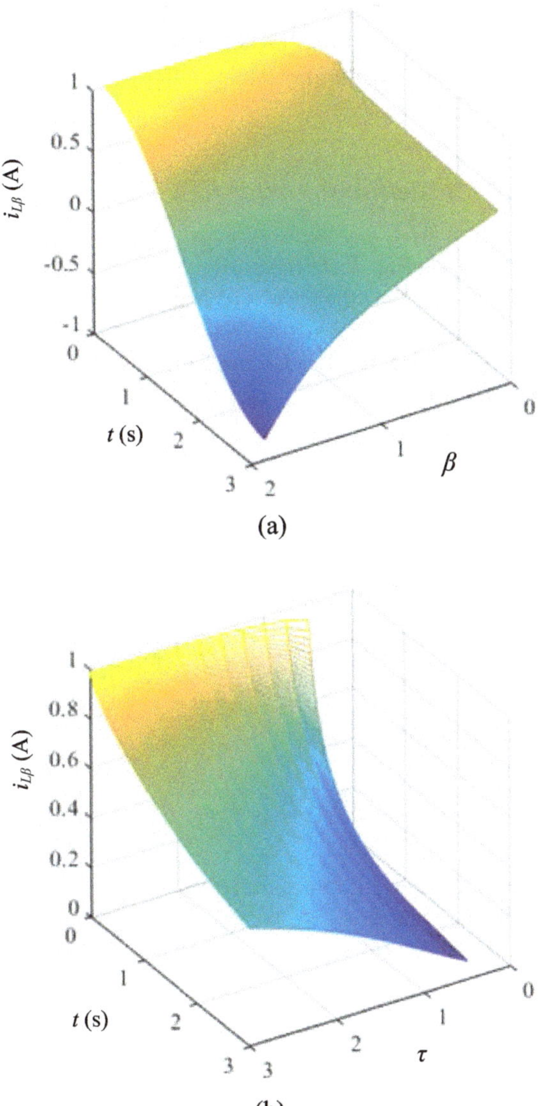

Fig. 4.15 The surfaces of the RL_β circuit discharging under Caputo definition for $I_0 = 1\text{A}$: **a** Relationship between discharging time and the order of inductor. **b** Relationship between discharging time and time constant

4.3 Zero-State Response of Fractional-Order RC_α and RL_β Circuits

The zero-state response refers to the response of the circuit in which the initial state of the energy storage component is equal to zero and only an external excitation source is working, which means that the initial voltage of the capacitor in the circuit is zero and the initial current of the inductor is zero. If initial conditions of fractional differential equations are equal to zero, the Laplace transform of the equation under R-L and Caputo definitions will lead to the same results. The transient responses of the RC_α and RL_β circuits have similar properties.

4.3.1 Fractional-Order RC_α Circuit

As shown in Fig. 4.16, the current source is i_s and switch S in the RC_α circuit turns off at $t = 0$, and then we will analyze the zero-input response of the circuit as $t > 0$. Actually, the zero-state response of the RC_α circuit represents the process of capacitor charging. By KCL and the property of components, the voltage of capacitor $u_{C\alpha}$ satisfies

$$\begin{cases} C_\alpha \frac{\mathrm{d}^\alpha u_{C\alpha}}{\mathrm{d}t^\alpha} + \frac{1}{R} u_{C\alpha} = i_s \\ u_{C\alpha}(0_+) = u_{C\alpha}(0_-) = 0 \end{cases} \tag{4.49}$$

where $t > 0$.

Taking Laplace transform of (4.49), it can be stated as

$$C_\alpha s^\alpha U_{C\alpha}(s) + \frac{1}{R} U_{C\alpha}(s) = I_s(s) \tag{4.50}$$

Then, there is

$$U_{C\alpha}(s) = \frac{1}{C_\alpha [s^\alpha + 1/(RC_\alpha)]} I_s(s) \tag{4.51}$$

By the inverse Laplace transform of (4.51), the analytical solution of voltage $u_{C\alpha}$ can be calculated as

$$u_{C\alpha}(t) = \mathcal{L}^{-1}[U_{C\alpha}(s)] = \frac{1}{C_\alpha} \int_0^t \tau^{\alpha-1} E_{\alpha,\alpha}\left(-\frac{1}{RC_\alpha}\tau^\alpha\right) i_s(t-\tau)\mathrm{d}\tau \, (t > 0) \tag{4.52}$$

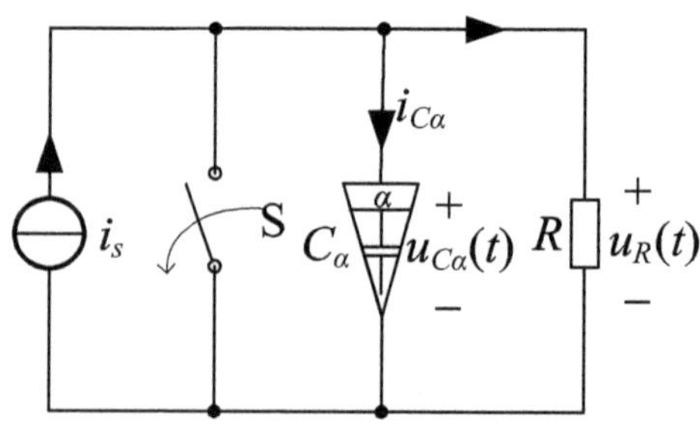

Fig. 4.16 Fractional-order RC_α circuit with zero state

Other currents and voltages can be derived as follows:
The voltage u_R across the resistor is

$$u_R(t) = u_{C\alpha}(t) \tag{4.53}$$

The current i_R flowing through the resistor is

$$i_R(t) = \frac{u_R(t)}{R} \tag{4.54}$$

The current $i_{C\alpha}$ of the capacitor is

$$i_{C\alpha}(t) = i_s(t) - i_R(t) = C_\alpha \frac{\mathrm{d}^\alpha u_{C\alpha}}{\mathrm{d}t^\alpha} \tag{4.55}$$

1. Response to DC Input

If the current source shown in Fig. 4.16 is DC current source (i.e., $i_s = I_0$) and its Laplace transform is $I_s(s) = I_0/s$, the voltage of the capacitor can be expressed as

$$U_{C\alpha}(s) = \frac{1}{C_\alpha[s^\alpha + 1/(RC_\alpha)]} \frac{I_0}{s} \tag{4.56}$$

The analytical solution of voltage across the capacitor can be determined by the inverse Laplace transform of (4.56) as

$$\begin{aligned} u_{C\alpha}(t) = \mathcal{L}^{-1}[U_{C\alpha}(s)] &= \frac{I_0}{C_\alpha} t^\alpha E_{\alpha,\alpha+1}\left(-\frac{1}{RC_\alpha} t^\alpha\right) \\ &= \frac{I_0}{C_\alpha} \sum_{k=0}^{\infty} \left(-\frac{1}{RC_\alpha}\right)^k \frac{t^{\alpha k+\alpha}}{\Gamma(\alpha k + \alpha + 1)} \end{aligned} \tag{4.57}$$

where $t > 0$.

Applying the unit step function $\varepsilon(t)$ to the equations mentioned above, the final expressions of currents and voltages for all time can be obtained as follows:

The voltage $u_{C\alpha}$ is

$$u_{C\alpha}(t) = \frac{I_0}{C_\alpha} t^\alpha E_{\alpha,\alpha+1}\left(-\frac{1}{RC_\alpha} t^\alpha\right)\varepsilon(t) \tag{4.58}$$

The voltage u_R is

$$u_R(t) = \frac{I_0}{C_\alpha} t^\alpha E_{\alpha,\alpha+1}\left(-\frac{1}{RC_\alpha} t^\alpha\right)\varepsilon(t) \tag{4.59}$$

The current i_R is

$$i_R(t) = \frac{I_0}{RC_\alpha} t^\alpha E_{\alpha,\alpha+1}\left(-\frac{1}{RC_\alpha} t^\alpha\right)\varepsilon(t) \tag{4.60}$$

The current $i_{C\alpha}$ is

$$u_{C\alpha}(t) = I_0\left[1 - \frac{t^\alpha}{RC_\alpha} E_{\alpha,\alpha+1}\left(-\frac{1}{RC_\alpha} t^\alpha\right)\right]\varepsilon(t) \tag{4.61}$$

The charging current and voltage of the fractional-order capacitor C_α are shown in Fig. 4.17, which implies that the circuit will finally reach the steady-state.

According to the final-value theorem of the Laplace transform, the steady-state of the voltage across the fractional-order capacitor can be calculated by

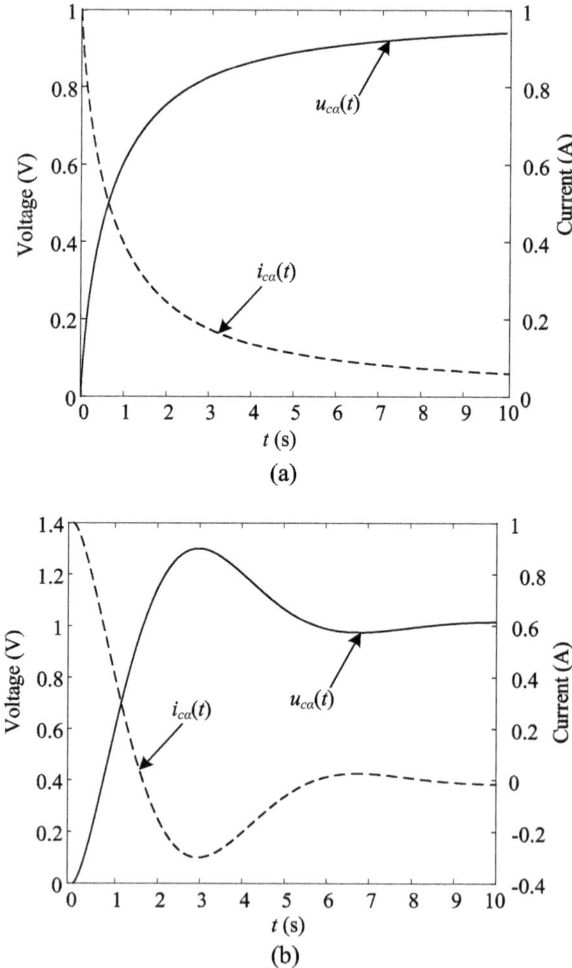

Fig. 4.17 RC_α circuit charges with DC current source for $R = 1\ \Omega$, $C_\alpha = 1$ F/s$^{1-\alpha}$, $I_0 = 1$ A: **a** $\alpha = 0.75$. **b** $\alpha = 1.5$

$$\lim_{t\to\infty} u_{C\alpha}(t) = \lim_{s\to 0} sU_{C\alpha}(s) = \lim_{s\to 0}\left(s\frac{I_0/C_\alpha}{s[s^\alpha + 1/(RC_\alpha)]}\right) = I_0 R \tag{4.62}$$

$$\lim_{t\to\infty} i_{C\alpha}(t) = I_0 - \lim_{t\to 0}\frac{u_{C\alpha}(t)}{R} = 0 \tag{4.63}$$

At $t = 0$, the voltage across the fractional-order capacitor C_α is $u_{C\alpha}(0_+) = u_{C\alpha}(0_-) = 0$, and the DC current I_0 starts to flow through capacitor C_α. With the increase of the voltage across capacitor C_α, the charging current of capacitor C_α decays to zero, and the current flowing through resistor R gradually increases. As $t \to \infty$, the charging current of the capacitor C_α is 0, and capacitor voltage reaches the steady-state that $u_{C\alpha} = RI_0$, which implies that the charging process is finished.

With the same charging current, the charging time of the RC_α circuit is influenced by the order α and time constant τ: The higher the order α is, the longer the charging time is, and the smaller the order α is, the shorter the charging time is. The surfaces of RC_α circuit charging with different parameters are shown in Fig. 4.18.

2. Response to AC Input

If the current source in Fig. 4.16 is an AC source and $i_s = I_m\cos(\omega t + \theta)$, the Laplace transform of the source is $I_s(s) = I_m\left(\frac{s\cos\theta - \omega\sin\theta}{s^2+\omega^2}\right)$. The switch S turns off at $t = 0$, and we will analyze the zero-state response with sinusoidal input.

The expression of capacitor voltage in the complex frequency domain is

$$U_{C\alpha}(s) = \frac{1}{C_\alpha[s^\alpha + 1/(RC_\alpha)]}\frac{I_m(s\cos\theta - \omega\sin\theta)}{(s^2+\omega^2)} \tag{4.64}$$

The analytical solution of the voltage across the fractional-order capacitor in the time domain can be obtained by the inverse Laplace transform of (4.64) as

$$u_{C\alpha}(t) = \mathcal{L}^{-1}[U_{C\alpha}(s)] = \frac{I_m}{C_\alpha}\int_0^t (t-\tau)^{\alpha-1}E_{\alpha,\alpha}\left[-\frac{1}{RC_\alpha}(t-\tau)^\alpha\right]\cos(\omega\tau + \theta)\mathrm{d}\tau \tag{4.65}$$

where $t > 0$.

The sinusoidal response of RC_α circuit is shown in Fig. 4.19, which illustrates that the current and voltage of fractional-order capacitors can reach the sinusoidal steady-state after a finite time. Obviously, the phase angle difference between capacitor voltage and current is determined by the order.

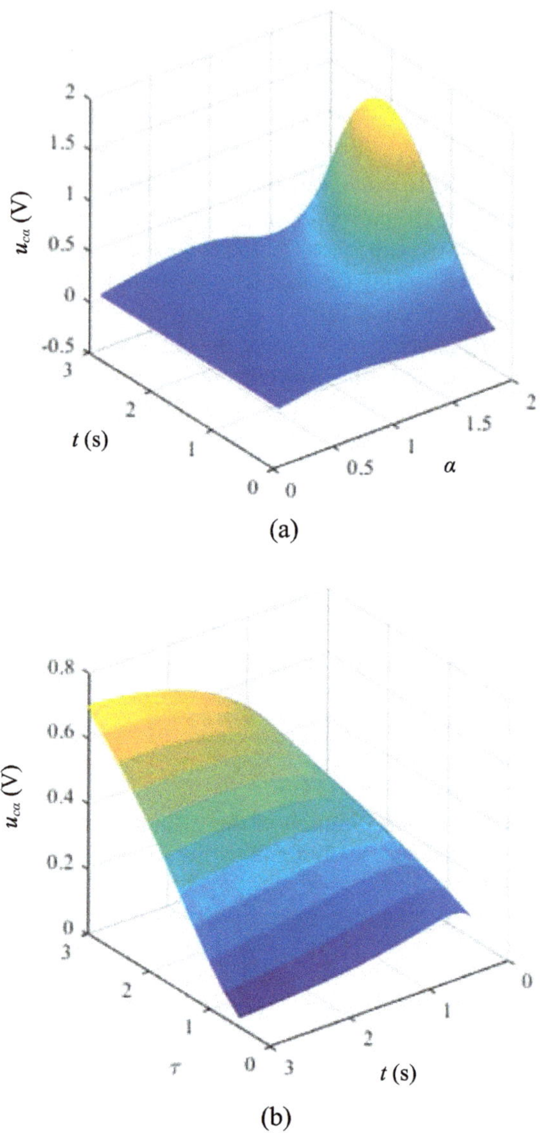

Fig. 4.18 The surfaces of the RC_α circuit charging with DC current source $I_0 =$ 1 A: **a** Responses of RC_α circuit with different orders. **b** Responses of RC_α circuit with different time constant

4.3.2 Fractional-Order RL_β Circuit

As shown in Fig. 4.20, the voltage source is u_s and switch S in the RL_β circuit turns off at $t = 0$, and then we will analyze the zero-input response of the circuit as $t > 0$. By KCL and the property of components, the inductor current $i_{L\beta}$ satisfies

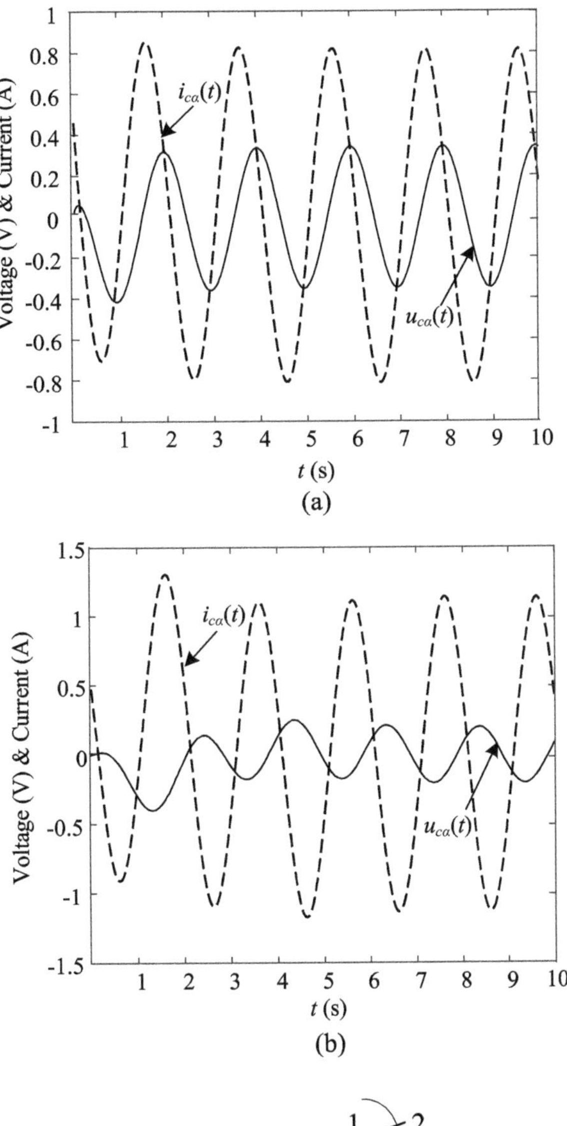

Fig. 4.19 RC_α circuit charging with sinusoidal current for $R = 1\ \Omega$, $C_\alpha = 1$ F/$s^{1-\alpha}$, $i_s = \cos(\pi t + \pi/3)$A: **a** $\alpha = 0.75$. **b** $\alpha = 1.5$

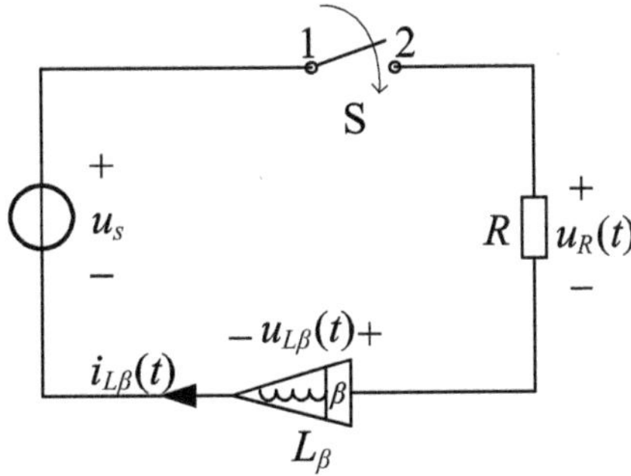

Fig. 4.20 Fractional-order RL_β circuit with zero state

$$\begin{cases} L_\beta \frac{\mathrm{d}^\beta i_{L\beta}}{\mathrm{d}t^\beta} + i_{L\beta}R = u_s \\ i_{L\beta}(0_+) = i_{L\beta}(0_-) = 0 \end{cases} \tag{4.66}$$

where $t > 0$.

Taking Laplace transform of (4.66), it can be derived as

$$L_\beta s^\beta I_{L\beta}(s) + RI_{L\beta}(s) = U_s(s) \tag{4.67}$$

Then, there is

$$I_{L\beta}(s) = \frac{1}{L_\beta\left[s^\beta + R/L_\beta\right]}U_s(s) \tag{4.68}$$

By the inverse Laplace transform of (4.68), the analytical solution of the inductor current $i_{L\beta}$ can be calculated as

$$i_{L\beta}(t) = \mathcal{L}^{-1}\left[I_{L\beta}(s)\right] = \frac{1}{L_\beta}\int_0^t \tau^{\beta-1}E_{\beta,\beta}\left(-\frac{R}{L_\beta}\tau^\alpha\right)u_s(t-\tau)\mathrm{d}\tau \ (t > 0) \tag{4.69}$$

Other currents and voltages can be derived as follows:

The resistor current i_R is

$$i_R(t) = i_{L\beta}(t) \tag{4.70}$$

The resistor voltage u_R is

$$u_R(t) = i_{L\beta}(t)R \tag{4.71}$$

The inductor voltage $u_{L\beta}$ is

$$u_{L\beta} = u_s(t) - u_R(t) \tag{4.72}$$

1. Response to DC Input

If the voltage source shown in Fig. 4.20 is DC source (i.e., $u_s = U_0$) and its Laplace transform is $U_s(s) = U_0/s$, the capacitor voltage can be expressed as

$$I_{L\beta}(s) = \frac{1}{L_\beta\left[s^\beta + R/L_\beta\right]}\frac{U_0}{s} \tag{4.73}$$

The analytical solution of inductor current can be determined by the inverse Laplace transform of (4.73) as

$$i_{L\beta}(t) = \mathcal{L}^{-1}[I_{L\beta}(s)] = \frac{U_0}{L_\beta} t^\beta E_{\beta,\beta+1}\left(-\frac{R}{L_\beta} t^\beta\right)$$
$$= \frac{U_0}{L_\beta} \sum_{k=0}^{\infty} \left(-\frac{R}{L_\beta}\right)^k \frac{t^{\beta k+\beta}}{\Gamma(\beta k+\beta+1)} \tag{4.74}$$

where $t > 0$.

Applying the unit step function $\varepsilon(t)$ to the equations mentioned above, the charging current and voltage of the fractional-order inductor L_β are shown in Fig. 4.21, which implies that the circuit will finally reach the steady-state, and the final expressions of currents and voltages for all time can be obtained as follows:

The inductor current $i_{L\beta}$ is

$$i_{L\beta}(t) = \frac{U_0}{L_\beta} t^\beta E_{\beta,\beta+1}\left(-\frac{R}{L_\beta} t^\beta\right)\varepsilon(t) \tag{4.75}$$

The resistor voltage u_R is

$$u_R(t) = \frac{R}{L_\beta} U_0 t^\beta E_{\beta,\beta+1}\left(-\frac{R}{L_\beta} t^\beta\right)\varepsilon(t) \tag{4.76}$$

The resistor current i_R is

$$i_{L\beta}(t) = \frac{U_0}{L_\beta} t^\beta E_{\beta,\beta+1}\left(-\frac{R}{L_\beta} t^\beta\right)\varepsilon(t) \tag{4.77}$$

The inductor voltage $u_{L\beta}$ is

$$u_{L\beta}(t) = U_0\left[1 - \frac{R}{L_\beta} t^\beta E_{\beta,\beta+1}\left(-\frac{R}{L_\beta} t^\beta\right)\right]\varepsilon(t) \tag{4.78}$$

2. Response to AC Input

If the voltage source shown in Fig. 4.20 is AC source and $u_s = U_m\cos(\omega t + \theta)$, the Laplace transform of the source is $U_s(s) = U_m\left(\frac{s\cos\theta - \omega\sin\theta}{s^2+\omega^2}\right)$. The switch S turns off at $t = 0$, and we will analyze the zero-state response with AC input. The sinusoidal response of $R_{L\beta}$ circuit is shown in Fig. 4.22, which illustrates that the current and voltage of the fractional-order inductor can reach the sinusoidal steady-state after a finite time.

The expression of inductor current in the complex frequency domain is

$$I_{L\beta}(s) = \frac{1}{L_\beta\left(s^\beta + R/L_\beta\right)} \frac{U_m(s\cos\theta - \omega\sin\theta)}{\left(s^2+\omega^2\right)} \tag{4.79}$$

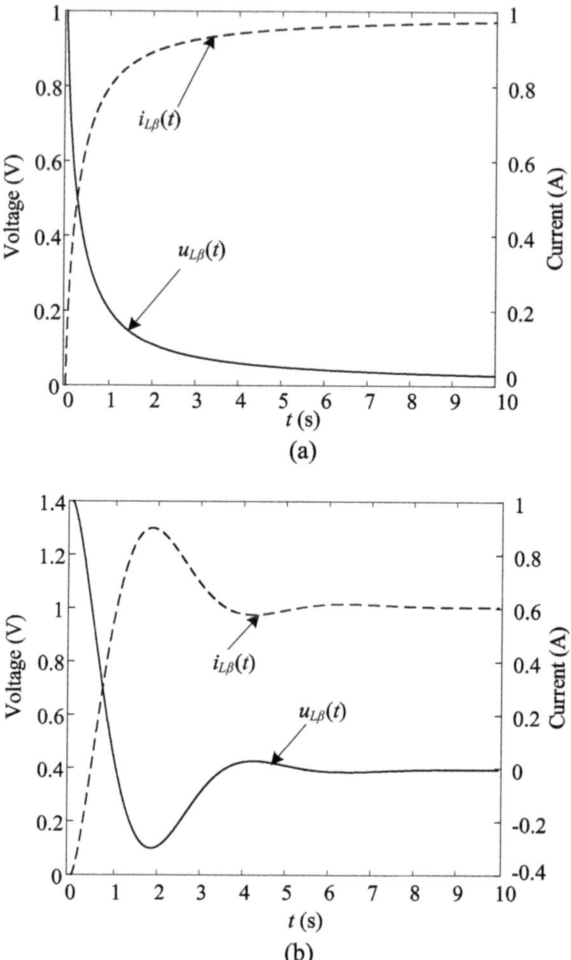

Fig. 4.21 RL_β circuit charging with DC voltage source for $R = 1\ \Omega$, $L_\beta = 0.5$ H/s$^{1-\beta}$, $U_0 = 1$ V: **a** $\beta = 0.75$. **b** $\beta = 1.5$

The analytical solution of the inductor current in the time domain can be obtained by the inverse Laplace transform of (4.79) as

$$i_{L\beta}(t) = \mathcal{L}^{-1}[I_{L\beta}(s)] = \frac{U_m}{L_\beta}\int_0^t (t-\tau)^{\beta-1}E_{\beta,\beta}\left(-\frac{R}{L_\beta}(t-\tau)^\beta\right)\cos(\omega\tau+\theta)\mathrm{d}\tau \tag{4.80}$$

where $t > 0$.

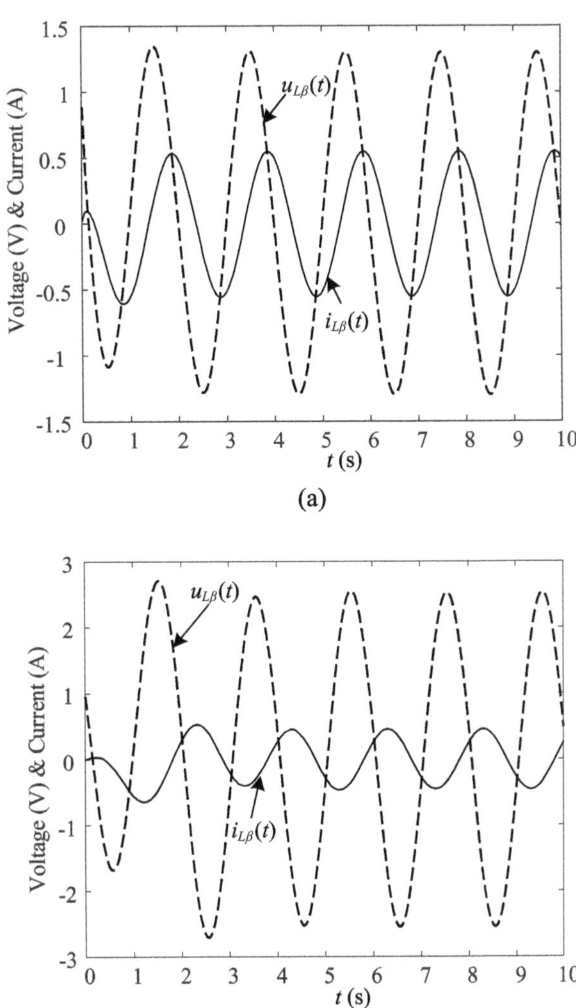

Fig. 4.22 RL_β circuit charging with sinusoidal voltage source for $R = 1\ \Omega$, $L_\beta = 0.5\ \text{H/s}^{1-\beta}$, $u_s = \cos(\pi t + \pi/3)$A: **a** $\beta = 0.75$, **b** $\beta = 1.5$

4.4 Complete Response of Fractional-Order RC_α and RL_β Circuits

4.4.1 Fractional-Order RC_α Circuit

The complete response of the circuit refers to the response with the initial state and external source, which is the sum of the zero-input and zero-state response of the fractional-order circuit. This section will introduce the solution of the complete

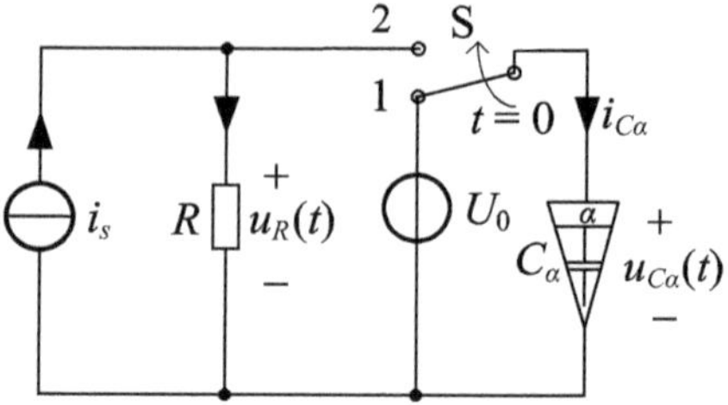

Fig. 4.23 Fractional-order RC_α circuit with initial state and a current source

response of the circuit with one fractional-order component. Here, we will only discuss the RC_α circuit and the analysis principle of other circuits are consistent.

As shown in Fig. 4.23, the original position of switch S is at "1", it changes to "2" at $t = 0$, and the input current source is i_s. Assuming that the initial voltage across the fractional-order capacitor is U_0, we will discuss the complete response of the circuit when $t > 0$. The voltage across the fractional-order capacitor $u_{C\alpha}$ satisfies

$$\begin{cases} C_\alpha \frac{\mathrm{d}^\alpha u_{C\alpha}}{\mathrm{d}t^\alpha} + \frac{1}{R} u_{C\alpha} = i_s \\ u_{C\alpha}(0_+) = u_{C\alpha}(0_-) = U_0 \\ Q_{C\alpha}(0_+) = Q_{C\alpha}(0_-) = Q_0 \end{cases} \tag{4.81}$$

where $t > 0$.

Taking Laplace transform of (4.81) under R-L definition, it can be stated as

$$s^\alpha U_{C\alpha}(s) - \left[{}_0D_t^{\alpha-1} u_{C\alpha}(t)\right]_{t=0} + \frac{1}{RC_\alpha} U_{C\alpha}(s) = \frac{I_s(s)}{C_\alpha} \tag{4.82}$$

The capacitor voltage in the complex frequency domain can be derived as

$$U_{C\alpha}(s) = \frac{I_s(s)/C_\alpha + \left[{}_0D_t^{\alpha-1} u_{C\alpha}(t)\right]_{t=0}}{s^\alpha + 1/(RC_\alpha)} = \underbrace{\frac{I_s(s)/C_\alpha}{s^\alpha + 1/(RC_\alpha)}}_{\text{zero-state response}} + \underbrace{\frac{\left[{}_0D_t^{\alpha-1} u_{C\alpha}(t)\right]_{t=0}}{s^\alpha + 1/(RC_\alpha)}}_{\text{zero-input response}} \tag{4.83}$$

Equation (4.83) shows that the complete response is a linear combination of zero-state and zero-input response. Hence, the analytical solution can be obtained by inverse Laplace transform as

$$u_{C\alpha}(t) = \mathcal{L}^{-1}[U_{C\alpha}(s)] = \mathcal{L}^{-1}\left[\frac{I_s(s)/C_\alpha}{s^\alpha + 1/(RC_\alpha)}\right] + \mathcal{L}^{-1}\left[\frac{\left[{}_0D_t^{\alpha-1} u_{C\alpha}(t)\right]_{t=0}}{s^\alpha + 1/(RC_\alpha)}\right]$$

$$= \underbrace{\int_0^t \frac{1}{C_\alpha}\tau^{\alpha-1}E_{\alpha,\alpha}\left(-\frac{1}{RC_\alpha}\tau^\alpha\right)i_s(t-\tau)\mathrm{d}\tau}_{\text{zero-state response}} + \underbrace{\frac{Q_0}{C_\alpha}t^{\alpha-1}E_{\alpha,\alpha}\left(-\frac{1}{RC_\alpha}t^\alpha\right)}_{\text{zero-input response}} \tag{4.84}$$

As for Caputo definition, applying Laplace transform to (4.81), there is

$$s^\alpha U_{C\alpha}(s) - s^{\alpha-1}u_{C\alpha}(0) + \frac{1}{RC_\alpha}U_{C\alpha}(s) = \frac{I_s(s)}{C_\alpha} \tag{4.85}$$

The capacitor voltage in the complex frequency domain can be derived as

$$U_{C\alpha}(s) = \frac{I_s(s)/C_\alpha + s^{\alpha-1}u_{C\alpha}(0)}{s^\alpha + 1/(RC_\alpha)} = \underbrace{\frac{I_s(s)/C_\alpha}{s^\alpha + 1/(RC_\alpha)}}_{\text{zero-state response}} + \underbrace{\frac{s^{\alpha-1}u_{C\alpha}(0)}{s^\alpha + 1/(RC_\alpha)}}_{\text{zero-input response}} \tag{4.86}$$

Hence, the analytical solution under Caputo definition is

$$\begin{aligned} u_{C\alpha}(t) &= \mathcal{L}^{-1}[U_{C\alpha}(s)] = \mathcal{L}^{-1}\left[\frac{I_s(s)/C_\alpha}{s^\alpha + 1/(RC_\alpha)}\right] + \mathcal{L}^{-1}\left[\frac{s^{\alpha-1}u_{C\alpha}(0)}{s^\alpha + 1/(RC_\alpha)}\right] \\ &= \underbrace{\int_0^t \frac{1}{C_\alpha}\tau^{\alpha-1}E_{\alpha,\alpha}\left(-\frac{1}{RC_\alpha}\tau^\alpha\right)i_s(t-\tau)\mathrm{d}\tau}_{\text{zero-state response}} + \underbrace{U_0 E_{\alpha,1}\left(-\frac{1}{RC_\alpha}t^\alpha\right)}_{\text{zero-input response}} \end{aligned} \tag{4.87}$$

Other physical variables in the RC_α circuit can be obtained by the same method.

1. Response to DC Input

If the current source shown in Fig. 4.23 is DC source and $i_s = I_0$, then the capacitor voltage can be expressed as follows:

(1) R-L fractional definition

$$\begin{aligned} u_{C\alpha}(t) &= \frac{I_0}{C_\alpha}t^\alpha E_{\alpha,\alpha+1}\left(-\frac{1}{RC_\alpha}t^\alpha\right) + \frac{Q_0}{C_\alpha}t^{\alpha-1}E_{\alpha,\alpha}\left(-\frac{1}{RC_\alpha}t^\alpha\right) \\ &= \underbrace{\frac{I_0}{C_\alpha}\sum_{k=0}^{\infty}\left(-\frac{1}{RC_\alpha}\right)^k \frac{t^{\alpha k+\alpha}}{\Gamma(\alpha k+\alpha+1)}}_{\text{zero-state response}} \\ &\quad + \underbrace{\frac{Q_0}{C_\alpha}\sum_{k=0}^{\infty}\left(-\frac{1}{RC_\alpha}\right)^k \frac{t^{\alpha k+\alpha-1}}{\Gamma(\alpha k+\alpha)}}_{\text{zero-input response}} \end{aligned} \tag{4.88}$$

(2) Caputo fractional definition

$$u_{C\alpha}(t) = \frac{I_0}{C_\alpha} t^\alpha E_{\alpha,\alpha+1}\left(-\frac{1}{RC_\alpha} t^\alpha\right) + U_0 E_{\alpha,1}\left(-\frac{1}{RC_\alpha} t^\alpha\right)$$
$$= \underbrace{\frac{I_0}{C_\alpha}\sum_{k=0}^{\infty}\left(-\frac{1}{RC_\alpha}\right)^k \frac{t^{\alpha k+\alpha}}{\Gamma(\alpha k+\alpha+1)}}_{\text{zero-state response}}$$
$$+ \underbrace{U_0 \sum_{k=0}^{\infty}\left(-\frac{1}{RC_\alpha}\right)^k \frac{t^{\alpha k}}{\Gamma(\alpha k+1)}}_{\text{zero-input response}} \tag{4.89}$$

Other voltages and currents can be derived from the KCL and KVL.

2. Response to AC Input

If the current source shown in Fig. 4.23 is an AC source and $i_s = I_m \cos(\omega t + \theta)$, then the capacitor voltage can be expressed as follows:

(1) R-L fractional definition

$$u_{C\alpha}(t) = \frac{I_m}{C_\alpha}\int_0^t (t-\tau)^{\alpha-1} E_{\alpha,\alpha}\left(-\frac{1}{RC_\alpha}(t-\tau)^\alpha\right)\cos(\omega\tau+\theta)\mathrm{d}\tau$$
$$+ \frac{Q_0}{C_\alpha} t^{\alpha-1} E_{\alpha,\alpha}\left(-\frac{1}{RC_\alpha} t^\alpha\right) \tag{4.90}$$

(2) Caputo fractional definition

$$u_{C\alpha}(t) = \frac{I_m}{C_\alpha}\int_0^t (t-\tau)^{\alpha-1} E_{\alpha,\alpha}\left(-\frac{1}{RC_\alpha}(t-\tau)^\alpha\right)\cos(\omega\tau+\theta)\mathrm{d}\tau$$
$$+ U_0 E_{\alpha,1}\left(-\frac{1}{RC_\alpha} t^\alpha\right) \tag{4.91}$$

Other voltages and currents can be derived from the KCL and KVL.

4.4.2 Fractional-Order RL$_\beta$ Circuit

As shown in Fig. 4.24, the original position of switch S is at "1", at $t = 0$, it changes to "2", and the input voltage source is u_s. Assuming that the initial current through the fractional-order capacitor is I_0, we will discuss the complete response of the circuit when $t > 0$. The voltage across the fractional-order inductor $i_{L\beta}$ satisfies

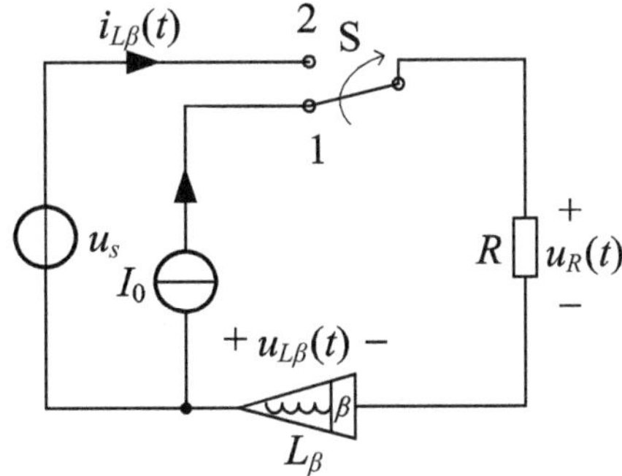

Fig. 4.24 Fractional-order RL_β circuit with initial state and a voltage source

$$\begin{cases} L_\beta \frac{\mathrm{d}^\beta i_{L\beta}}{\mathrm{d}t^\beta} + Ri_{L\beta} = u_s \\ i_{L\beta}(0_+) = i_{L\beta}(0_-) = I_0 \\ \psi_{L\beta}(0_+) = \psi_{L\beta}(0_-) = \psi_0 \end{cases} \tag{4.92}$$

where $t > 0$.

Taking Laplace transform of (4.92) under R-L definition, it satisfies

$$s^\beta I_{L\beta}(s) - \left[{}_0D_t^{\beta-1} i_{L\beta}(t)\right]_{t=0} + \frac{R}{L_\beta} I_{L\beta}(s) = \frac{U_s(s)}{L_\beta} \tag{4.93}$$

The inductor current in the complex frequency domain can be derived as

$$I_{L\beta}(s) = \frac{U_s(s)/L_\beta + \left[{}_0D_t^{\beta-1} i_{L\beta}(t)\right]_{t=0}}{s^\beta + R/L_\beta} = \underbrace{\frac{U_s(s)/L_\beta}{s^\beta + R/L_\beta}}_{\text{zero-state response}} + \underbrace{\frac{\left[{}_0D_t^{\beta-1} i_{L\beta}(t)\right]_{t=0}}{s^\beta + R/L_\beta}}_{\text{zero-input response}} \tag{4.94}$$

Equation (4.94) shows that the complete response is a linear combination of zero-state and zero-input response. Hence, the analytical solution can be obtained by inverse Laplace transform as

$$\begin{aligned} i_{L\beta}(t) &= \mathcal{L}^{-1}\left[I_{L\beta}(s)\right] = \mathcal{L}^{-1}\left[\frac{U_s(s)/L_\beta}{s^\beta + R/L_\beta}\right] + \mathcal{L}^{-1}\left[\frac{\left[{}_0D_t^{\beta-1} i_{L\beta}(t)\right]_{t=0}}{s^\beta + R/L_\beta}\right] \\ &= \underbrace{\int_0^t \frac{1}{L_\beta}\tau^{\beta-1} E_{\beta,\beta}\left(-\frac{R}{L_\beta}\tau^\beta\right) u_s(t-\tau)\mathrm{d}\tau}_{\text{zero-state response}} + \underbrace{\frac{\psi_0}{L_\beta} t^{\beta-1} E_{\beta,\beta}\left(-\frac{R}{L_\beta} t^\beta\right)}_{\text{zero-input response}} \end{aligned} \tag{4.95}$$

As for Caputo definition, applying Laplace transform to (4.92), there is

$$s^{\beta}I_{L\beta}(s)-s^{\beta-1}i_{L\beta}(0)+\frac{R}{L_{\beta}}I_{L\beta}(s)=\frac{U_s(s)}{L_{\beta}} \tag{4.96}$$

The inductor current in the complex frequency domain can be derived as

$$I_{L\beta}(s)=\frac{U_s(s)/L_{\beta}+s^{\beta-1}i_{L\beta}(0)}{s^{\beta}+R/L_{\beta}}=\underbrace{\frac{U_s(s)/L_{\beta}}{s^{\beta}+R/L_{\beta}}}_{\text{zero-state response}}+\underbrace{\frac{s^{\beta-1}i_{L\beta}(0)}{s^{\alpha}+R/L_{\beta}}}_{\text{zero-input response}} \tag{4.97}$$

Hence, the analytical solution under Caputo definition is

$$\begin{aligned}i_{L\beta}(t)&=\mathcal{L}^{-1}\left[I_{L\beta}(s)\right]=\mathcal{L}^{-1}\left[\frac{U_s(s)/L_{\beta}}{s^{\alpha}+R/L_{\beta}}\right]+\mathcal{L}^{-1}\left[\frac{s^{\beta-1}i_{L\beta}(0)}{s^{\beta}+R/L_{\beta}}\right]\\&=\underbrace{\int_0^t\frac{1}{L_{\beta}}\tau^{\beta-1}E_{\beta,\beta}\left(-\frac{R}{L_{\beta}}\tau^{\beta}\right)u_s(t-\tau)\mathrm{d}\tau}_{\text{zero-state response}}+\underbrace{I_0E_{\beta,1}\left(-\frac{R}{L_{\beta}}t^{\beta}\right)}_{\text{zero-input response}}\end{aligned} \tag{4.98}$$

Other physical variables in the RL_{β} circuit can be obtained by the same method.

1. Response to DC Input

If the voltage source shown in Fig. 4.24 is DC source and $u_s=U_0$, then the inductor current can be expressed as follows:

(1) R-L fractional definition

$$\begin{aligned}i_{L\beta}(t)&=\frac{U_0}{L_{\beta}}t^{\beta}E_{\beta,\beta+1}\left(-\frac{R}{L_{\beta}}t^{\beta}\right)+\frac{\psi_0}{L_{\beta}}t^{\beta-1}E_{\beta,\beta}\left(-\frac{R}{L_{\beta}}t^{\beta}\right)\\&=\underbrace{\frac{U_0}{L_{\beta}}\sum_{k=0}^{\infty}\left(-\frac{R}{L_{\beta}}\right)^k\frac{t^{\beta k+\beta}}{\Gamma(\beta k+\beta+1)}}_{\text{zero-state response}}+\underbrace{\frac{\psi_0}{L_{\beta}}\sum_{k=0}^{\infty}\left(-\frac{R}{L_{\beta}}\right)^k\frac{t^{\beta k+\beta-1}}{\Gamma(\beta k+\beta)}}_{\text{zero-input response}}\end{aligned} \tag{4.99}$$

(2) Caputo fractional definition

$$i_{L\beta}(t)=\frac{U_0}{L_{\beta}}t^{\beta}E_{\beta,\beta+1}\left(-\frac{R}{L_{\beta}}t^{\beta}\right)+U_0E_{\beta,1}\left(-\frac{R}{L_{\beta}}t^{\beta}\right)$$

$$= \underbrace{\frac{U_0}{L_\beta}\sum_{k=0}^{\infty}\left(-\frac{R}{L_\beta}\right)^k \frac{t^{\beta k+\beta}}{\Gamma(\beta k+\beta+1)}}_{\text{zero-state response}} + \underbrace{U_0\sum_{k=0}^{\infty}\left(-\frac{R}{L_\beta}\right)^k \frac{t^{\beta k}}{\Gamma(\beta k+1)}}_{\text{zero-input response}} \tag{4.100}$$

Other voltages and currents can be derived from the KCL and KVL.

2. Response to AC Input

If the voltage source shown in Fig. 4.24 is an AC source and $u_s = U_m\cos(\omega t+\theta)$, then the capacitor voltage can be expressed as follows:

(1) R-L fractional definition

$$i_{L\beta}(t) = \frac{U_m}{L_\beta}\int_0^t (t-\tau)^{\beta-1}E_{\beta,\beta}\left(-\frac{R}{L_\beta}(t-\tau)^\beta\right)\cos(\omega\tau+\theta)\mathrm{d}\tau + \frac{\psi_0}{L_\beta}t^{\beta-1}E_{\beta,\beta}\left(-\frac{R}{C_\alpha}t^\beta\right) \tag{4.101}$$

(2) Caputo fractional definition

$$i_{L\beta}(t) = \frac{U_m}{L_\beta}\int_0^t (t-\tau)^{\beta-1}E_{\beta,\beta}\left(-\frac{R}{L_\beta}(t-\tau)^\beta\right)\cos(\omega\tau+\theta)\mathrm{d}\tau + I_0E_{\beta,1}\left(-\frac{R}{L_\beta}t^\beta\right) \tag{4.102}$$

Other voltages and currents can be derived from the KCL and KVL.

4.5 Step and Pulse Response of Fractional-Order RC_α and RL_β Circuits

The step response of the circuit refers to the zero-state response generated by the unit step excitation, which is denoted by $s(t)$. The pulse response of the circuit refers to the zero-state response generated by the impulse excitation, which is denoted by $h(t)$. This section will discuss the time-domain response of the circuit containing one fractional-order component with unit step response and impulse response, respectively.

4.5.1 Step Response

1. Fractional-order RC_α Circuit

As shown in Fig. 4.25, assuming that the external excitation is a unit step current source and $i_s = \varepsilon(t)$, the step response of the RC_α circuit will be discussed later.

According to KCL and the property of components, the capacitor voltage $u_{C\alpha}$ satisfies

$$\begin{cases} C_\alpha \frac{\mathrm{d}^\alpha u_{C\alpha}}{\mathrm{d}t^\alpha} + \frac{1}{R} u_{C_\alpha} = \varepsilon(t) \\ u_{C\alpha}(0_-) = 0 \end{cases} \tag{4.103}$$

Taking Laplace transform of (4.103), it can be states as

$$C_\alpha s^\alpha U_{C\alpha}(s) + \frac{1}{R} U_{C\alpha}(s) = \frac{1}{s} \tag{4.104}$$

The capacitor voltage in the complex frequency domain can be derived as

$$U_{C\alpha}(s) = \frac{1}{C_\alpha [s^\alpha + 1/(RC_\alpha)]} \frac{1}{s} \tag{4.105}$$

Taking inverse Laplace transform of (4.105), the analytical solution of the unit step response is

$$\begin{aligned} u_{C\alpha}(t) &= \mathcal{L}^{-1}[U_{C\alpha}(s)] = \frac{1}{C_\alpha} t^\alpha E_{\alpha,\alpha+1}\left(-\frac{1}{RC_\alpha} t^\alpha\right)\varepsilon(t) \\ &= \frac{1}{C_\alpha} \sum_{k=0}^{\infty} \left(-\frac{1}{RC_\alpha}\right)^k \frac{t^{\alpha k+\alpha}}{\Gamma(\alpha k+\alpha+1)} \varepsilon(t) \end{aligned} \tag{4.106}$$

From (4.106), it can be concluded that the step response of the RC_α circuit is similar to the zero-state response since the ideal switching action can be represented by the unit step function.

2. Fractional-order RL_β Circuit

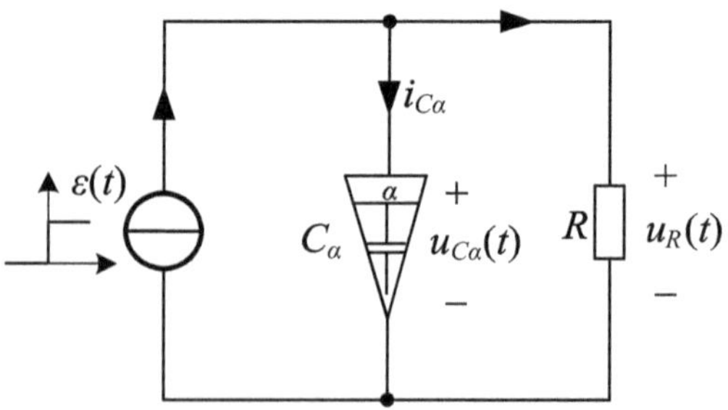

Fig. 4.25 RC_α circuit with step excitation

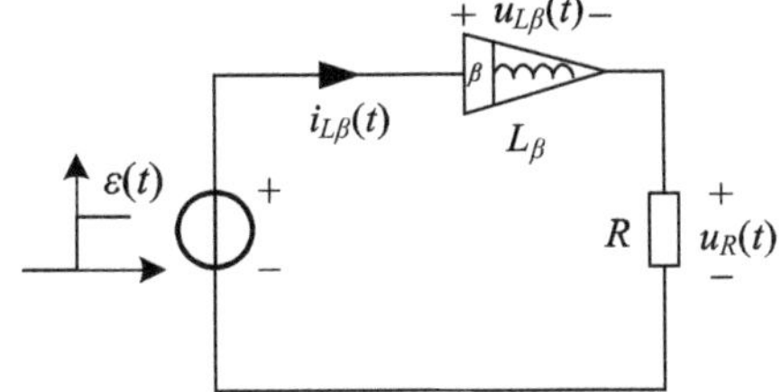

Fig. 4.26 RL_β circuit with step excitation

As shown in Fig. 4.26, assuming that the external excitation is a unit step voltage source and $u_s = \varepsilon(t)$, the step response of the RL_β circuit will be analyzed later.

Based on KVL and the property of components, the inductor current $i_{L\beta}$ satisfies

$$\begin{cases} L_\beta \frac{\mathrm{d}^\beta i_{L\beta}}{\mathrm{d}t^\beta} + i_{L\beta} R = \varepsilon(t) \\ i_{L\beta}(0_-) = 0 \end{cases} \tag{4.107}$$

Applying Laplace transform to (4.107), there is

$$L_\beta s^\beta I_{L\beta}(s) + R I_{L\beta}(s) = \frac{1}{s} \tag{4.108}$$

The inductor current in the complex frequency domain can be derived as

$$I_{L\beta}(s) = \frac{1}{L_\beta\left(s^\beta + R/L_\beta\right)} \frac{1}{s} \tag{4.109}$$

Using inverse Laplace transform to (4.109), the analytical solution of the step response can be derived as

$$\begin{aligned} i_{L\beta}(t) = \mathcal{L}^{-1}[I_{L\beta}(s)] &= \frac{1}{L_\beta} t^\beta E_{\beta,\beta+1}\left(-\frac{R}{L_\beta} t^\beta\right)\varepsilon(t) \\ &= \frac{1}{L_\beta} \sum_{k=0}^{\infty} \left(-\frac{R}{L_\beta}\right)^k \frac{t^{\beta k+\beta}}{\Gamma(\beta k + \beta + 1)} \varepsilon(t) \end{aligned} \tag{4.110}$$

4.5.2 Pulse Response

1. Fractional-order RC_α Circuit

As shown in Fig. 4.27, supposing that the external excitation is impulse current source and $i_s = \delta(t)$, the impulse response of the RC_α circuit will be analyzed later.

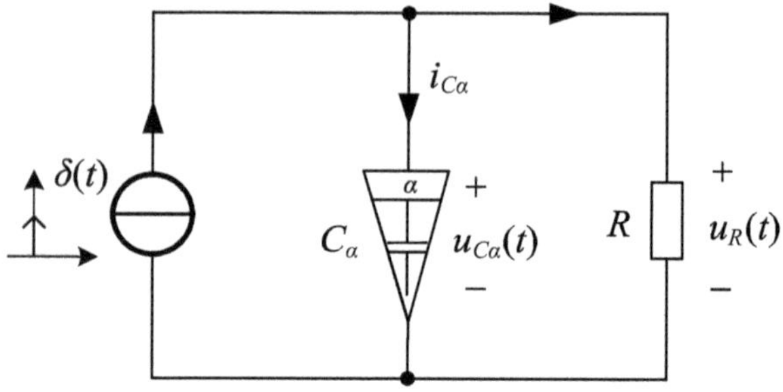

Fig. 4.27 RC_α circuit with pulse excitation

Based on KCL and the property of components, the capacitor voltage $u_{C\alpha}$ can be derived from

$$\begin{cases} C_\alpha \frac{d^\alpha u_{C\alpha}}{dt^\alpha} + \frac{1}{R} u_{C\alpha} = \delta(t) \\ u_{C\alpha}(0_-) = 0 \end{cases} \tag{4.111}$$

Taking Laplace transform of (4.111), there is

$$C_\alpha s^\alpha U_{C\alpha}(s) + \frac{1}{R} U_{C\alpha}(s) = 1 \tag{4.112}$$

The capacitor voltage in the complex frequency domain can be derived as

$$U_{C\alpha}(s) = \frac{1}{C_\alpha[s^\alpha + 1/(RC_\alpha)]} \tag{4.113}$$

Using inverse Laplace transform to (4.113), we obtain the analytical solution of the impulse response as

$$\begin{aligned} u_{C\alpha}(t) &= \mathcal{L}^{-1}[U_{C\alpha}(s)] = \frac{1}{C_\alpha} t^{\alpha-1} E_{\alpha,\alpha}\left(-\frac{1}{RC_\alpha} t^\alpha\right) \\ &= \frac{1}{C_\alpha} \sum_{k=0}^{\infty} \left(-\frac{1}{RC_\alpha}\right)^k \frac{t^{\alpha k+\alpha-1}}{\Gamma(\alpha k+\alpha)} \end{aligned} \tag{4.114}$$

From (4.114), it can be concluded that the impulse response of the RC_α circuit is similar to the zero-input response. Since the energy of impulse input is infinite, the capacitor can be charged to steady-state instantaneously, and as $t > 0$, the circuit starts to discharge like the zero-input response.

2. Fractional-order RL_β Circuit

As shown in Fig. 4.28, assuming that the external excitation is the unit impulse voltage source and $u_s = \delta(t)$, we will analyze the impulse response of the RL_β circuit.

Based on KVL, the inductor current $i_{L\beta}$ satisfies

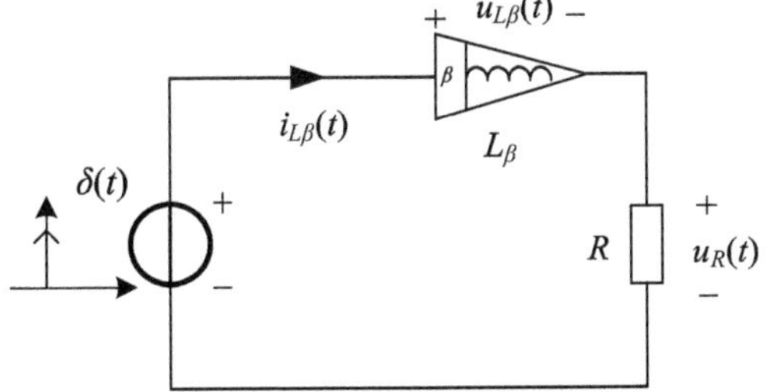

Fig. 4.28 RL_β circuit with pulse excitation

$$\begin{cases} L_\beta \frac{\mathrm{d}^\beta i_{L\beta}}{\mathrm{d}t^\beta} + i_{L\beta} R = \delta(t) \\ i_{L\beta}(0_-) = 0 \end{cases} \tag{4.115}$$

Taking Laplace transform of (4.115), there is

$$L_\beta s^\beta I_{L\beta}(s) + R I_{L\beta}(s) = 1 \tag{4.116}$$

from which we obtain

$$I_{L\beta}(s) = \frac{1}{L_\beta\left(s^\beta + R/L_\beta\right)} \tag{4.117}$$

The analytical solution of impulse response can be calculated by the inverse Laplace transform of (4.117) as

$$i_{L\beta}(t) = \mathcal{L}^{-1}[I_{L\beta}(s)] = \frac{1}{L_\beta} t^{\beta-1} E_{\beta,\beta}\left(-\frac{R}{L_\beta} t^\beta\right) = \frac{1}{L_\beta} \sum_{k=0}^{\infty} \left(-\frac{R}{L_\beta}\right)^k \frac{t^{\beta k+\beta-1}}{\Gamma(\beta k+\beta)} \tag{4.118}$$

4.6 Zero-Input Response of Fractional-Order $RL_\beta C_\alpha$ Circuits

The circuit containing two fractional-order components will be discussed in this section, including series and parallel $RL_\beta C_\alpha$ circuits with non-zero initial states. The zero-input response in the time domain is only determined by the initial energy and circuit topology since there is no external source.

4.6.1 Series Fractional-Order $RL_\beta C_\alpha$ Circuit

As shown in Fig. 4.29, the initial voltage across the capacitor is $u_{C\alpha}(0_-) = U_0$ at $t = 0_-$, and the initial current flowing through the inductor L_β is $i_{L\beta}(0_-) = I_0$. Since the branch current in the series circuit is equal everywhere, we can assume that $i_{C\alpha} = i_{L\beta} = i_R = i$. According to KVL, the voltage sum of the resistor, capacitor, and inductor is equal to zero, which means that $u_{C\alpha} + u_{L\beta} + u_R = 0$. Hence, the fractional differential equation and its initial conditions can be stated as

$$\begin{cases} \frac{1}{C_\alpha} {}_0D_t^{-\alpha} i + L_\beta {}_0D_t^\beta i + iR = 0 \\ u_{C\alpha}(0_+) = u_{C\alpha}(0_-) = U_0 \\ i_{L\beta}(0_+) = i_{L\beta}(0_-) = I_0 \\ Q_{C\alpha}(0_+) = Q_{C\alpha}(0_-) = Q_0 \\ \psi_{L\beta}(0_+) = \psi_{L\beta}(0_-) = \psi_0 \end{cases} \tag{4.119}$$

According to the relationship between capacitor voltage and current, the fractional integral operator can be eliminated, and then the voltage of capacitor can be derived from (4.119) as

$$L_\beta C_\alpha \, {}_0D_t^\beta \left({}_0D_t^\alpha u_{C\alpha} \right) + RC_\alpha \, {}_0D_t^\alpha u_{C\alpha} + u_{C\alpha} = 0 \tag{4.120}$$

At $t = 0_+$, the initial conditions satisfy

$$i(0_+) = C_\alpha \left[{}_0D_t^\alpha u_{C\alpha}(t) \right]_{t=0} = I_0 \tag{4.121}$$

3. R-L Definition

If we apply the property of R-L definition to (4.120), it can be simplified as

$$L_\beta C_\alpha \left[{}_0D_t^{\alpha+\beta} u_{C\alpha} - \left[{}_0D_t^{\alpha-1} u_{C\alpha}(t) \right]_{t=0} \frac{t^{-\beta-1}}{\Gamma(-\beta)} \right] + RC_\alpha {}_0D_t^\alpha u_{C\alpha} + u_{C\alpha} = 0 \tag{4.122}$$

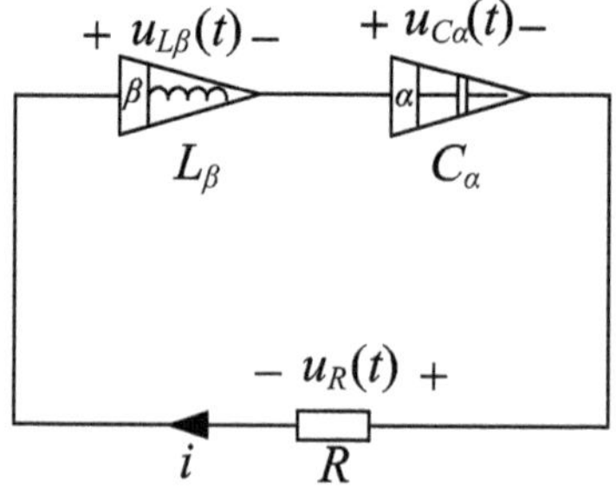

Fig. 4.29 Series $RL_\beta C_\alpha$ circuit with zero input

Applying Laplace transform to (4.119) and (4.122), there is

$$\frac{1}{C_\alpha}s^{-\alpha}I(s)+L_\beta\Big[s^\beta I(s)-\Big[{}_0D_t^{\beta-1}i_{L\beta}(t)\Big]_{t=0}\Big]+RI(s)=0 \tag{4.123}$$

$$L_\beta C_\alpha\Big[s^{\alpha+\beta}U_{C\alpha}(s)-\Big[{}_0D_t^{\beta-1}\big({}_0D_t^\alpha u_{C\alpha}(t)\big)\Big]_{t=0}-\big[{}_0D_t^{\alpha-1}u_{C\alpha}(t)\big]_{t=0}s^\beta\Big] \\ +RC_\alpha\big(s^\alpha U_{C\alpha}(s)-\big[{}_0D_t^{\alpha-1}u_{C\alpha}(t)\big]_{t=0}\big)+U_{C\alpha}(s)=0 \tag{4.124}$$

Then, the inductor current can be derived as

$$I_{L\beta}(s)=\frac{\Big[{}_0D_t^{\beta-1}i_{L\beta}(t)\Big]_{t=0}s^\alpha}{s^{\alpha+\beta}+\frac{R}{L_\beta}s^\alpha+\frac{1}{L_\beta C_\alpha}} \tag{4.125}$$

The capacitor voltage is

$$U_{C\alpha}(s)=\frac{\Big(\frac{R}{L_\beta}+s^\beta\Big)\big[{}_0D_t^{\alpha-1}u_{C\alpha}(t)\big]_{t=0}+\Big[{}_0D_t^{\beta-1}\big({}_0D_t^\alpha u_{C\alpha}(t)\big)\Big]_{t=0}}{s^{\alpha+\beta}+\frac{R}{L_\beta}s^\alpha+\frac{1}{L_\beta C_\alpha}} \tag{4.126}$$

The initial conditions can be denoted by

$$\Big[{}_0D_t^{\beta-1}\big({}_0D_t^\alpha u_{C\alpha}(t)\big)\Big]_{t=0}=\Big[{}_0D_t^{\beta-1}\frac{I_0}{C_\alpha}\Big]_{t=0}=\Big[\frac{I_0}{C_\alpha}\frac{t^{1-\beta}}{\Gamma(2-\beta)}\Big]_{t=0}=0 \tag{4.127}$$

where $\big[{}_0D_t^{\alpha-1}u_{C\alpha}(t)\big]_{t=0}=\frac{Q_0}{C_\alpha}$, $\Big[{}_0D_t^{\beta-1}i_{L\beta}(t)\Big]_{t=0}=\frac{\psi_0}{L_\beta}$.

Therefore, the analytical solutions in the time domain can be derived from the inverse Laplace transform as follows:

The inductor current $i_{L\beta}$ is

$$\begin{aligned} i_{L\beta}(t)&=\mathcal{L}^{-1}\big[I_{L\beta}(s)\big] \\ &=\Big[{}_0D_t^{\beta-1}i_{L\beta}(t)\Big]_{t=0}\sum_{r=0}^{\infty}\Big(-\frac{R}{L_\beta}\Big)^r t^{\alpha r+\beta-1}E_{\alpha+\beta,\beta+\alpha r}^{r+1}\Big(-\frac{1}{L_\beta C_\alpha}t^{\alpha+\beta}\Big) \end{aligned} \tag{4.128}$$

The capacitor voltage $u_{C\alpha}$ is

$$\begin{aligned} u_{C\alpha}(t)&=\mathcal{L}^{-1}[U_{C\alpha}(s)] \\ &=\frac{R}{L_\beta}\big[{}_0D_t^{\alpha-1}u_{C\alpha}(t)\big]_{t=0} \\ &\quad\times\sum_{r=0}^{\infty}\Big(-\frac{R}{L_\beta}\Big)^r t^{\alpha r+\alpha+\beta-1}E_{\alpha+\beta,\alpha+\beta+\alpha r}^{r+1}\Big(-\frac{1}{L_\beta C_\alpha}t^{\alpha+\beta}\Big) \end{aligned}$$

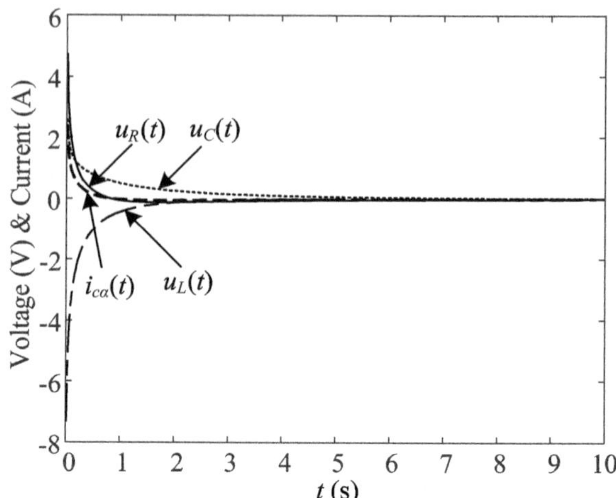

Fig. 4.30 Zero-input response of series $RL_\beta C_\alpha$ circuit under R-L definition for $\alpha = 0.75$, $\beta = 0.75$, $R = 2\ \Omega$, $C_\alpha = 1\ \text{F/s}^{1-\alpha}$, $L_\beta = 1\ \text{H/s}^{1-\beta}$, $I_0 = 1$ A, $U_0 = 1$ V

$$+\left[{}_0D_t^{\alpha-1}u_{C\alpha}(t)\right]_{t=0}\sum_{r=0}^{\infty}\left(-\frac{R}{L_\beta}\right)^r t^{\alpha r+\alpha-1}E_{\alpha+\beta,\alpha+\alpha r}^{r+1}\left(-\frac{1}{L_\beta C_\alpha}t^{\alpha+\beta}\right) \tag{4.129}$$

where $E_{\alpha+\beta,\alpha+\alpha r}^{r+1}\left(-\frac{1}{L_\beta C_\alpha}t^{\alpha+\beta}\right)$ is the three-parameter Mittag–Leffler function [6].

Other currents and voltages of the circuit can be derived from the above equations by the characteristics of the series circuit. The zero-input response under the R-L definition of series $RL_\beta C_\alpha$ circuit is shown in Fig. 4.30, which implies that the currents and voltages finally decay to zero.

4. Caputo Definition

Applying the property of R-L definition to (4.120), it can be simplified as

$$L_\beta C_{\alpha 0}D_t^{\alpha+\beta}u_{C\alpha} + RC_{\alpha 0}D_t^{\alpha}u_{C\alpha} + u_{C\alpha} = 0 \tag{4.130}$$

Taking Laplace transform of (4.119) and (4.130), there is

$$\frac{1}{C_\alpha}s^{-\alpha}I(s) + L_\beta\left[s^\beta I(s) - s^{\beta-1}i_{L\beta}(0)\right] + RI(s) = 0 \tag{4.131}$$

$$\begin{aligned}&L_\beta C_\alpha\left[s^{\alpha+\beta}U_{C\alpha}(s) - s^{\alpha+\beta-1}u_{C\alpha}(0)\right] + RC_\alpha\left[s^\alpha U_{C\alpha}(s) - s^{\alpha-1}u_{C\alpha}(0)\right]\\&+U_{C\alpha}(s) = 0\end{aligned} \tag{4.132}$$

Then, the inductor current can be deduced as

$$I_{L\beta}(s) = \frac{i_{L\beta}(0)s^{\alpha+\beta-1}}{s^{\alpha+\beta} + \frac{R}{L_\beta}s^\alpha + \frac{1}{L_\beta C_\alpha}} \tag{4.133}$$

The capacitor voltage is

$$U_{C\alpha}(s) = \frac{u_{C\alpha}(0)\left(\frac{R}{L_\beta}s^{\alpha-1} + s^{\alpha+\beta-1}\right)}{s^{\alpha+\beta} + \frac{R}{L_\beta}s^{\alpha} + \frac{1}{L_\beta C_\alpha}} \tag{4.134}$$

The analytical solutions in the time domain can be derived from the above equations by inverse Laplace transform, which means:

The inductor current $i_{L\beta}$ is

$$\begin{aligned} i_{L\beta}(t) &= \mathcal{L}^{-1}\left[I_{L\beta}(s)\right] \\ &= I_0 \sum_{r=0}^{\infty} \left(-\frac{R}{L_\beta}\right)^r t^{\beta r} E_{\alpha+\beta,1+\beta r}^{r+1}\left(-\frac{1}{L_\beta C_\alpha}t^{\alpha+\beta}\right) \end{aligned} \tag{4.135}$$

and the capacitor voltage $u_{C\alpha}$ is

$$\begin{aligned} u_{C\alpha}(t) &= \mathcal{L}^{-1}[U_{C\alpha}(s)] \\ &= \frac{R}{L_\beta}U_0 \sum_{r=0}^{\infty} \left(-\frac{R}{L_\beta}\right)^r t^{\beta r+\beta} E_{\alpha+\beta,\beta+1+\beta r}^{r+1}\left(-\frac{1}{L_\beta C_\alpha}t^{\alpha+\beta}\right) \\ &\quad + U_0 \sum_{r=0}^{\infty} \left(-\frac{R}{L_\beta}\right)^r t^{\beta r} E_{\alpha+\beta,1+\beta r}^{r+1}\left(-\frac{1}{L_\beta C_\alpha}t^{\alpha+\beta}\right) \end{aligned} \tag{4.136}$$

Other currents and voltages of the circuit can be derived from the above equations by the characteristics of the series circuit. Hence, the zero-input response under Caputo definition of series $RL_\beta C_\alpha$ circuit is shown in Fig. 4.31, which implies that the currents and voltages finally decay to zero.

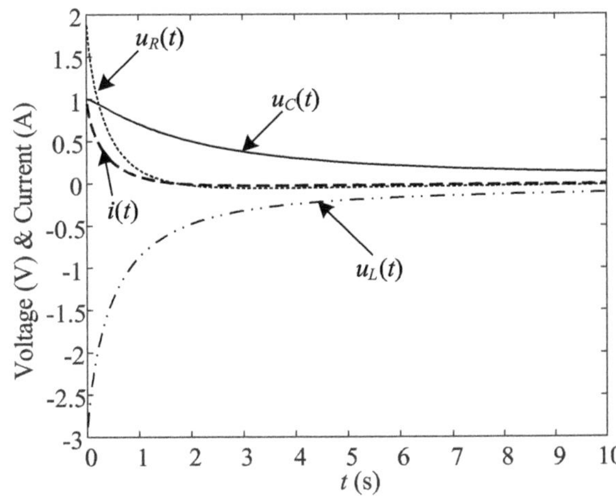

Fig. 4.31 Zero-input response of series $RL_\beta C_\alpha$ circuit under Caputo definition for $\alpha = 0.75$, $\beta = 0.75$, $R = 2\ \Omega$, $C_\alpha = 1$ F/s$^{1-\alpha}$, $L_\beta = 1$ H/s$^{1-\beta}$, $I_0 = 1$ A, $U_0 = 1$ V

4.6.2 Parallel Fractional-Order $RL_\beta C_\alpha$ Circuit

As shown in Fig. 4.32, when $t = 0_-$, the initial voltage across the capacitor is $u_{C\alpha}(0_-) = U_0$ and the initial current flowing through the inductor L_β is $i_{L\beta}(0_-) = I_0$. Since the component voltages in the parallel circuit are the same, we can assume that $u_{C\alpha} = u_{L\beta} = u_R = u$. According to KCL, the current sum of the resistor, capacitor, and inductor is equal to zero, which means that $i_{C\alpha} + i_{L\beta} + i_R = 0$. Then, the fractional differential equation and its initial conditions can be stated as

$$\begin{cases} \frac{1}{L_\beta}{}_0D_t^{-\beta}u + C_\alpha {}_0D_t^\alpha u + \frac{u}{R} = 0 \\ u_{C\alpha}(0_+) = u_{C\alpha}(0_-) = U_0 \\ i_{L\beta}(0_+) = i_{L\beta}(0_-) = I_0 \\ Q_{C\alpha}(0_+) = Q_{C\alpha}(0_-) = Q_0 \\ \psi_{L\beta}(0_+) = \psi_{L\beta}(0_-) = \psi_0 \end{cases} \tag{4.137}$$

From the voltage-current relationship of the inductor, we can eliminate the fractional integral operator, then the current of the inductor can be derived from (4.137) as

$$L_\beta C_\alpha {}_0D_t^\alpha\left({}_0D_t^\beta i_{L\beta}\right) + \frac{L_\beta}{R}{}_0D_t^\beta i_{L\beta} + i_{L\beta} = 0 \tag{4.138}$$

At $t = 0_+$, the initial conditions satisfy

$$u(0_+) = L_\beta\left[{}_0D_t^\beta i_{L\beta}(t)\right]_{t=0} = U_0 \tag{4.139}$$

5. R-L Definition

Under R-L definition, (4.138) can be simplified as

$$L_\beta C_\alpha\left[{}_0D_t^{\alpha+\beta} i_{L\beta} - \left[{}_0D_t^{\beta-1} i_{L\beta}(t)\right]_{t=0}\frac{t^{-\alpha-1}}{\Gamma(-\alpha)}\right] + \frac{L_\beta}{R}{}_0D_t^\beta i_{L\beta} + i_{L\beta} = 0 \tag{4.140}$$

Using Laplace transform to (4.137) and (4.140), there are

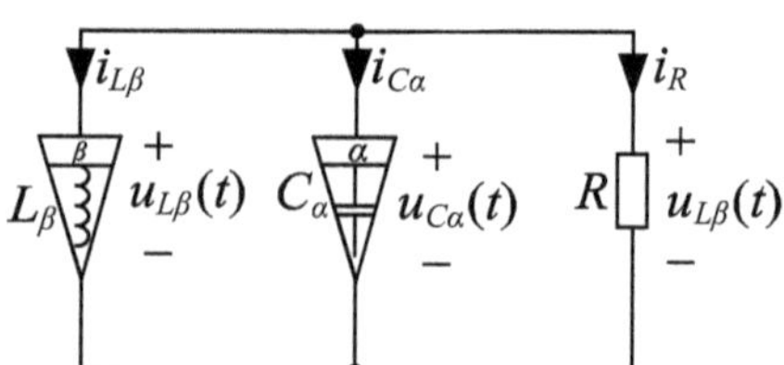

Fig. 4.32 Parallel $RL_\beta C_\alpha$ circuit with zero input

$$\frac{1}{L_\beta}s^{-\beta}U(s)+C_\alpha\left[s^\alpha U(s)-\left[{}_0D_t^{\alpha-1}u(t)\right]_{t=0}\right]+\frac{1}{R}U(s)=0 \tag{4.141}$$

$$L_\beta C_\alpha\left[s^{\alpha+\beta}I_{L\beta}(s)-\left[{}_0D_t^{\alpha-1}\left({}_0D_t^\beta i_{L\beta}(t)\right)\right]_{t=0}-\left[{}_0D_t^{\beta-1}i_{L\beta}(t)\right]_{t=0}s^\alpha\right] \\ +\frac{L_\beta}{R}\left[s^\beta I_{L\beta}(s)-\left[{}_0D_t^{\beta-1}i_{L\beta}(t)\right]_{t=0}\right]+I_{L\beta}(s)=0 \tag{4.142}$$

the capacitor voltage in the complex frequency can be obtained as

$$U_{C\alpha}(s)=\frac{\left[{}_0D_t^{\alpha-1}u_{C\alpha}(t)\right]_{t=0}s^\beta}{s^{\alpha+\beta}+\frac{1}{RC_\alpha}s^\beta+\frac{1}{L_\beta C_\alpha}} \tag{4.143}$$

the inductor current is

$$I_{L\beta}(s)=\frac{\left[{}_0D_t^{\beta-1}i_{L\beta}(t)\right]_{t=0}\left(s^\alpha+\frac{1}{RC_\alpha}\right)+\left[{}_0D_t^{\alpha-1}\left({}_0D_t^\beta i_{L\beta}(t)\right)\right]_{t=0}}{s^{\alpha+\beta}+\frac{1}{RC_\alpha}s^\beta+\frac{1}{L_\beta C_\alpha}} \tag{4.144}$$

and the initial conditions can be denoted by

$$\begin{cases}\left[{}_0D_t^{\alpha-1}\left({}_0D_t^\beta i_{L\beta}(t)\right)\right]_{t=0}=\left[{}_0D_t^{\alpha-1}\frac{U_0}{L_\beta}\right]_{t=0}=\left[\frac{U_0}{L_\beta}\frac{t^{1-\alpha}}{\Gamma(2-\alpha)}\right]_{t=0}=0\\ \left[{}_0D_t^{\alpha-1}u_{C\alpha}(t)\right]_{t=0}=\frac{Q_0}{C_\alpha}\\ \left[{}_0D_t^{\beta-1}i_{L\beta}(t)\right]_{t=0}=\frac{\psi_0}{L_\beta}\end{cases} \tag{4.145}$$

Thus, the analytical solutions in the time domain can be derived from the inverse Laplace transform, the inductor current $i_{L\beta}$ is deduced as

$$\begin{aligned} i_{L\beta}(t)&=\mathcal{L}^{-1}\left[i_{L\beta}(s)\right]\\ &=\left[{}_0D_t^{\beta-1}i_{L\beta}(t)\right]_{t=0}\sum_{r=0}^{\infty}\left(-\frac{1}{RC_\alpha}\right)^r t^{\alpha r+\beta-1}E_{\alpha+\beta,\beta+\alpha r}^{r+1}\left(-\frac{1}{L_\beta C_\alpha}t^{\alpha+\beta}\right)\\ &\quad+\left[\left[{}_0D_t^{\alpha+\beta-1}i_{L\beta}(t)\right]_{t=0}+\frac{1}{RC_\alpha}\left[{}_0D_t^{\beta-1}i_{L\beta}(t)\right]_{t=0}\right]\\ &\quad\times\sum_{r=0}^{\infty}\left(-\frac{1}{RC_\alpha}\right)^r t^{\alpha r+\alpha+\beta-1}E_{\alpha+\beta,\alpha+\beta+\alpha r}^{r+1}\left(-\frac{1}{L_\beta C_\alpha}t^{\alpha+\beta}\right) \end{aligned} \tag{4.146}$$

and the capacitor voltage $u_{C\alpha}$ is

$$u_{C\alpha}(t)=\mathcal{L}^{-1}[U_{C\alpha}(s)]$$

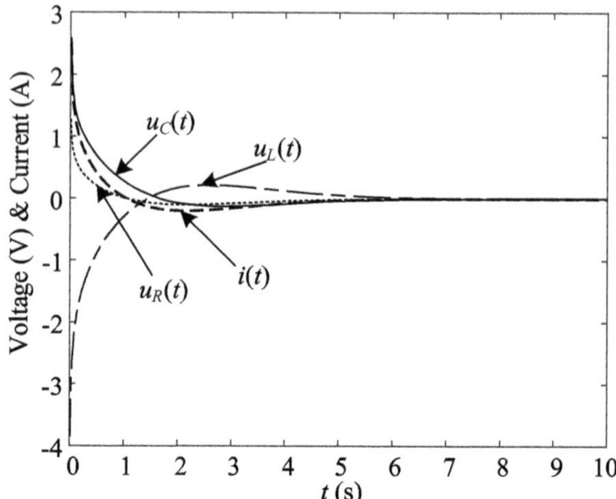

Fig. 4.33 Zero-input response of parallel $RL_\beta C_\alpha$ circuit under R-L definition for $\alpha = 0.75$, $\beta = 0.75$, $R = 2\Omega$, $C_\alpha = 1$ F/s$^{1-\alpha}$, $L_\beta = 1$ H/s$^{1-\beta}$, $I_0 = 1$ A, $U_0 = 1$ V

$$= \left[{}_0D_t^{\alpha-1}u_{C\alpha}(t)\right]_{t=0}\sum_{r=0}^{\infty}\left(-\frac{1}{RC_\alpha}\right)^r t^{\alpha r+\alpha-1}E_{\alpha+\beta,\alpha+\alpha r}^{r+1}\left(-\frac{1}{L_\beta C_\alpha}t^{\alpha+\beta}\right) \tag{4.147}$$

Other currents and voltages of the circuit can be derived from the above equations by the characteristics of the parallel circuit. Hence, the zero-input response under the R-L definition of parallel $RL_\beta C_\alpha$ circuit is shown in Fig. 4.33, which implies that the currents and voltages finally decay to zero.

6. Caputo Definition

If we apply the property of Caputo definition to (4.138), it can be simplified as

$$L_\beta C_\alpha\, {}_0D_t^{\alpha+\beta}i_{L\beta} + \frac{L_\beta}{R}{}_0D_t^\beta i_{L\beta} + i_{L\beta} = 0 \tag{4.148}$$

Taking Laplace transform of (4.137) and (4.148), there are

$$\frac{1}{L_\beta}s^{-\beta}U(s) + C_\alpha\left[s^\alpha U(s) - s^{\alpha-1}u_{C\alpha}(0)\right] + \frac{1}{R}U(s) = 0 \tag{4.149}$$

$$L_\beta C_\alpha\left[s^{\alpha+\beta}I_{L\beta}(s) - s^{\alpha+\beta-1}i_{L\beta}(0)\right] + \frac{L_\beta}{R}\left[s^\beta I_{L\beta}(s) - s^{\beta-1}i_{L\beta}(0)\right] + I_{L\beta}(s) = 0 \tag{4.150}$$

Then, the capacitor voltage can be derived as

$$U_{C\alpha}(s) = \frac{s^{\alpha+\beta-1}u_{C\alpha}(0)}{s^{\alpha+\beta} + \frac{1}{RC_\alpha}s^\beta + \frac{1}{L_\beta C_\alpha}} \tag{4.151}$$

and inductor current is

$$I_{L\beta}(s) = \frac{\left(\frac{1}{RC_\alpha}s^{\beta-1} + s^{\alpha+\beta-1}\right)i_{L\beta}(0)}{s^{\alpha+\beta} + \frac{1}{RC_\alpha}s^\beta + \frac{1}{L_\beta C_\alpha}} \tag{4.152}$$

Therefore, the analytical solutions in the time domain can be derived from the inverse Laplace transform, the capacitor voltage $u_{C\alpha}$ is obtained as

$$\begin{aligned} u_{C\alpha}(t) &= \mathcal{L}^{-1}[U_{C\alpha}(s)] \\ &= U_0 \sum_{r=0}^{\infty}\left(-\frac{1}{RC_\alpha}\right)^r t^{\alpha r} E_{\alpha+\beta,1+\alpha r}^{r+1}\left(-\frac{1}{L_\beta C_\alpha}t^{\alpha+\beta}\right) \end{aligned} \tag{4.153}$$

and inductor current $i_{L\beta}$ is

$$\begin{aligned} i_{L\beta}(t) &= \mathcal{L}^{-1}\left[I_{L\beta}(s)\right] \\ &= \frac{I_0}{RC_\alpha} \sum_{r=0}^{\infty}\left(-\frac{1}{RC_\alpha}\right)^r t^{\alpha r+\alpha} E_{\alpha+\beta,\alpha+1+\alpha r}^{r+1}\left(-\frac{1}{L_\beta C_\alpha}t^{\alpha+\beta}\right) \\ &\quad + I_0 \sum_{r=0}^{\infty}\left(-\frac{1}{RC_\alpha}\right)^r t^{\alpha r} E_{\alpha+\beta,1+\alpha r}^{r+1}\left(-\frac{1}{L_\beta C_\alpha}t^{\alpha+\beta}\right) \end{aligned} \tag{4.154}$$

Other currents and voltages of the circuit can be derived from the above equations by the characteristics of the parallel circuit. The zero-input response under Caputo definition of parallel $RL_\beta C_\alpha$ circuit is shown in Fig. 4.34, which implies that the currents and voltages finally decay to zero.

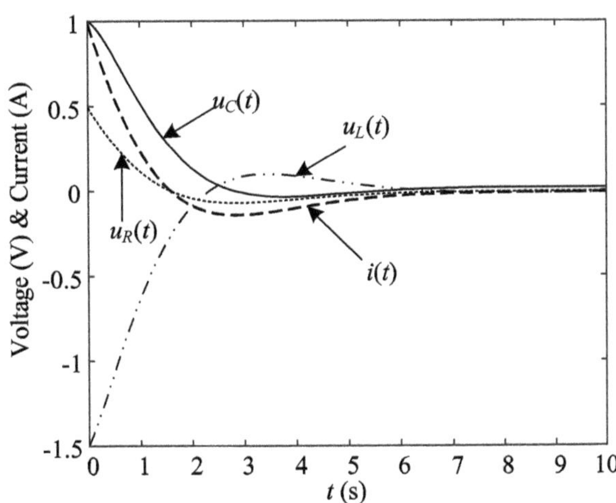

Fig. 4.34 Zero-input response of parallel $RL_\beta C_\alpha$ circuit under Caputo definition for $\alpha = 0.75$, $\beta = 0.75$, $R = 2\ \Omega$, $C_\alpha = 1$ F/$s^{1-\alpha}$, $L_\beta = 1$ H/$s^{1-\beta}$, $I_0 = 1$ A, $U_0 = 1$ V

4.7 Zero-State Response of Fractional-Order $RL_\beta C_\alpha$ Circuits

In this section, the time-domain response of a circuit containing two fractional-order components with zero initial states will be discussed, which means that there is no initial energy stored in the circuit. The zero-state response of series and parallel $RL_\beta C_\alpha$ circuit will be analyzed with DC source and AC source, respectively.

4.7.1 Series Fractional-Order $RL_\beta C_\alpha$ Circuits

As shown in Fig. 4.35, at $t = 0_-$, all the initial states of the circuit component are 0, and the excitation source is u_s, which means that there is no initial energy. According to KVL, we can obtain the voltage sum of the resistor, capacitor, and inductor is equal to the excitation voltage, which means that $u_{C\alpha} + u_{L\beta} + u_R = u_s$. Since the current of the series circuit is equal everywhere, we can assume that $i_{C\alpha} = i_{L\beta} = i_R = I$, then, the fractional differential equation and its initial conditions can be stated as

$$\begin{cases} \frac{1}{C_\alpha} {}_0D_t^{-\alpha} i + L_\beta {}_0D_t^\beta i + iR = u_s \\ u_{C\alpha}(0_+) = u_{C\alpha}(0_-) = 0 \\ i_{L\beta}(0_+) = i_{L\beta}(0_-) = 0 \end{cases} \tag{4.155}$$

Based on the voltage-current relationship of the capacitor, the integral operator can be eliminated, and the voltage of the capacitor can be derived as

$$L_\beta C_\alpha \, {}_0D_t^\beta \left({}_0D_t^\alpha u_{C\alpha} \right) + RC_\alpha \, {}_0D_t^\alpha u_{C\alpha} + u_{C\alpha} = u_s \tag{4.156}$$

Applying Laplace transform to (4.155) and (4.156), there are

$$\frac{1}{C_\alpha} s^{-\alpha} I(s) + L_\beta s^\beta I(s) + RI(s) = U_s(s) \tag{4.157}$$

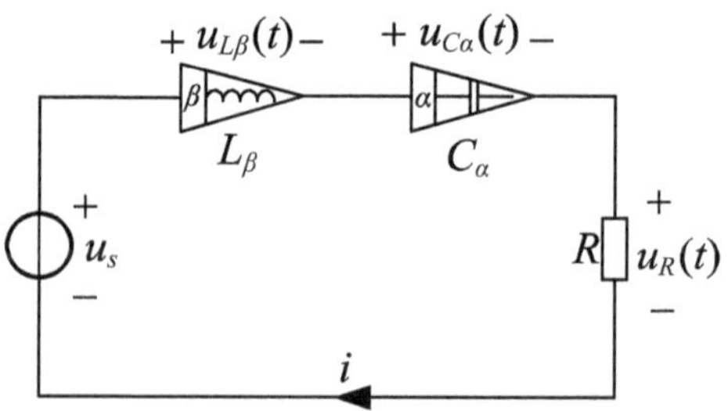

Fig. 4.35 Series $RL_\beta C_\alpha$ circuit with zero state

$$L_\beta C_\alpha s^{\alpha+\beta} U_{C\alpha}(s) + RC_\alpha s^\alpha U_{C\alpha}(s) + U_{C\alpha}(s) = U_s(s) \tag{4.158}$$

The inductor current can be derived as

$$I_{L\beta}(s) = \frac{\frac{1}{L_\beta} s^\alpha}{s^{\alpha+\beta} + \frac{R}{L_\beta} s^\alpha + \frac{1}{L_\beta C_\alpha}} U(s) \tag{4.159}$$

and capacitor voltage is

$$U_{C\alpha}(s) = \frac{\frac{1}{L_\beta C_\alpha}}{s^{\alpha+\beta} + \frac{R}{L_\beta} s^\alpha + \frac{1}{L_\beta C_\alpha}} U_s(s) \tag{4.160}$$

Then, the analytical solution in the time domain can be derived from the inverse Laplace transform, the inductor current $i_{L\beta}$ can be deduced as

$$\begin{aligned} i(t) &= i_{L\beta}(t) = \mathcal{L}^{-1}\big[I_{L\beta}(s)\big] \\ &= \frac{1}{L_\beta} \int_0^t \sum_{r=0}^{\infty} \left(-\frac{R}{L_\beta}\right)^r \tau^{\alpha r+\beta-1} E^{r+1}_{\alpha+\beta,\alpha r+\beta}\left(-\frac{1}{L_\beta C_\alpha}\tau^{\alpha+\beta}\right) u_s(t-\tau)\mathrm{d}\tau \end{aligned} \tag{4.161}$$

and capacitor voltage $u_{C\alpha}$ is

$$\begin{aligned} u_{C\alpha}(t) &= L^{-1}[U_{C\alpha}(s)] \\ &= \frac{1}{L_\beta C_\alpha} \int_0^t \sum_{r=0}^{\infty} \left(-\frac{R}{L_\beta}\right)^r \tau^{\alpha r+\alpha+\beta-1} \\ &\quad \times E^{r+1}_{\alpha+\beta,\alpha+\beta+\alpha r}\left(-\frac{1}{L_\beta C_\alpha}\tau^{\alpha+\beta}\right) u_s(t-\tau) d\tau \end{aligned} \tag{4.162}$$

Other currents and voltages of the circuit can be derived from the above equations by the characteristics of the parallel circuit, then the resistor voltage u_R is

$$u_R(t) = i(t)R \tag{4.163}$$

and the inductor voltage $u_{L\beta}$ is

$$u_{L\beta}(t) = L_\beta \frac{\mathrm{d}^\beta i_{L\beta}}{\mathrm{d}t^\beta} = u_s(t) - u_{C\alpha}(t) - u_R(t) \tag{4.164}$$

7. Reponse to DC Source

If the voltage source in Fig. 4.35 is DC source and $u_s = U_0$, the analytical solution can be obtained by inverse Laplace transform as follows:

The inductor current $i_{L\beta}$ is

$$\begin{aligned} i_{L\beta}(t) &= \mathcal{L}^{-1}\left[i_{L\beta}(s)\right] = \mathcal{L}^{-1}\left[\frac{\frac{1}{L_\beta}s^\alpha}{s^{\alpha+\beta}+\frac{R}{L_\beta}s^\alpha+\frac{1}{L_\beta C_\alpha}}\frac{U_0}{s}\right] \\ &= \frac{U_0}{L_\beta}\sum_{r=0}^{\infty}\left(-\frac{R}{L_\beta}\right)^r t^{\alpha r+\beta}E^{r+1}_{\alpha+\beta,\alpha r+\beta+1}\left(-\frac{1}{L_\beta C_\alpha}t^{\alpha+\beta}\right) \end{aligned} \tag{4.165}$$

The capacitor voltage $u_{C\alpha}$ is

$$\begin{aligned} u_{C\alpha}(t) &= \mathcal{L}^{-1}[U_{C\alpha}(s)] = \mathcal{L}^{-1}\left[\frac{\frac{1}{L_\beta C_\alpha}}{s^{\alpha+\beta}+\frac{R}{L_\beta}s^\alpha+\frac{1}{L_\beta C_\alpha}}\frac{U_0}{s}\right] \\ &= \frac{U_0}{L_\beta C_\alpha}\sum_{r=0}^{\infty}\left(-\frac{R}{L_\beta}\right)^r t^{\alpha r+\alpha+\beta}E^{r+1}_{\alpha+\beta,\alpha r+\alpha+\beta+1}\left(-\frac{1}{L_\beta C_\alpha}t^{\alpha+\beta}\right) \end{aligned} \tag{4.166}$$

Thus, the zero-state response of the series $RL_\beta C_\alpha$ circuit under DC source is shown in Fig. 4.36. The behaviors of the transition state are influenced by the orders of fractional-order components, which illustrates that the oscillation of currents and voltages will increase as the orders increase.

8. Response to AC Source

If the voltage source in Fig. 4.35 is an AC source and $u_s = U_m\cos(\omega t + \theta)$, the analytical solution can be obtained by inverse Laplace transform as follows:

The inductor current $i_{L\beta}$ is

$$\begin{aligned} i_{L\beta}(t) &= \mathcal{L}^{-1}\left[I_{L\beta}(s)\right] \\ &= \mathcal{L}^{-1}\left[\frac{s^\alpha/L_\beta}{s^{\alpha+\beta}+R/L_\beta s^\alpha+1/\left(L_\beta C_\alpha\right)}\frac{U_m(s\cos\theta-\omega\sin\theta)}{\left(s^2+\omega^2\right)}\right] \\ &= \frac{U_m}{L_\beta}\int_0^t\sum_{r=0}^{\infty}\left(-\frac{R}{L_\beta}\right)^r \tau^{\alpha r+\beta-1}E^{r+1}_{\alpha+\beta,\alpha r+\beta}\left(-\frac{1}{L_\beta C_\alpha}\tau^{\alpha+\beta}\right) \\ &\quad \times \cos[\omega(t-\tau)+\theta]\mathrm{d}\tau \end{aligned} \tag{4.167}$$

The capacitor voltage $u_{C\alpha}$ is

$$\begin{aligned} u_{C\alpha}(t) &= \mathcal{L}^{-1}[U_{C\alpha}(s)] \\ &= \mathcal{L}^{-1}\left[\frac{1/\left(L_\beta C_\alpha\right)}{s^{\alpha+\beta}+1/(RC_\alpha)s^\beta+1/\left(L_\beta C_\alpha\right)}\frac{U_m(s\cos\theta-\omega\sin\theta)}{\left(s^2+\omega^2\right)}\right] \end{aligned}$$

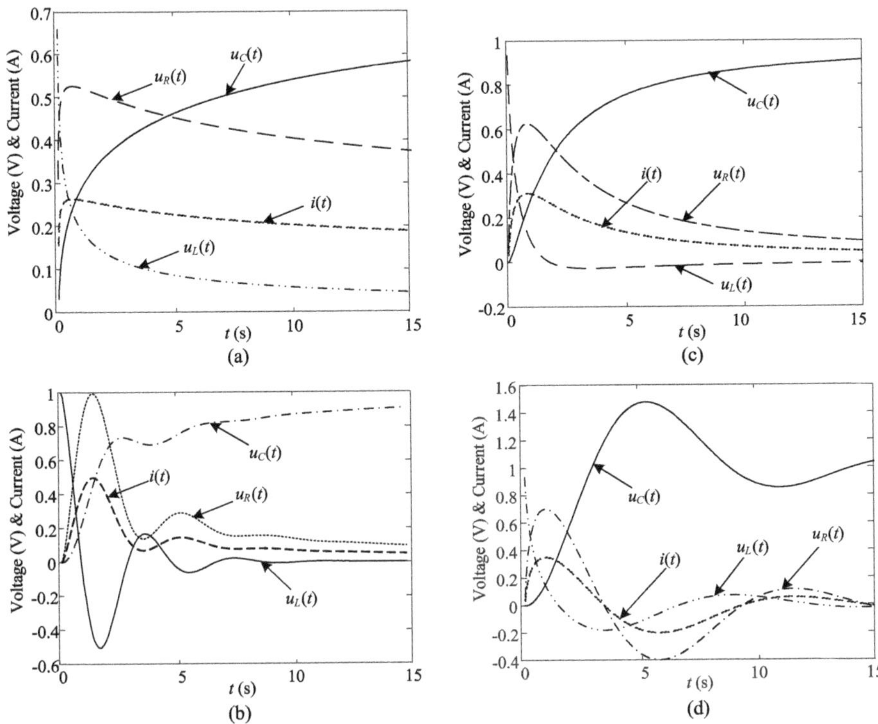

Fig. 4.36 Zero-state response of series $RL_\beta C_\alpha$ with DC voltage source for $R = 2\ \Omega$, $C_\alpha = 1\ \text{F/s}^{1-\alpha}$, $L_\beta = 1\ \text{H/s}^{1-\beta}$, $U_0 = 1$ V: **a** $\alpha = 0.35$, $\beta = 0.35$. **b** $\alpha = 0.75$, $\beta = 1.5$. **c** $\alpha = 0.75$, $\beta = 0.75$. **d** $\alpha = 1.5$, $\beta = 0.75$

$$= \frac{U_m}{L_\beta C_\alpha} \int_0^t \sum_{r=0}^{\infty} \left(-\frac{R}{L_\beta}\right)^r \tau^{\alpha r+\alpha+\beta-1} E^{r+1}_{\alpha+\beta,\alpha+\beta+\alpha r}\left(-\frac{1}{L_\beta C_\alpha}\tau^{\alpha+\beta}\right) \times \cos[\omega(t-\tau)+\theta]\mathrm{d}\tau \tag{4.168}$$

Then, the zero-state response of the series $RL_\beta C_\alpha$ circuit under AC source is shown in Fig. 4.37, which implies that the circuit voltage and current will finally reach the steady-state, and the phase angle difference is determined by the orders of the fractional-order components.

4.7.2 Parallel Fractional-Order $RL_\beta C_\alpha$ Circuits

As shown in Fig. 4.38, all the initial states of the circuit components are equal to zero at $t = 0_-$, which means that there is no initial energy, and the excitation current source is i_s. Since the component voltages of the parallel circuit are the same, we

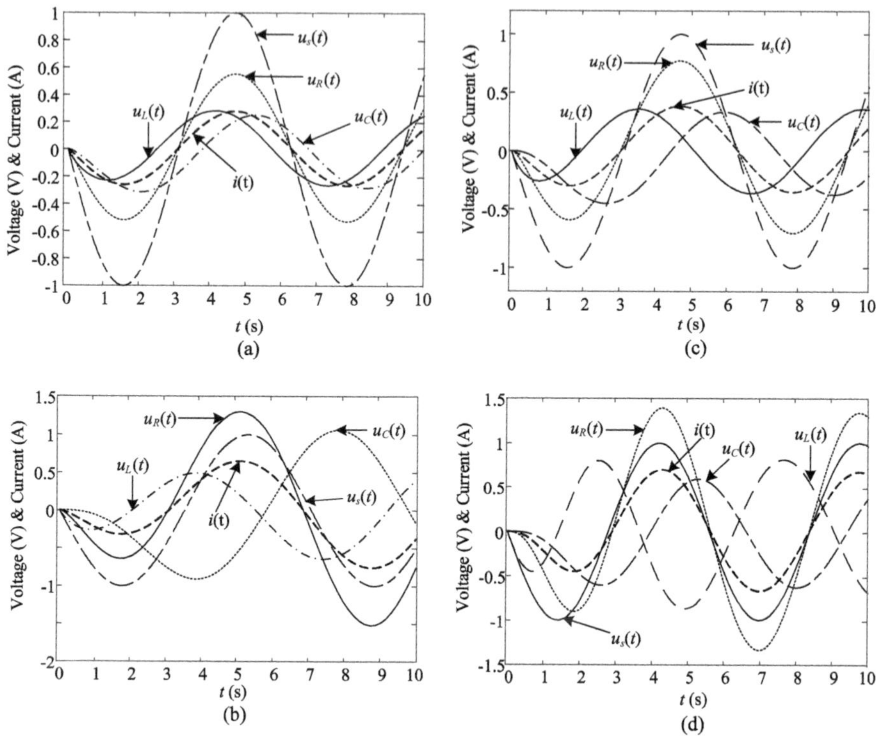

Fig. 4.37 Zero-state response of series $RL_{\beta}C_{\alpha}$ with sinusoidal voltage source for $R = 2\ \Omega$, $C_{\alpha} = 1\ \text{F/s}^{1-\alpha}$, $L_{\beta} = 1\ \text{H/s}^{1-\beta}$, $U_{\text{m}} = 1$ V: **a** $\alpha = 0.35$, $\beta = 0.35$. **b** $\alpha = 0.75$, $\beta = 1.5$. **c** $\alpha = 0.75$, $\beta = 0.75$. **d** $\alpha = 1.5$, $\beta = 0.75$

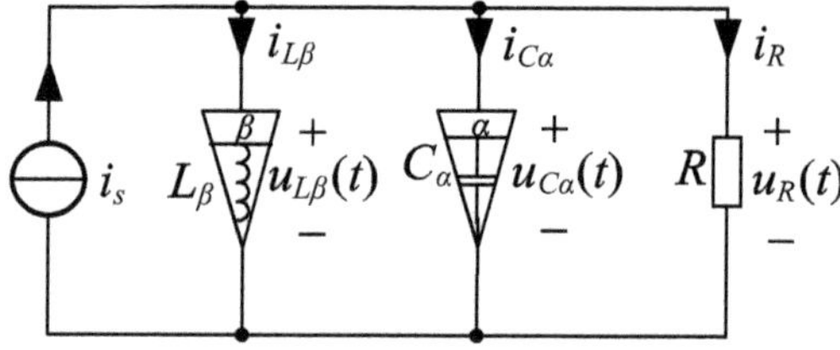

Fig. 4.38 Parallel $RL_{\beta}C_{\alpha}$ circuit with zero state

can assume that $u_{C\alpha} = u_{L\beta} = u_R = u$. According to KCL, we can obtain that the current sum of the resistor, capacitor, and inductor is equal to the excitation current, which means that $i_{C\alpha} + i_{L\beta} + i_R = i_s$. Then, the fractional differential equation and its initial conditions can be stated as

$$\begin{cases} \frac{1}{L_\beta} {}_0D_t^{-\beta}u + C_\alpha {}_0D_t^{\alpha}u + \frac{u}{R} = i_s \\ u_{C\alpha}(0_+) = u_{C\alpha}(0_-) = 0 \\ i_{L\beta}(0_+) = i_{L\beta}(0_-) = 0 \end{cases} \tag{4.169}$$

Based on the voltage-current relationship of the inductor, the fractional integral operator can be eliminated, and the current of the inductor can be derived from (4.169) as

$$L_\beta C_\alpha \, {}_0D_t^\alpha\left({}_0D_t^\beta i_{L\beta}\right) + \frac{L_\beta}{R}{}_0D_t^\beta i_{L\beta} + i_{L\beta} = i_s \quad (t > 0) \tag{4.170}$$

Using Laplace transform to (4.169) and (4.170), there is

$$\frac{1}{L_\beta}s^{-\beta}U(s) + C_\alpha s^\alpha U(s) + \frac{1}{R}U(s) = I_s(s) \tag{4.171}$$

$$L_\beta C_\alpha s^{\alpha+\beta} I_{L\beta}(s) + \frac{L_\beta}{R}s^\beta I_{L\beta}(s) + I_{L\beta}(s) = I_s(s) \tag{4.172}$$

the capacitor voltage can be written as

$$U_{C\alpha}(s) = \frac{s^\beta / C_\alpha}{s^{\alpha+\beta} + \frac{1}{RC_\alpha}s^\beta + \frac{1}{L_\beta C_\alpha}} I_s(s) \tag{4.173}$$

and the inductor current is

$$I_{L\beta}(s) = \frac{\frac{1}{L_\beta C_\alpha}}{s^{\alpha+\beta} + \frac{1}{RC_\alpha}s^\beta + \frac{1}{L_\beta C_\alpha}} I_s(s) \tag{4.174}$$

Then, the analytical solutions in the time domain can be derived from the inverse Laplace transform as follows:

The capacitor voltage $u_{C\alpha}$ is

$$\begin{aligned} u(t) &= u_{C\alpha}(t) = \mathcal{L}^{-1}[U_{C\alpha}(s)] \\ &= \frac{1}{C_\alpha}\int_0^t \sum_{r=0}^{\infty}\left(-\frac{1}{RC_\alpha}\right)^r \tau^{\alpha r+\alpha-1} E_{\alpha+\beta,\alpha+\alpha r}^{r+1}\left(-\frac{1}{L_\beta C_\alpha}\tau^{\alpha+\beta}\right) i_s(t-\tau)\mathrm{d}\tau \end{aligned} \tag{4.175}$$

and inductor current $i_{L\beta}$ is

$$\begin{aligned} i_{L\beta}(t) &= \mathcal{L}^{-1}\left[I_{L\beta}(s)\right] \\ &= \frac{1}{L_\beta C_\alpha}\int_0^t \sum_{r=0}^{\infty}\left(-\frac{1}{RC_\alpha}\right)^r \tau^{\alpha r+\alpha+\beta-1} E_{\alpha+\beta,\alpha+\beta+\alpha r}^{r+1}\left(-\frac{1}{L_\beta C_\alpha}\tau^{\alpha+\beta}\right) \\ &\quad \times i_s(t-\tau)\mathrm{d}\tau \end{aligned} \tag{4.176}$$

Other currents and voltages of the circuit can be derived from the above equations by the characteristics of the parallel circuit, the resistor current i_R can be represented by

$$i_R(t) = \frac{u(t)}{R} \tag{4.177}$$

and capacitor current $i_{C\alpha}$ is

$$i_{C\alpha}(t) = C_\alpha \frac{\mathrm{d}^\alpha u_{C\alpha}}{\mathrm{d}t^\alpha} = i_s(t) - i_{L\beta}(t) - i_R(t) \tag{4.178}$$

1. Response to DC Source

If the current source shown in Fig. 4.38 is DC source and $i_s = I_0$, the analytical solution can be obtained by inverse Laplace transform as follows:

The capacitor voltage $u_{C\alpha}$ is

$$\begin{aligned} u_{C\alpha}(t) &= \mathcal{L}^{-1}[U_{C\alpha}(s)] = \mathcal{L}^{-1}\left[\frac{s^\beta/C_\alpha}{s^{\alpha+\beta} + \frac{1}{RC_\alpha}s^\beta + \frac{1}{L_\beta C_\alpha}}\frac{I_0}{s}\right] \\ &= \frac{I_0}{C_\alpha}\sum_{r=0}^{\infty}\left(-\frac{1}{RC_\alpha}\right)^r t^{\alpha r+\alpha} E^{r+1}_{\alpha+\beta,\alpha+\alpha r+1}\left(-\frac{1}{L_\beta C_\alpha}t^{\alpha+\beta}\right) \end{aligned} \tag{4.179}$$

and inductor current $i_{L\beta}$ is

$$\begin{aligned} i_{L\beta}(t) &= \mathcal{L}^{-1}\left[I_{L\beta}(s)\right] = \mathcal{L}^{-1}\left[\frac{\frac{1}{L_\beta C_\alpha}}{s^{\alpha+\beta} + \frac{1}{RC_\alpha}s^\beta + \frac{1}{L_\beta C_\alpha}}\frac{I_0}{s}\right] \\ &= \frac{I_0}{L_\beta C_\alpha}\sum_{r=0}^{\infty}\left(-\frac{1}{RC_\alpha}\right)^r (t-\tau)^{\alpha r+\alpha+\beta} E^{r+1}_{\alpha+\beta,\alpha r+\alpha+\beta+1}\left(-\frac{1}{L_\beta C_\alpha}(t-\tau)^{\alpha+\beta}\right) \end{aligned} \tag{4.180}$$

Then, the zero-state response of parallel $RL_\beta C_\alpha$ circuit under DC source is shown in Fig. 4.39, which implies that the inductor current will increase to steady-state and the circuit voltage will finally decay to zero.

2. Response to AC Source

If the current source shown in Fig. 4.38 is an AC source and $i_s = I_m\cos(\omega t + \theta)$, the analytical solutions can be obtained by inverse Laplace transform, the capacitor voltage $u_{C\alpha}$ can be derived as

$$u_{C\alpha}(t) = \mathcal{L}^{-1}[U_{C\alpha}(s)] = \mathcal{L}^{-1}\left[\frac{s^\beta/C_\alpha}{s^{\alpha+\beta} + \frac{1}{RC_\alpha}s^\beta + \frac{1}{L_\beta C_\alpha}} I_m \frac{(s\cos\theta - \omega\sin\theta)}{\left(s^2+\omega^2\right)}\right]$$

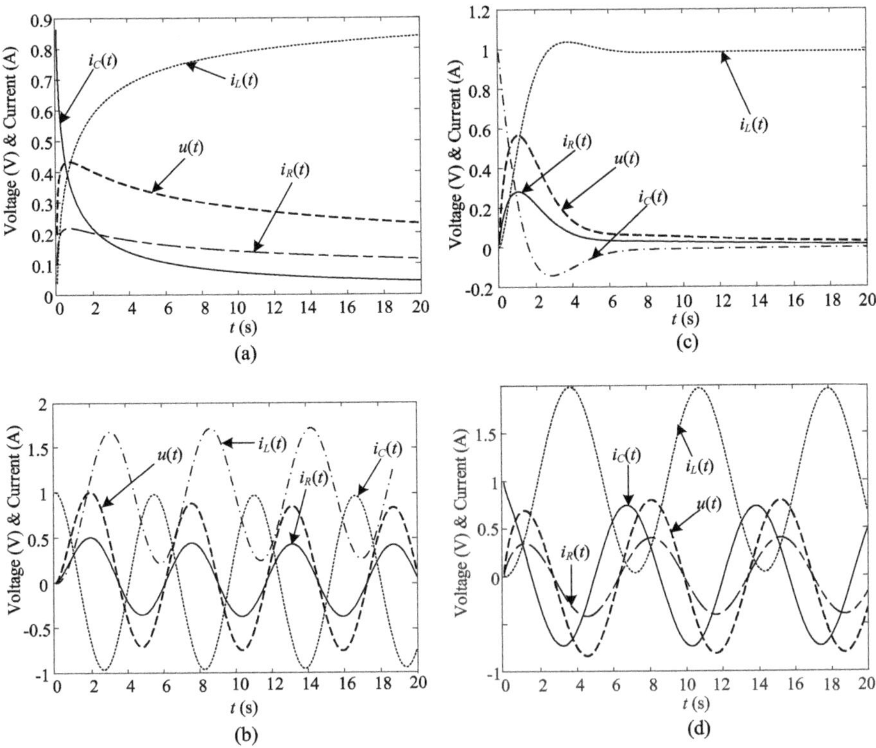

Fig. 4.39 Zero-state response of parallel $RL_\beta C_\alpha$ with DC current source for $R = 2\ \Omega$, $C_\alpha = 1$ F/$s^{1-\alpha}$, $L_\beta = 1$ H/$s^{1-\beta}$, $I_0 = 1$ A: **a** $\alpha = 0.35$, $\beta = 0.35$. **b** $\alpha = 0.75$, $\beta = 1.5$. **c** $\alpha = 0.75$, $\beta = 0.75$. **d** $\alpha = 1.5$, $\beta = 0.75$

$$= \frac{I_m}{C_\alpha}\int\limits_0^t \sum_{r=0}^{\infty}\left(-\frac{1}{RC_\alpha}\right)^r (t-\tau)^{\alpha r+\alpha-1} E_{\alpha+\beta,\alpha+\alpha r}^{r+1}\left(-\frac{1}{L_\beta C_\alpha}(t-\tau)^{\alpha+\beta}\right) \times \cos(\omega\tau+\theta)\mathrm{d}\tau \tag{4.181}$$

and inductor current $i_{L\beta}$ is

$$i_{L\beta}(t) = \mathcal{L}^{-1}\left[I_{L\beta}(s)\right] = \mathcal{L}^{-1}\left[\frac{\frac{1}{L_\beta C_\alpha}}{s^{\alpha+\beta}+\frac{1}{RC_\alpha}s^\beta+\frac{1}{L_\beta C_\alpha}}\frac{I_m(s\cos\theta-\omega\sin\theta)}{(s^2+\omega^2)}\right]$$
$$= \frac{I_0}{L_\beta C_\alpha}\int\limits_0^t \sum_{r=0}^{\infty}\left(-\frac{1}{RC_\alpha}\right)^r (t-\tau)^{\alpha r+\alpha+\beta-1} E_{\alpha+\beta,\alpha+\beta+\alpha r}^{r+1}\left(-\frac{1}{L_\beta C_\alpha}(t-\tau)^{\alpha+\beta}\right) \times \cos(\omega\tau+\theta)\mathrm{d}\tau \tag{4.182}$$

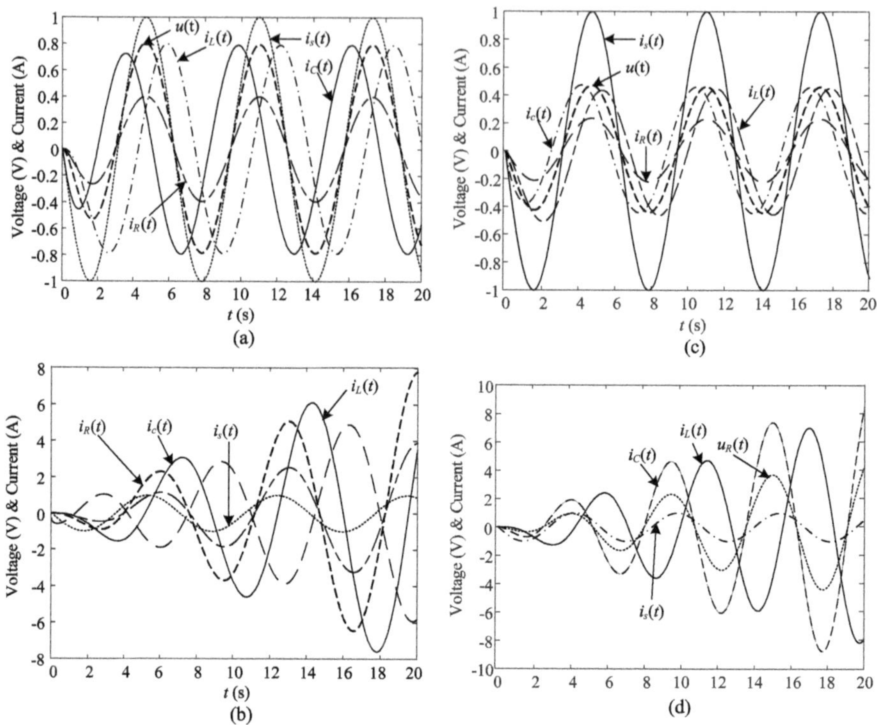

Fig. 4.40 Zero-state response of parallel $RL_\beta C_\alpha$ with sinusoidal current source for $R = 2\ \Omega$, $C_\alpha = 1\ \text{F}/s^{1-\alpha}$, $L_\beta = 1\ \text{H}/s^{1-\beta}$, $I_0 = 1$ A: **a** $\alpha = 0.35$, $\beta = 0.35$. **b** $\alpha = 0.75$, $\beta = 1.5$. **c** $\alpha = 0.75$, $\beta = 0.75$. **d** $\alpha = 1.5$, $\beta = 0.75$

Thus, the zero-state response of the parallel $RL_\beta C_\alpha$ circuit under an AC source is shown in Fig. 4.40, which implies that the currents and voltages will finally reach the sinusoidal steady-state.

4.8 Complete Response of Fractional-Order $RL_\beta C_\alpha$ Circuits

This section will introduce the solution of the complete response of the circuit with two fractional-order components, including series and parallel $RL_\beta C_\alpha$ circuits.

4.8.1 Series Fractional-Order $RL_\beta C_\alpha$ Circuits

As shown in Fig. 4.35, the external source is u_s, supposing that the initial voltage across the fractional-order capacitor is U_0, and the current flowing through the capacitor is I_0. We will discuss the complete response of the circuit when $t > 0$. According to KVL, the voltage sum of the resistor, capacitor, and inductor is equal to the voltage of the source, which means that $u_{C\alpha} + u_{L\beta} + u = u_s$. Then, the fractional differential equation and its initial conditions can be stated as

$$\begin{cases} \frac{1}{C_\alpha} {}_0D_t^{-\alpha} i + L_\beta \, {}_0D_t^{\beta} i + iR = u_s \\ u_{C\alpha}(0^+) = u_{C\alpha}(0^-) = U_0 \\ i_{L\beta}(0^+) = i_{L\beta}(0^-) = I_0 \end{cases} \tag{4.183}$$

The capacitor voltage satisfies

$$L_\beta C_\alpha \, {}_0D_t^{\beta}\left({}_0D_t^{\alpha} u_{C\alpha}\right) + RC_\alpha \, {}_0D_t^{\alpha} u_{C\alpha} + u_{C\alpha} = u_s \tag{4.184}$$

1. R-L Definition

Under R-L definition, applying the Laplace transform to (4.183) and (4.184), the inductor current and capacitor voltage satisfy

$$\frac{1}{C_\alpha} s^{-\alpha} I(s) + L_\beta\left[s^\beta I(s) - \left[{}_0D_t^{\beta-1} i_{L\beta}(t)\right]_{t=0}\right] + RI(s) = U_s(s) \tag{4.185}$$

$$\begin{aligned} & L_\beta C_\alpha\left[s^{\alpha+\beta} U_{C\alpha}(s) - \left[{}_0D_t^{\alpha+\beta-1} u_{C\alpha}(t)\right]_{t=0} - \left[{}_0D_t^{\alpha-1} u_{C\alpha}(t)\right]_{t=0} s^\beta\right] \\ & + RC_\alpha\left[s^\alpha U_{C\alpha}(s) - \left[{}_0D_t^{\alpha-1} u_{C\alpha}(t)\right]_{t=0}\right] + U_{C\alpha}(s) = U_s(s) \end{aligned} \tag{4.186}$$

the inductor current in the complex frequency domain is

$$I_{L\beta}(s) = \underbrace{\frac{\frac{1}{L_\beta} s^\alpha U_s(s)}{s^{\alpha+\beta} + \frac{R}{L_\beta} s^\alpha + \frac{1}{L_\beta C_\alpha}}}_{\text{zero-state response}} + \underbrace{\frac{\left[{}_0D_t^{\beta-1} i_{L\beta}(t)\right]_{t=0} s^\alpha}{s^{\alpha+\beta} + \frac{R}{L_\beta} s^\alpha + \frac{1}{L_\beta C_\alpha}}}_{\text{zero-input response}} \tag{4.187}$$

and capacitor voltage in the complex frequency domain is

$$U_{C\alpha}(s) = \underbrace{\frac{\frac{1}{L_\beta C_\alpha} U(s)}{s^{\alpha+\beta} + \frac{R}{L_\beta} s^\alpha + \frac{1}{L_\beta C_\alpha}}}_{\text{zero-state response}}$$

$$+\underbrace{\frac{\left(\frac{R}{L_\beta}+s^\beta\right)\left[{}_0D_t^{\alpha-1}u_{C\alpha}(t)\right]_{t=0}+\left[{}_0D_t^{\alpha+\beta-1}u_{C\alpha}(t)\right]_{t=0}}{s^{\alpha+\beta}+\frac{R}{L_\beta}s^\alpha+\frac{1}{L_\beta C_\alpha}}}_{\text{zero-input response}} \tag{4.188}$$

Then, the analytical solution in the time domain can be derived from the inverse Laplace transform, the inductor current $i_{L\beta}$ is

$$\begin{aligned} i_{L\beta}(t) &= \mathcal{L}^{-1}\left[I_{L\beta}(s)\right] \\ &= \left[{}_0D_t^{\beta-1}i_{L\beta}(t)\right]_{t=0}\sum_{r=0}^{\infty}\left(-\frac{R}{L_\beta}\right)^r t^{\alpha r+\beta-1}E_{\alpha+\beta,\beta+\alpha r}^{r+1}\left(-\frac{1}{L_\beta C_\alpha}t^{\alpha+\beta}\right) \\ &\quad+\frac{1}{L_\beta}\int_0^t u_s(t-\tau)\sum_{r=0}^{\infty}\left(-\frac{R}{L_\beta}\right)^r \tau^{\alpha r+\beta}E_{\alpha+\beta,\alpha r+\beta+1}^{r+1}\left(-\frac{1}{L_\beta C_\alpha}\tau^{\alpha+\beta}\right)\mathrm{d}\tau \end{aligned} \tag{4.189}$$

and the capacitor voltage $u_{C\alpha}$ is

$$\begin{aligned} u_{C\alpha}(t) &= \mathcal{L}^{-1}[U_{C\alpha}(s)] \\ &= \left(\left[{}_0D_t^{\alpha+\beta-1}u_{C\alpha}(t)\right]_{t=0}+\frac{R}{L_\beta}\left[{}_0D_t^{\alpha-1}u_{C\alpha}(t)\right]_{t=0}\right) \\ &\quad\times\sum_{r=0}^{\infty}\left(-\frac{R}{L_\beta}\right)^r t^{\alpha r+\alpha+\beta-1}E_{\alpha+\beta,\alpha+\beta+\alpha r}^{r+1}\left(-\frac{1}{L_\beta C_\alpha}t^{\alpha+\beta}\right) \\ &\quad+\left[{}_0D_t^{\alpha-1}u_{C\alpha}(t)\right]_{t=0}\sum_{r=0}^{\infty}\left(-\frac{R}{L_\beta}\right)^r t^{\alpha r+\alpha-1}E_{\alpha+\beta,\alpha+\alpha r}^{r+1}\left(-\frac{1}{L_\beta C_\alpha}t^{\alpha+\beta}\right) \\ &\quad+\frac{1}{L_\beta C_\alpha}\int_0^t\sum_{r=0}^{\infty}\left(-\frac{R}{L_\beta}\right)^r \tau^{\alpha r+\alpha+\beta-1}E_{\alpha+\beta,\alpha+\beta+\alpha r}^{r+1}\left(-\frac{1}{L_\beta C_\alpha}\tau^{\alpha+\beta}\right) \\ &\quad\times u_s(t-\tau)\mathrm{d}\tau \end{aligned} \tag{4.190}$$

Other currents and voltages of the circuit can be derived from the above equations by the characteristics of the series circuit.

2. Caputo Definition

Under Caputo definition, taking Laplace transform of (4.183) and (4.184), there are

$$\frac{1}{C_\alpha}s^{-\alpha}I(s)+L_\beta\left[s^\beta I(s)-s^{\beta-1}i_{L\beta}(0)\right]+RI(s)=U_s(s) \tag{4.191}$$

$$\begin{aligned} &L_\beta C_\alpha\left[s^{\alpha+\beta}U_{C\alpha}(s)-s^{\alpha+\beta-1}u_{C\alpha}(0)\right]+RC_\alpha\left[s^\alpha U_{C\alpha}(s)-s^{\alpha-1}u_{C\alpha}(0)\right] \\ &\quad+U_{C\alpha}(s)=U_s(s) \end{aligned} \tag{4.192}$$

from which the inductor current is

$$I_{L\beta}(s)=\underbrace{\frac{\frac{1}{L_\beta}s^\alpha U(s)}{s^{\alpha+\beta}+\frac{R}{L_\beta}s^\alpha+\frac{1}{L_\beta C_\alpha}}}_{\text{zero-state response}}+\underbrace{\frac{i_{L\beta}(0)s^{\alpha+\beta-1}}{s^{\alpha+\beta}+\frac{R}{L_\beta}s^\alpha+\frac{1}{L_\beta C_\alpha}}}_{\text{zero-input response}} \tag{4.193}$$

and the capacitor voltage is

$$U_{C\alpha}(s)=\underbrace{\frac{\frac{1}{L_\beta C_\alpha}U(s)}{s^{\alpha+\beta}+\frac{R}{L_\beta}s^\alpha+\frac{1}{L_\beta C_\alpha}}}_{\text{zero-state response}}+\underbrace{\frac{u_{C\alpha}(0)\left(\frac{R}{L_\beta}s^{\alpha-1}+s^{\alpha+\beta-1}\right)}{s^{\alpha+\beta}+\frac{R}{L_\beta}s^\alpha+\frac{1}{L_\beta C_\alpha}}}_{\text{zero-input response}} \tag{4.194}$$

Then, the analytical solution in the time domain can be derived from the inverse Laplace transform, the inductor current $i_{L\beta}$ can be derived as

$$\begin{aligned} i_{L\beta}(t)&=\mathcal{L}^{-1}\left[i_{L\beta}(s)\right]\\ &=I_0\sum_{r=0}^{\infty}\left(-\frac{R}{L_\beta}\right)^r t^{\beta r}E^{r+1}_{\alpha+\beta,1+\beta r}\left(-\frac{1}{L_\beta C_\alpha}t^{\alpha+\beta}\right)\\ &\quad+\frac{1}{L_\beta}\int_0^t\sum_{r=0}^{\infty}\left(-\frac{R}{L_\beta}\right)^r\tau^{\alpha r+\beta}E^{r+1}_{\alpha+\beta,\alpha r+\beta+1}\left(-\frac{1}{L_\beta C_\alpha}\tau^{\alpha+\beta}\right)u_s(t-\tau)\mathrm{d}\tau \end{aligned} \tag{4.195}$$

and capacitor voltage $u_{C\alpha}$ is

$$\begin{aligned} u_{C\alpha}(t)&=\mathcal{L}^{-1}[U_{C\alpha}(s)]\\ &=\frac{R}{L_\beta}U_0\sum_{r=0}^{\infty}\left(-\frac{R}{L_\beta}\right)^r t^{\beta r+\beta}E^{r+1}_{\alpha+\beta,\beta+1+\beta r}\left(-\frac{1}{L_\beta C_\alpha}t^{\alpha+\beta}\right)\\ &\quad+U_0\sum_{r=0}^{\infty}\left(-\frac{R}{L_\beta}\right)^r t^{\beta r}E^{r+1}_{\alpha+\beta,1+\beta r}\left(-\frac{1}{L_\beta C_\alpha}t^{\alpha+\beta}\right)\\ &\quad+\frac{1}{L_\beta C_\alpha}\int_0^t\sum_{r=0}^{\infty}\left(-\frac{R}{L_\beta}\right)^r\tau^{\alpha r+\alpha+\beta-1}E^{r+1}_{\alpha+\beta,\alpha+\beta+\alpha r}\left(-\frac{1}{L_\beta C_\alpha}\tau^{\alpha+\beta}\right)\\ &\quad\times u_s(t-\tau)\mathrm{d}\tau \end{aligned} \tag{4.196}$$

Other currents and voltages of the circuit can be derived from the above equations by the characteristics of the series circuit.

4.8.2 Parallel Fractional-Order $RL_\beta C_\alpha$ Circuits

As shown in Fig. 4.38, the external source is i_s, supposing that the initial voltage across the fractional-order capacitor is U_0, and the current flowing through the capacitor is I_0. We will discuss the complete response of the circuit when $t > 0$. According to KVL, the current sum of the resistor, capacitor, and inductor is equal to the current source, which means that $i_{C\alpha} + i_{L\beta} + i_R = i_s$. Then, the fractional differential equation and its initial conditions can be stated as

$$\begin{cases} C_\alpha\, {}_0D_t^\alpha u + \frac{1}{L_\beta}{}_0D_t^{-\beta} u + \frac{u}{R} = i_s \\ u_{C\alpha}(0^+) = u_{C\alpha}(0^-) = U_0 \\ i_{L\beta}(0^+) = i_{L\beta}(0^-) = I_0 \end{cases} \tag{4.197}$$

The inductor current satisfies

$$L_\beta C_\alpha\, {}_0D_t^\alpha\left({}_0D_t^\beta i_{L\beta}\right) + \frac{L_\beta}{R}{}_0D_t^\beta i_{L\beta} + i_{L\beta} = i_s \tag{4.198}$$

1. R-L Definition

Under R-L definition, applying the Laplace transform to (4.197) and (4.198), we have

$$\frac{1}{L_\beta}s^{-\beta}U(s) + C_\alpha\left[s^\alpha U(s) - \left[{}_0D_t^{\alpha-1}u_{C\alpha}(t)\right]_{t=0}\right] + U(s)/R = I_s(s) \tag{4.199}$$

$$\begin{aligned} &L_\beta C_\alpha\left[s^{\alpha+\beta}I_{L\beta}(s) - \left[{}_0D_t^{\alpha+\beta-1}i_{L\beta}(t)\right]_{t=0} - \left[{}_0D_t^{\beta-1}i_{L\beta}(t)\right]_{t=0}s^\alpha\right] \\ &+ \frac{L_\beta}{R}\left(s^\beta I_{L\beta}(s) - \left[{}_0D_t^{\beta-1}i_{L\beta}(t)\right]_{t=0}\right) + I_{L\beta}(s) = I_s(s) \end{aligned} \tag{4.200}$$

from which the inductor current is

$$\begin{aligned} I_{L\beta}(s) = &\underbrace{\frac{\frac{1}{L_\beta C_\alpha}I_s(s)}{s^{\alpha+\beta} + \frac{1}{RC_\alpha}s^\beta + \frac{1}{L_\beta C_\alpha}}}_{\text{zero - stateresponse}} \\ &+ \underbrace{\frac{\left(\frac{1}{RC_\alpha} + s^\alpha\right)\left[{}_0D_t^{\beta-1}i_{L\beta}(t)\right]_{t=0} + \left[{}_0D_t^{\alpha+\beta-1}i_{L\beta}(t)\right]_{t=0}}{s^{\alpha+\beta} + \frac{1}{RC_\alpha}s^\beta + \frac{1}{L_\beta C_\alpha}}}_{\text{zero}-\text{input response}} \end{aligned} \tag{4.201}$$

and capacitor voltage is

$$U_{C\alpha}(s) = \underbrace{\frac{\frac{1}{C_\alpha}s^\beta I_s(s)}{s^{\alpha+\beta} + \frac{1}{RC_\alpha}s^\beta + \frac{1}{L_\beta C_\alpha}}}_{\text{zero-state response}} + \underbrace{\frac{\left[{}_0D_t^{\alpha-1}u_{C\alpha}(t)\right]_{t=0}s^\beta}{s^{\alpha+\beta} + \frac{1}{RC_\alpha}s^\beta + \frac{1}{L_\beta C_\alpha}}}_{\text{zero-input response}} \tag{4.202}$$

Then, the analytical solution in the time domain can be derived from the inverse Laplace transform, the inductor current $i_{L\beta}$ is

$$\begin{aligned} i_{L\beta}(t) &= \mathcal{L}^{-1}\left[I_{L\beta}(s)\right] \\ &= \left[{}_0D_t^{\beta-1}i_{L\beta}(t)\right]_{t=0}\sum_{r=0}^{\infty}\left(-\frac{1}{RC_\alpha}\right)^r t^{\alpha r+\beta-1}E_{\alpha+\beta,\beta+\alpha r}^{r+1}\left(-\frac{1}{L_\beta C_\alpha}t^{\alpha+\beta}\right) \\ &+ \left(\left[{}_0D_t^{\alpha+\beta-1}i_{L\beta}(t)\right]_{t=0} + \frac{1}{RC_\alpha}\left[{}_0D_t^{\beta-1}i_{L\beta}(t)\right]_{t=0}\right) \\ &\times \sum_{r=0}^{\infty}\left(-\frac{1}{RC_\alpha}\right)^r t^{\alpha r+\alpha+\beta-1}E_{\alpha+\beta,\alpha+\beta+\alpha r}^{r+1}\left(-\frac{1}{L_\beta C_\alpha}t^{\alpha+\beta}\right) \\ &+ \frac{1}{L_\beta C_\alpha}\int_0^t\sum_{r=0}^{\infty}\left(-\frac{1}{RC_\alpha}\right)^r \tau^{\alpha r+\alpha+\beta-1}E_{\alpha+\beta,\alpha+\beta+\alpha r}^{r+1}\left(-\frac{1}{L_\beta C_\alpha}\tau^{\alpha+\beta}\right) \\ &\times i_s(t-\tau)\mathrm{d}\tau \end{aligned} \tag{4.203}$$

and the capacitor voltage $u_{C\alpha}$ is

$$\begin{aligned} u_{C\alpha}(t) &= \mathcal{L}^{-1}[U_{C\alpha}(s)] \\ &= \frac{1}{C_\alpha}\int_0^t\sum_{r=0}^{\infty}\left(-\frac{1}{RC_\alpha}\right)^r \tau^{\alpha r+\alpha-1}E_{\alpha+\beta,\alpha+\alpha r}^{r+1}\left(-\frac{1}{L_\beta C_\alpha}\tau^{\alpha+\beta}\right)i_s(t-\tau)\mathrm{d}\tau \\ &+ \left[{}_0D_t^{\alpha-1}u_{C\alpha}(t)\right]_{t=0}\sum_{r=0}^{\infty}\left(-\frac{1}{RC_\alpha}\right)^r t^{\alpha r+\alpha-1}E_{\alpha+\beta,\alpha+\alpha r}^{r+1}\left(-\frac{1}{L_\beta C_\alpha}t^{\alpha+\beta}\right) \end{aligned} \tag{4.204}$$

Other currents and voltages of the circuit can be derived from the above equations by the characteristics of the parallel circuit.

2. Caputo Definition

Under Caputo definition, taking Laplace transform of (4.197) and (4.198), there are

$$\frac{1}{L_\beta}s^{-\beta}U(s) + C_\alpha\left(s^\alpha U(s) - s^{\alpha-1}u_{C\alpha}(0)\right) + \frac{1}{R}U(s) = I_s(s) \tag{4.205}$$

$$L_\beta C_\alpha\left[s^{\alpha+\beta}I_{L\beta}(s) - s^{\alpha+\beta-1}i_{L\beta}(0)\right] + \frac{L_\beta}{R}\left[s^\beta I_{L\beta}(s) - s^{\beta-1}i_{L\beta}(0)\right]$$

$$+ I_{L\beta}(s) = I_s(s) \tag{4.206}$$

from which the inductor current is

$$I_{L\beta}(s) = \underbrace{\frac{\frac{1}{L_\beta C_\alpha} I_s(s)}{s^{\alpha+\beta} + \frac{1}{RC_\alpha} s^\beta + \frac{1}{L_\beta C_\alpha}}}_{\text{zero-state response}} + \underbrace{\frac{\left(\frac{1}{RC_\alpha} s^{\beta-1} + s^{\alpha+\beta-1}\right) i_{L\beta}(0)}{s^{\alpha+\beta} + \frac{1}{RC_\alpha} s^\beta + \frac{1}{L_\beta C_\alpha}}}_{\text{zero-input response}} \tag{4.207}$$

and capacitor voltage is

$$U_{C\alpha}(s) = \underbrace{\frac{\frac{1}{C_\alpha} s^\beta I_s(s)}{s^{\alpha+\beta} + \frac{1}{RC_\alpha} s^\beta + \frac{1}{L_\beta C_\alpha}}}_{\text{zero-state response}} + \underbrace{\frac{s^{\alpha+\beta-1} u_{C\alpha}(0)}{s^{\alpha+\beta} + \frac{1}{RC_\alpha} s^\beta + \frac{1}{L_\beta C_\alpha}}}_{\text{zero-input response}} \tag{4.208}$$

Then, the analytical solution in the time domain can be derived from the inverse Laplace transform, the inductor current $i_{L\beta}$ can be derived as

$$\begin{aligned} i_{L\beta}(t) &= \mathcal{L}^{-1}\left[i_{L\beta}(s)\right] \\ &= \frac{I_0}{RC_\alpha} \sum_{r=0}^{\infty} \left(-\frac{1}{RC_\alpha}\right)^r t^{\alpha r+\alpha} E_{\alpha+\beta,\alpha+1+\alpha r}^{r+1}\left(-\frac{1}{L_\beta C_\alpha} t^{\alpha+\beta}\right) \\ &+ I_0 \sum_{r=0}^{\infty} \left(-\frac{1}{RC_\alpha}\right)^r t^{\alpha r} E_{\alpha+\beta,1+\alpha r}^{r+1}\left(-\frac{1}{L_\beta C_\alpha} t^{\alpha+\beta}\right) \\ &+ \frac{1}{L_\beta C_\alpha} \int_0^t \sum_{r=0}^{\infty} \left(-\frac{1}{RC_\alpha}\right)^r \tau^{\alpha r+\alpha+\beta-1} E_{\alpha+\beta,\alpha+\beta+\alpha r}^{r+1}\left(-\frac{1}{L_\beta C_\alpha} \tau^{\alpha+\beta}\right) \\ &\times i_s(t-\tau)\mathrm{d}\tau \end{aligned} \tag{4.209}$$

and capacitor voltage $u_{C\alpha}$ is

$$\begin{aligned} u_{C\alpha}(t) &= \mathcal{L}^{-1}[U_{C\alpha}(s)] \\ &= U_0 \sum_{r=0}^{\infty} \left(-\frac{1}{RC_\alpha}\right)^r t^{\alpha r} E_{\alpha+\beta,1+\alpha r}^{r+1}\left(-\frac{1}{L_\beta C_\alpha} t^{\alpha+\beta}\right) \\ &\frac{1}{C_\alpha} \int_0^t \sum_{r=0}^{\infty} \left(-\frac{1}{RC_\alpha}\right)^r \tau^{\alpha r+\alpha-1} E_{\alpha+\beta,\alpha+\alpha r}^{r+1}\left(-\frac{1}{L_\beta C_\alpha} \tau^{\alpha+\beta}\right) i_s(t-\tau)\mathrm{d}\tau \end{aligned} \tag{4.210}$$

Other currents and voltages of the circuit can be derived from the above equations by the characteristics of the parallel circuit.

Otherwise, there is another method to solve fractional differential equations. Time-domain solutions of the response of the $RL_{\beta}C_{\alpha}$ circuit can be obtained by applying the method of decomposing rational function into partial fraction function [11]. If the general rational function form has only a single pole, it can be expressed as a sum of partial fractions by introducing a new variable $\omega = s^{1/n}$, which means

$$H(\omega) = \frac{A\omega^{q} + B}{\omega^{n_1} + a\omega^{m} + b} = \sum_{i=1}^{k_1} \frac{R_i}{\omega - \omega_i} + \sum_{j=1}^{k_2} \left(\frac{C_j}{\omega - \omega_j} + \frac{C_j^*}{\omega - \omega_j^*} \right) \tag{4.211}$$

where ω_i are real poles, ω_j and ω_j^* are complex conjugate poles, R_i, C_j and C_j^* are decomposition coefficients (residues).

Assuming that there are k_1 real poles and $2k_2$ complex conjugate poles, E(s) is the Laplace transform of the excitation source, we will discuss the solutions of different poles locations:

Supposing that the function $H(\omega)$ has only k_1 real poles, the Laplace transform of inductor current or capacitor voltage is given by

$$Y(s)_{\text{real}} = E(s) \sum_{i=1}^{k_1} \frac{R_i}{s^{\lambda} - \omega_i} \tag{4.212}$$

where $\lambda = 1/n$, R_i are appropriate real decomposition coefficients (residues).

Then, the circuit response $y(t)$ in the time domain is

$$y(t) = \mathcal{L}^{-1}[Y(s)_{\text{real}}] = \sum_{i=1}^{k_1} R_i \int_0^t e(t-\tau)\tau^{\lambda-1} E_{\lambda,\lambda}\left(\omega_i \tau^{\lambda}\right) \mathrm{d}\tau \tag{4.213}$$

If the function $H(\omega)$ has only $2k_2$ complex conjugate poles, the Laplace transform of inductor current or capacitor voltage can be described as

$$Y(s)_{\text{imaginary}} = E(s) \sum_{j=1}^{k_2} \left(\frac{C_j}{s^{\lambda} - \omega_j} + \frac{C_j^*}{s^{\lambda} - \omega_j^*} \right) \tag{4.214}$$

Using the representation of the Mittag–Leffler function as a series, the voltage or current is defined by

$$\begin{aligned} y(t) &= \mathcal{L}^{-1}\left[Y(s)_{\text{imaginary}}\right] \\ &= \sum_{i=1}^{k_2} \sum_{k=0}^{\infty} \frac{2|C_j||\omega_j|^k \cos(k\upsilon + \varphi)}{\Gamma(\lambda k + \lambda)} \int_0^t e(t-\tau)\tau^{\lambda(k+1)-1} \mathrm{d}\tau \end{aligned} \tag{4.215}$$

where

$$\varphi = \arctan\left(\frac{\mathrm{Im}\{C_j\}}{\mathrm{Re}\{C_j\}}\right),\ \upsilon = \arctan\left(\frac{\mathrm{Im}\{\omega_j\}}{\mathrm{Re}\{\omega_j\}}\right) \tag{4.216}$$

$\lambda = 1/n$, C_j and C_j^* are appropriate complex partial fraction decomposition coefficients (residues), φ and υ are arguments of the complex coefficients.

If the function $H(\omega)$ contains real and complex conjugate poles, and then the current and voltages are a sum of the relations derived in (4.212) and (4.214), there is

$$Y(s)_{\text{complex}} = E(s)\left\{\sum_{i=1}^{k_1}\frac{R_i}{s^\lambda - \omega_i} + \sum_{j=1}^{k_2}\left(\frac{C_j}{s^\lambda - \omega_j} + \frac{C_j^*}{s^\lambda - \omega_j^*}\right)\right\} \tag{4.217}$$

Then, the voltage or current is derived as

$$\begin{aligned} y(t) = \mathcal{L}^{-1}\left[Y(s)_{\text{complex}}\right] &= \sum_{i=1}^{k_1} R_i \int_0^t e(t-\tau)\tau^{\lambda-1}E_{\lambda,\lambda}\left(\omega_i\tau^\lambda\right)\mathrm{d}\tau \\ &+ \sum_{i=1}^{k_2}\sum_{k=0}^{\infty}\frac{2|C_j||\omega_j|^k\cos(k\upsilon+\varphi)}{\Gamma(\lambda k+\lambda)}\int_0^t e(t-\tau)\tau^{\lambda(k+1)-1}\mathrm{d}\tau \end{aligned} \tag{4.218}$$

where

$$\varphi = \arctan\left(\frac{\mathrm{Im}\{C_j\}}{\mathrm{Re}\{C_j\}}\right),\ \upsilon = \arctan\left(\frac{\mathrm{Im}\{\omega_j\}}{\mathrm{Re}\{\omega_j\}}\right) \tag{4.219}$$

$\lambda = 1/n$, R_i are appropriate real decomposition coefficients (residues), C_j and C_j^* are appropriate complex partial fraction decomposition coefficients (residues), φ and υ are arguments of the complex coefficients.

4.9 Step and Pulse Response of Fractional-Order $RL_\beta C_\alpha$ Circuits

4.9.1 Step Response

1. Series Fractional-order $RL_\beta C_\alpha$ Circuit

As shown in Fig. 4.41, at $t = 0_-$, the voltage across the fractional-order capacitor C_α is $u_{C\alpha}(0_-) = 0$, and the current flowing through the fractional-order inductor L_β is $i_{L\beta}(0_-) = 0$. Let the external excitation be unit step voltage source and $u_s = \varepsilon(t)$, the step response of the series $RL_\beta C_\alpha$ circuit can be analyzed. The fractional differential equation and its initial conditions can be stated by KVL as

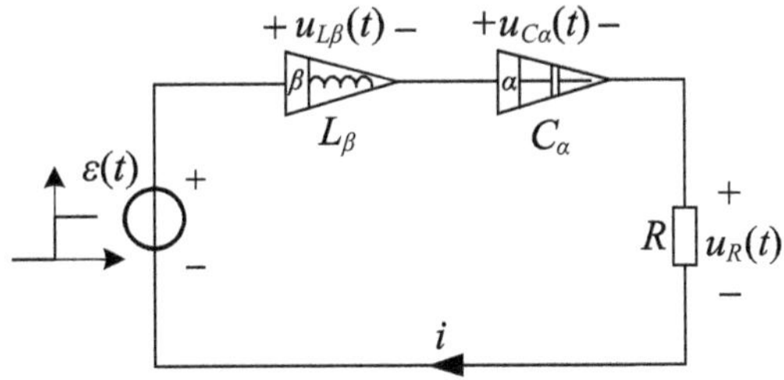

Fig. 4.41 Series $RL_\beta C_\alpha$ circuit with step excitation

$$\begin{cases} \frac{1}{C_\alpha}{}_0D_t^{-\alpha}i + L_\beta\,{}_0D_t^{\beta}i + iR = \varepsilon(t) \\ u_{C\alpha}(0^-) = 0 \\ i_{L\beta}(0^-) = 0 \end{cases} \tag{4.220}$$

From the voltage-current relationship of the capacitor, the integral operator can be eliminated, the voltage of the capacitor can be derived from (4.220) as

$$L_\beta C_\alpha\,{}_0D_t^{\beta}\left({}_0D_t^{\alpha}u_{C\alpha}\right) + RC_\alpha\,{}_0D_t^{\alpha}u_{C\alpha} + u_{C\alpha} = \varepsilon(t) \tag{4.221}$$

Applying Laplace transform to (4.220) and (4.221), there are

$$\frac{1}{C_\alpha}s^{-\alpha}I(s) + L_\beta s^{\beta}I(s) + RI(s) = \frac{1}{s} \tag{4.222}$$

$$L_\beta C_\alpha s^{\alpha+\beta}U_{C\alpha}(s) + RC_\alpha s^{\alpha}U_{C\alpha}(s) + U_{C\alpha}(s) = \frac{1}{s} \tag{4.223}$$

the inductor current can be deduced as

$$I_{L\beta}(s) = \frac{\frac{1}{L_\beta}s^{\alpha}}{s^{\alpha+\beta} + \frac{R}{L_\beta}s^{\alpha} + \frac{1}{L_\beta C_\alpha}}\frac{1}{s} \tag{4.224}$$

and the capacitor voltage is

$$U_{C\alpha}(s) = \frac{\frac{1}{L_\beta C_\alpha}}{s^{\alpha+\beta} + \frac{R}{L_\beta}s^{\alpha} + \frac{1}{L_\beta C_\alpha}}\frac{1}{s} \tag{4.225}$$

Then, the analytical solutions in the time domain can be derived from the inverse Laplace transform and the inductor current $i_{L\beta}$ can be obtained as

$$i_{L\beta}(t) = \mathcal{L}^{-1}\left[i_{L\beta}(s)\right] = \frac{1}{L_\beta}\sum_{r=0}^{\infty}\left(-\frac{R}{L_\beta}\right)^{r} t^{\alpha r+\beta}E_{\alpha+\beta,\alpha r+\beta+1}^{r+1}\left(-\frac{1}{L_\beta C_\alpha}t^{\alpha+\beta}\right) \tag{4.226}$$

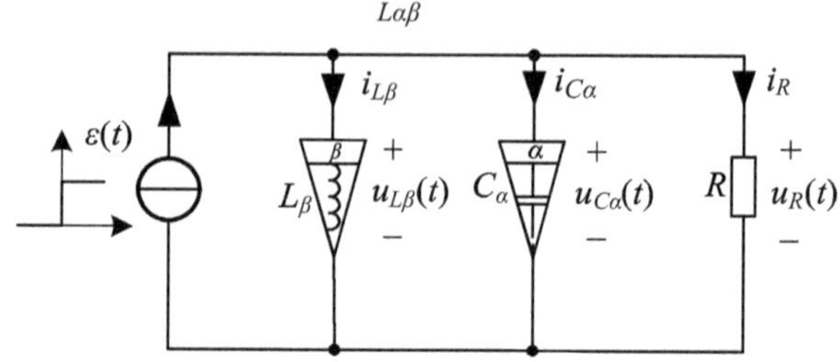

Fig. 4.42 Parallel $RL_\beta C_\alpha$ circuit with step excitation

the capacitor voltage $u_{C\alpha}$ is

$$\begin{aligned} u_{C\alpha}(t) &= \mathcal{L}^{-1}[U_{C\alpha}(s)] \\ &= \frac{1}{L_\beta C_\alpha}\sum_{r=0}^{\infty}\left(-\frac{R}{L_\beta}\right)^r t^{\alpha r+\alpha+\beta}E^{r+1}_{\alpha+\beta,\alpha r+\alpha+\beta+1}\left(-\frac{1}{L_\beta C_\alpha}t^{\alpha+\beta}\right) \end{aligned} \tag{4.227}$$

Other currents and voltages of the circuit can be derived from the above equations by the characteristics of the series circuit.

2. Parallel Fractional-order $RL_\beta C_\alpha$ Circuit

As shown in Fig. 4.42, at $t = 0_-$, the voltage across the capacitor C_α is $u_{C\alpha}(0_-) = 0$, and the current flowing through the inductor L_β is $i_{L\beta}(0_-) = 0$. Supposing that the external excitation is unit step voltage source and $u_s = \varepsilon(t)$, we will analyze the step response of the parallel $RL_\beta C_\alpha$ circuit. The fractional differential equation and its initial conditions can be stated by KVL as.

$$\begin{cases} \frac{1}{L_\beta}{}_0D_t^{-\beta}u + C_\alpha\,{}_0D_t^\alpha u + \frac{u}{R} = \varepsilon(t) \\ u_{C\alpha}(0^-) = 0 \\ i_{L\beta}(0^-) = 0 \end{cases} \tag{4.228}$$

By the relationship between the voltage and current of the inductor, we can eliminate the fractional integral operator, then the current of the inductor can be derived from (4.228) as

$$L_\beta C_\alpha\,{}_0D_t^\alpha\left({}_0D_t^\beta i_{L\beta}\right) + \frac{L_\beta}{R}{}_0D_t^\beta i_{L\beta} + i_{L\beta} = \varepsilon(t) \tag{4.229}$$

Taking Laplace transform of (4.228) and (4.229), there are

$$\frac{1}{L_\beta}s^{-\beta}U(s) + C_\alpha s^\alpha U(s) + \frac{1}{R}U(s) = \frac{1}{s} \tag{4.230}$$

$$L_\beta C_\alpha s^{\alpha+\beta}I_{L\beta}(s) + \frac{L_\beta}{R}s^\beta I_{L\beta}(s) + I_{L\beta}(s) = \frac{1}{s} \tag{4.231}$$

the capacitor voltage can be written as

$$U_{C\alpha}(s) = \frac{\frac{1}{C_\alpha}s^\beta}{s^{\alpha+\beta} + \frac{1}{RC_\alpha}s^\beta + \frac{1}{L_\beta C_\alpha}} \frac{1}{s} \tag{4.232}$$

and the inductor current is

$$I_{L\beta}(s) = \frac{\frac{1}{L_\beta C_\alpha}}{s^{\alpha+\beta} + \frac{1}{RC_\alpha}s^\beta + \frac{1}{L_\beta C_\alpha}} \frac{1}{s} \tag{4.233}$$

Then, the analytical solutions in the time domain can be derived from the inverse Laplace transform, the capacitor voltage $u_{C\alpha}$ is determined by

$$u_{C\alpha}(t) = \mathcal{L}^{-1}[U_{C\alpha}(s)] = \frac{1}{C_\alpha}\sum_{r=0}^{\infty}\left(-\frac{1}{RC_\alpha}\right)^r t^{\alpha r+\alpha} E^{r+1}_{\alpha+\beta,\alpha r+\alpha+1}\left(-\frac{1}{L_\beta C_\alpha}t^{\alpha+\beta}\right) \tag{4.234}$$

and the inductor current $i_{L\beta}$ is

$$\begin{aligned} i_{L\beta}(t) &= \mathcal{L}^{-1}\left[I_{L\beta}(s)\right] \\ &= \frac{1}{L_\beta C_\alpha}\sum_{r=0}^{\infty}\left(-\frac{1}{RC_\alpha}\right)^r t^{\alpha r+\alpha+\beta} E^{r+1}_{\alpha+\beta,\alpha r+\alpha+\beta+1}\left(-\frac{1}{L_\beta C_\alpha}t^{\alpha+\beta}\right) \end{aligned} \tag{4.235}$$

Other currents and voltages of the circuit can be derived from the above equations by the characteristics of the parallel circuit.

4.9.2 Pulse Response

1. Series Fractional-order $RL_\beta C_\alpha$ Circuit

As shown in Fig. 4.43, at $t = 0_-$, the voltage across the fractional-order capacitor C_α is $u_{C\alpha}(0_-) = 0$, and the current flowing through the fractional-order inductor L_β is $i_{L\beta}(0_-) = 0$. Considering that the external excitation is an impulse voltage source

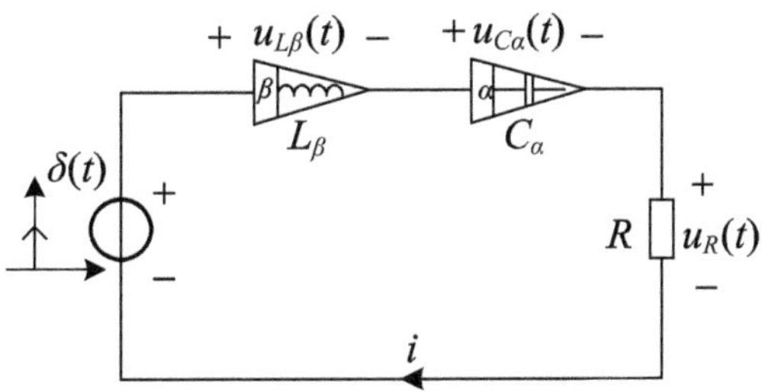

Fig. 4.43 Series $RL_\beta C_\alpha$ circuit with pulse excitation

and $u_s = \delta(t)$, we will analyze the impulse response of the series $RL_\beta C_\alpha$ circuit. According to KVL, the fractional differential equation and its initial conditions can be stated as

$$\begin{cases} \frac{1}{C_\alpha}{}_0D_t^{-\alpha}i + L_\beta\,{}_0D_t^{\beta}i + iR = \delta(t) \\ u_{C\alpha}(0^-) = 0 \\ i_{L\beta}(0^-) = 0 \end{cases} \tag{4.236}$$

From the voltage-current relationship of the capacitor, we can eliminate the fractional integral operator, and then the voltage of the capacitor can be derived from (4.237) as

$$L_\beta C_\alpha\,{}_0D_t^{\beta}\left({}_0D_t^{\alpha}u_{C\alpha}\right) + RC_\alpha\,{}_0D_t^{\alpha}u_{C\alpha} + u_{C\alpha} = \delta(t) \tag{4.237}$$

Applying Laplace transform to (4.236) and (4.237), there are

$$\frac{1}{C_\alpha}s^{-\alpha}I(s) + L_\beta s^{\beta}I(s) + RI(s) = 1 \tag{4.238}$$

$$L_\beta C_\alpha s^{\alpha+\beta}U_{C\alpha}(s) + RC_\alpha s^{\alpha}U_{C\alpha}(s) + U_{C\alpha}(s) = 1 \tag{4.239}$$

The inductor current can be obtained as

$$I_{L\beta}(s) = \frac{\frac{1}{L_\beta}s^{\alpha}}{s^{\alpha+\beta} + \frac{R}{L_\beta}s^{\alpha} + \frac{1}{L_\beta C_\alpha}} \tag{4.240}$$

and the capacitor voltage is

$$U_{C\alpha}(s) = \frac{\frac{1}{L_\beta C_\alpha}}{s^{\alpha+\beta} + \frac{R}{L_\beta}s^{\alpha} + \frac{1}{L_\beta C_\alpha}} \tag{4.241}$$

Then, the analytical solutions in the time domain can be derived from the inverse Laplace transform, the inductor current $i_{L\beta}$ is

$$i_{L\beta}(t) = \mathcal{L}^{-1}\left[i_{L\beta}(s)\right] = \frac{1}{L_\beta}\sum_{r=0}^{\infty}\left(-\frac{R}{L_\beta}\right)^{r} t^{\alpha r+\beta-1}E^{r+1}_{\alpha+\beta,\beta+\alpha r}\left(-\frac{1}{L_\beta C_\alpha}t^{\alpha+\beta}\right) \tag{4.242}$$

and the capacitor voltage $u_{C\alpha}$ is

$$\begin{aligned} u_{C\alpha}(t) &= \mathcal{L}^{-1}[U_{C\alpha}(s)] \\ &= \frac{1}{L_\beta C_\alpha}\sum_{r=0}^{\infty}\left(-\frac{R}{L_\beta}\right)^{r} t^{\alpha r+\alpha+\beta-1}E^{r+1}_{\alpha+\beta,\alpha+\beta+\alpha r}\left(-\frac{1}{L_\beta C_\alpha}t^{\alpha+\beta}\right) \end{aligned} \tag{4.243}$$

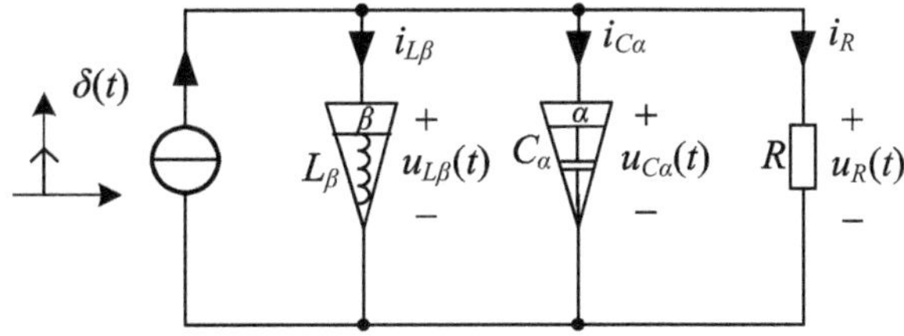

Fig. 4.44 Parallel $RL_\beta C_\alpha$ circuit with pulse excitation

Other currents and voltages of the circuit can be derived from the above equations by the characteristics of the series circuit.

2. Parallel Fractional-order $RL_\beta C_\alpha$ Circuit

As shown in Fig. 4.44, at $t = 0_-$, the voltage across the fractional-order capacitor C_α is $u_{C\alpha}(0_-) = 0$, and the current flowing through the fractional-order inductor L_β is $i_{L\beta}(0_-) = 0$. Supposing that the external excitation is impulse current source and $i_s = \delta(t)$, we will analyze the impulse response of the parallel $RL_\beta C_\alpha$ circuit. Based on KCL, the fractional differential equation and its initial conditions can be described as

$$\begin{cases} \frac{1}{L_\beta} {}_0D_t^{-\beta} u + C_\alpha \, {}_0D_t^\alpha u + \frac{u}{R} = \delta(t) \\ u_{C\alpha}(0^-) = 0 \\ i_{L\beta}(0^-) = 0 \end{cases} \tag{4.244}$$

Based on the relationship between the voltage and current of the inductor, we can eliminate the fractional integral operator, the current of the inductor can be derived from (4.244)

$$L_\beta C_\alpha \, {}_0D_t^\alpha \left({}_0D_t^\beta i_{L\beta} \right) + \frac{L_\beta}{R} {}_0D_t^\beta i_{L\beta} + i_{L\beta} = \delta(t) \tag{4.245}$$

Using Laplace transform to (4.244) and (4.245), there are

$$\frac{1}{L_\beta} s^{-\beta} U(s) + C_\alpha s^\alpha U(s) + \frac{1}{R} U(s) = 1 \tag{4.246}$$

$$L_\beta C_\alpha s^{\alpha+\beta} I_{L\beta}(s) + \frac{L_\beta}{R} s^\beta I_{L\beta}(s) + I_{L\beta}(s) = 1 \tag{4.247}$$

Then, the capacitor voltage can be obtained as

$$U_{C\alpha}(s) = \frac{s^\beta / C_\alpha}{s^{\alpha+\beta} + \frac{1}{RC_\alpha} s^\beta + \frac{1}{L_\beta C_\alpha}} \tag{4.248}$$

and the inductor current is

$$I_{L\beta}(s) = \frac{\frac{1}{L_\beta C_\alpha}}{s^{\alpha+\beta} + \frac{1}{RC_\alpha}s^\beta + \frac{1}{L_\beta C_\alpha}} \tag{4.249}$$

Hence, the analytical solutions in the time domain can be derived from the inverse Laplace transform, and then the capacitor voltage $u_{C\alpha}$ can be written as

$$u_{C\alpha}(t) = \mathcal{L}^{-1}[U_{C\alpha}(s)] = \frac{1}{C_\alpha}\sum_{r=0}^{\infty}\left(-\frac{1}{RC_\alpha}\right)^r t^{\alpha r+\alpha-1}E^{r+1}_{\alpha+\beta,\alpha+\alpha r}\left(-\frac{1}{L_\beta C_\alpha}t^{\alpha+\beta}\right) \tag{4.250}$$

and the inductor current $i_{L\beta}$ is

$$\begin{aligned} i_{L\beta}(t) &= \mathcal{L}^{-1}\left[I_{L\beta}(s)\right] \\ &= \frac{1}{L_\beta C_\alpha}\sum_{r=0}^{\infty}\left(-\frac{1}{RC_\alpha}\right)^r t^{\alpha r+\alpha+\beta-1}E^{r+1}_{\alpha+\beta,\alpha+\beta+\alpha r}\left(-\frac{1}{L_\beta C_\alpha}t^{\alpha+\beta}\right) \end{aligned} \tag{4.251}$$

Other currents and voltages of the circuit can be derived from the above equations by the characteristics of the parallel circuit.

4.10 Response to Arbitrary Excitations

There is a linear relationship between the zero-state response of the linear circuit and the excitation source because the operators of fractional and integer-order calculus satisfy the properties of homogeneity and additivity. If H_{00} represents the zero-state response operator, and the excitation sources are i_{s1}, i_{s2}, $i_{s1} + i_{s2}$, and Ki_{s1}, the corresponding zero-state responses can be expressed as $H_{00}[i_{s1}]$, $H_{00}[i_{s2}]$, $H_{00}[i_{s1} + i_{s2}]$ and $H_{00}[Ki_{s1}]$, which implies that.

Additivity:

$$H_{00}[i_{s1} + i_{s2}] = H_{00}[i_{s1}] + H_{00}[i_{s2}] \tag{4.252}$$

Homogeneity:

$$H_{00}[Ki_{s1}] = KH_{00}[i_{s1}] \tag{4.253}$$

It can be concluded that if the excitation source is the superposition of two excitation sources, the zero-state response is the superposition of the zero-state response when the two excitation sources are input separately. If the amplitude of the excitation source is expanded to K times, the zero-state response will also increase to K times. Therefore, the physical quantities in the linear circuit satisfy the linear relationship.

In a linear circuit, the complete response is the superposition of the zero-input and the zero-state responses. Supposing that $H_{0i}(t)$ represents the zero-input response and $H_{00}(t)$ denotes the zero-state response, the complete response $H(t)$ can be calculated by the linear combination of $H_{0i}(t)$ and $H_{00}(t)$ as

$$H(t) = H_{0i}(t) + H_{00}(t) \tag{4.254}$$

Decomposing the complete response into the sum of zero-input and zero-state responses is an important theorem of the linear circuit, which plays an important role in circuit analysis. The effects of the external excitation and the initial energy of the circuit are decoupled, which means that they can be analyzed separately. Hence, we can perform the linear superposition to obtain the solutions, which can simplify the circuit analysis.

For each segment of arbitrary waveforms, it can be approximated as the superposition of a series of coefficients $f(t_k)\cdot\Delta$and a delay rectangle pulse $p_\Delta(t-t_k)$with a width of Δ. The zero-state response of each segment of the waveform excitation is the superposition of the zero-state response generated by the series of rectangular pulse functions $f(t_k)\Delta f(t_k)p_\Delta(t-t_k)\cdot\Delta$. Assume that the zero-state response generated by the delay rectangle pulse $p_\Delta(t-t_k)$is $h_\Delta(t-t_k)$, the waveform excitation expression can be expressed as

$$y_o(t) \approx \sum_{k=0}^{n-1} f(t_k)h_\Delta(t-t_k)\cdot\Delta \tag{4.255}$$

As $n\to\infty$, $\Delta\to 0$, we can obtain the following conclusion:

(1) The discrete variable t_k is equal to continuous variable τ.
(2) The response of delay rectangle pulse $h_\Delta(t-t_k)$represents the impulse response $h(t-\tau)$.
(3) The increment Δ of discrete variable t_k changes to the increment $\mathrm{d}\tau$ of continuous variable.
(4) The infinite series sum changes to the integral, which means

$$y_o(t) = \int_{t_0}^{t_n} f(\tau)h(t-\tau)\mathrm{d}\tau \tag{4.256}$$

where $f(t)$ is the excitation function, and $h(t)$ is the impulse response function of the circuit. (4.256) can be used to calculate the zero-state response under arbitrary excitation, which is related to the convolution integral.

For a circuit network with non-zero states, the complete response is the superposition of the zero-state and zero-input responses, which means

$$y_o(t) = H_{0i}(t) + H_{00}(t) = H_{0i}(t) + \int_{t_0}^{t_n} f(\tau)h(t-\tau)\mathrm{d}\tau \tag{4.257}$$

4.11 Summary

This chapter introduces the derivation and solving methods of fractional differential equations for circuits with one and two fractional-order components, respectively. The zero-state response, zero-input response, and complete response of series and parallel RC_α, RL_β, $RL_\beta C_\alpha$ circuits are analyzed. The influence of circuit parameters on the time-domain response is discussed. When the initial energy in the circuit is equal to zero, since the initial value of the Laplace transform is not involved, the solution of the fractional differential equation under R-L definition is consistent with the result under Caputo definition. When the initial energy of fractional-order components in the circuit is not equal to zero, the transient state of physical quantities of the circuit described by R-L and Caputo definitions are quite different. The transient state under the R-L definition decays more rapidly, and the voltage or current curves have an impulse at the moment of state change. As for the Caputo definition, there is no impulse in the voltage or current curves at the moment of state changing, but the transient state decays slowly. All the differences originate from the definition of the initial conditions required for solving fractional differential equations under R-L and Caputo definitions. Meanwhile, the physical meaning of initial conditions in the circuit still needs to be further researched.

References

1. Podlubny I (1998) Fractional differential equations: an introduction to fractional derivatives, fractional differential equations, to methods of their solution and some of their applications. Academic, San Diego, CA, USA
2. Abate J, Whitt W (2006) A unified framework for numerically inverting Laplace transforms. INFORMS J Comput 18(4):408–421
3. Mittag-Leffler Function (2012) MATLAB central file exchange [Online]. Available: https://www.mathworks.com/matlabcentral/fileexchange/8738-mittag-leffler-function
4. FOTF Toolbox (2017) MATLAB central file exchange [Online]. Available: https://www.mathworks.com/matlabcentral/fileexchange/60874-fotf-toolbox
5. Lin SD, Lu CH (2013) Laplace transform for solving some families of fractional differential equations and its applications. Adv Differ Equ-Ny 2013 (in English)
6. Kochubei A, Gorenflo R, Mainardi F, Rogosin S, Luchko Y (2019) Mittag-Leffler function: properties and applications. In: Basic theory, pp 269–296
7. Tomovski Ž, Hilfer R, Srivastava HM (2010) Fractional and operational calculus with generalized fractional derivative operators and Mittag-Leffler type functions. Integral Transform Spec Funct 21(11):797–814
8. AbdelAty AM, Radwan AG, Ahmed WA, Faied M (2016) Charging and discharging RC α circuit under Riemann-Liouville and Caputo fractional derivatives. In: 2016 13th International conference on electrical engineering/electronics, computer, telecommunications and information technology (ECTI-CON). IEEE, pp 1–4
9. Jiang YW, Zhang B (2020) Comparative study of Riemann-Liouville and Caputo derivative definitions in time-domain analysis of fractional-order capacitor (in English). IEEE T Circuits-II 67(10):2184–2188
10. Guía M, Gómez F, Rosales J (2013) Analysis on the time and frequency domain for the RC electric circuit of fractional order. Open Phys 11(10)
11. Jakubowska A, Walczak J (2016) Analysis of the transient state in a series circuit of the class $RL_\beta C_\alpha$. Circuits Syst Signal Process 35(6):1831–1853

Chapter 5
Sinusoidal Steady-State Analysis of Fractional-Order Circuits

In actual production and engineering applications, sinusoidal AC power supplies have been widely used, so many practical circuits operate in sinusoidal steady state, such as most circuits in power systems. Therefore, the sinusoidal steady-state analysis of the circuit is very important.

Sinusoidal steady-state analysis is often used to study and discuss the steady-state response of a linear time-invariant circuit under sinusoidal power supplies with the same frequency. The phasor method is a convenient method to analyze the sinusoidal steady-state circuit. It uses complex numbers called phasors to represent the sinusoidal quantities and transforms the differential or integral equations describing the sinusoidal steady-state circuits into complex algebra equations, thus simplifying the analysis and calculation of the circuit.

This chapter adopts the phasor method to analyze the sinusoidal steady-state response of fractional-order circuits. Firstly, the fundamental concepts of sinusoidal waveforms are introduced, including the phase difference, the effective value of the sinusoidal quantity, and its phasor representation method. Then, the phasor representations of fractional-order capacitors, fractional-order capacitors, and fractional-order mutual inductors are given, and the phasor representation and analysis methods of fractional-order circuits are also presented. Finally, in order to illustrate the application of the phasor method in the sinusoidal steady state, several typical fractional-order circuits are introduced as examples for analysis, such as fractional-order resonant circuits, fractional-order mutual inductance coupling circuits, and fractional-order filter circuits [1].

B. Zhang and X. Shu, *Fractional-Order Electrical Circuit Theory*, CPSS Power Electronics Series, https://doi.org/10.1007/978-981-16-2822-1_5

5.1 Preliminaries

5.1.1 Phase Difference

For sinusoidal waveforms, there are two different instantaneous expressions, namely "sin" and "cos". In this book, "sin" is used as the standard form of instantaneous expression. For example, the instantaneous expression of sinusoidal voltage $u(t)$ is written as

$$u(t) = U_m \sin(\omega t + \theta) \tag{5.1}$$

where U_m is the amplitude, θ is the initial phase angle, and ω is the angular frequency. U_m, θ, and ω are three essential variables of sinusoidal quantities. Only when these three variables are fixed, the sinusoidal waveform can be uniquely determined.

In addition, the subtraction of initial phase angles of two sinusoidal waveforms with the same frequency is termed as the phase difference between them, which is a constant that does not vary with time. When the two sinusoidal waveforms have the same frequency, the phase difference is valid and fixed. Supposing that there are two sinusoidal waveforms, which are expressed as

$$\begin{cases} u_1(t) = U_m \sin(\omega t + \theta_1) \\ u_2(t) = U_m \sin(\omega t + \theta_2) \end{cases} \tag{5.2}$$

where $\varphi_1 = \omega t + \theta_1$ and $\varphi_2 = \omega t + \theta_2$ are the phase angles of u_1 and u_2, respectively. The phase difference of the two waveforms can be obtained as $\theta = \varphi_1 - \varphi_2 = \theta_1 - \theta_2$. It is noted that all expressions must be written in a unified standard form like "sin" or "cos" before calculating the phase difference.

If $\theta > 0$ ($\theta_1 > \theta_2$), it implies that $u_1(t)$ leads $u_2(t)$. If $\theta < 0$ ($\theta_1 < \theta_2$), it denotes that $u_1(t)$ lags $u_2(t)$. If $\theta = 0$ ($\theta_1 = \theta_2$), it represents that $u_1(t)$ and $u_2(t)$ are in phase. Figure 5.1 shows the waveforms of two sinusoidal waveforms with different initial phase angles. The lead-lag relationship between these two phasors can be judged by the peaks or the zero-crossing points of the waveforms. As shown in Fig. 5.1, $u_1(t)$

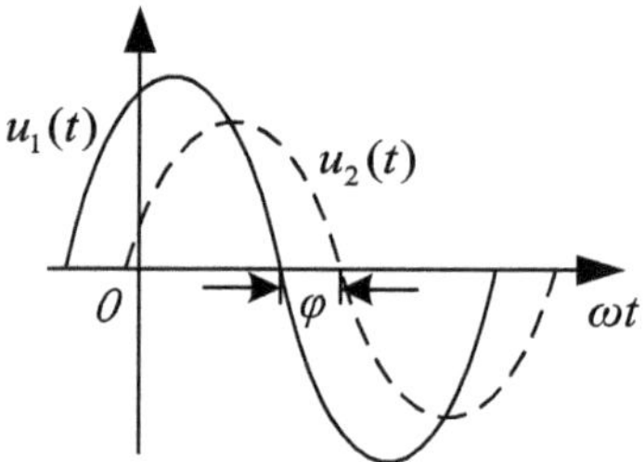

Fig. 5.1 Waveforms of $u_1(t)$ and $u_2(t)$

always reaches its peaks or zero-crossing points before $u_2(t)$, which means that $u_1(t)$ leads $u_2(t)$, and the phase difference is φ.

5.1.2 Effective Values

The effective value of the sinusoidal waveform is defined by the root-mean-square (RMS) value of the waveform in one period. For instance, if the sinusoidal current is described as $i(t) = I_m \sin(\omega t + \theta)$, its effective value I can be defined as

$$I = \sqrt{\frac{1}{T}\int_0^T I_m^2 \sin^2(\omega t + \theta)\mathrm{d}t} = \frac{I_m}{\sqrt{2}} \tag{5.3}$$

where I_m is the amplitude or maximum value of the current.

Similarly, the effective value of the voltage can be determined by

$$U = \frac{U_m}{\sqrt{2}} \tag{5.4}$$

where U_m is the amplitude of the voltage.

5.1.3 Phasor Representation

The phasor method plays an important role in the sinusoidal steady-state analysis of fractional-order circuits. Taking voltage $u(t) = \sqrt{2}U \sin(\omega t + \phi_u)$ and current $i(t) = \sqrt{2}I \sin(\omega t + \phi_i)$ as examples, based on Euler's identity, all sinusoidal variables with the same frequency at the steady state can be expressed as

$$\begin{cases} \dot{U} = \sqrt{2}U\mathrm{e}^{j\phi_u} = \sqrt{2}U\angle\phi_u \\ \dot{I} = \sqrt{2}I\mathrm{e}^{j\phi_i} = \sqrt{2}I\angle\phi_i \end{cases} \tag{5.5}$$

where U and I are effective values of $u(t)$ and $i(t)$, ϕ_u and ϕ_i are initial phase angles of $u(t)$ and $i(t)$, $\dot{U}$ and $\dot{I}$ are phasor forms of $u(t)$ and $i(t)$, respectively. Hence, the imaginary part of the phasor can represent the sinusoidal waveform, which is convenient for the sinusoidal steady-state analysis of a fractional-order circuit.

5.2 Phasor Representation of Fractional-Order Components and Circuits

5.2.1 Fractional-Order Capacitor

Based on the introduction of fractional-order components in Chap. 2, by applying Laplace transform to (2.2), the impedance of a fractional-order capacitor can be represented as $Z_\alpha(s) = U_C(s)/I_C(s) = 1/(C_\alpha s^\alpha)$. Replacing s with $j\omega$, the impedance of the fractional-order capacitor can be written as

$$Z_\alpha = \frac{1}{(j\omega)^\alpha C_\alpha} = \frac{1}{\omega^\alpha C_\alpha}\mathrm{e}^{-j\frac{\pi\alpha}{2}} = C_\alpha\frac{1}{\omega^\alpha C_\alpha}\cos\left(\frac{\pi}{2}\alpha\right) - j\frac{1}{\omega^\alpha C_\alpha}\sin\left(\frac{\pi}{2}\alpha\right) \quad (5.6)$$

Similarly, the impedance of the fractional-order capacitor can be represented as

$$Y_\alpha = C_\alpha(j\omega)^\alpha = C_\alpha\omega^\alpha \mathrm{e}^{j\frac{\pi\alpha}{2}} = C_\alpha\omega^\alpha\left[\cos\left(\frac{\pi}{2}\alpha\right) + j\sin\left(\frac{\pi}{2}\alpha\right)\right] \quad (5.7)$$

It can be seen from (4.245) and (5.7) that the impedance and admittance of the fractional-order capacitor depends not only on the pseudo-capacitance value C_α and angular frequency ω, but also on the fractional order α. Besides, it can be noted that the impedance and admittance contain real and imaginary parts, which are completely different from the traditional integer-order capacitors. In particular, when $\alpha < 1$, the real part of the impedance is positive. When $\alpha > 1$, the real part of the impedance is negative.

If the voltage across the fractional-order capacitor is a sinusoidal waveform, i.e., $u_C(t) = U_{Cm}\sin(\omega t + \theta_1)$, its corresponding phasor form $\dot{U}_C$ can be written as

$$\dot{U}_C = U_{Cm}\angle\theta_1 = \sqrt{2}U_C\angle\theta_1 \quad (5.8)$$

where U_{Cm} and U_C are the amplitude and effective values of the voltage $u_C(t)$.

The phasor form $\dot{I}_C$ of the current flowing through the fractional-order capacitor can be deduced as

$$\dot{I}_C = \dot{U}_C Y_\alpha = \omega^\alpha C_\alpha U_{Cm}\angle\left(\theta_1 + \frac{\pi\alpha}{2}\right) \quad (5.9)$$

Extracting the imaginary part of (5.9), the sinusoidal expression of the current can be expressed as

$$i_C(t) = \omega^\alpha C_\alpha U_{Cm}\sin\left(\omega t + \theta_1 + \frac{\pi\alpha}{2}\right) \quad (5.10)$$

From (5.8), (5.9) and (5.10), the phasors diagram and time-domain waveform of the current and voltage of the fractional-order capacitor can be drawn as shown in Fig. 5.2. $\varphi_C = \pi\alpha/2$ is the phase difference between voltage and current waveforms

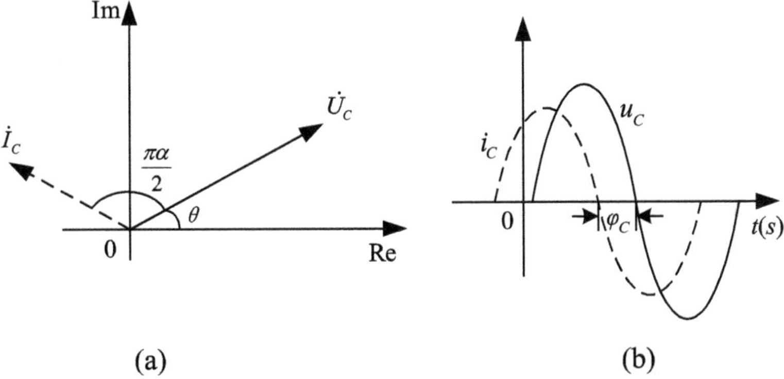

Fig. 5.2 Voltage and current of fractional-order capacitor: **a** Phasor diagram. **b** Time-domain waveforms

of the fractional-order capacitor, which is only determined by α and not influenced by the operating frequency, so the fractional-order capacitor is also called the constant phase element [2, 3].

Without loss of generality, assuming that the initial phase angle of the voltage across the fractional-order capacitor is zero, the voltage can be described as

$$u_C(t) = U_{Cm}\sin(\omega t) \tag{5.11}$$

Then the current of the fractional-order capacitor can be obtained as

$$i_C(t) = \omega^\alpha C_\alpha U_{Cm}\sin\left(\omega t + \frac{\pi\alpha}{2}\right) \tag{5.12}$$

The instantaneous power $p_{C\alpha}$ of the fractional-order capacitor can be calculated as

$$\begin{aligned} p_{C\alpha}(t) &= u_C(t)i_C(t) \\ &= \frac{\omega^\alpha C_\alpha U_{Cm}^2}{2}\left\{\cos\left(\frac{\pi}{2}\alpha\right)[1-\cos(2\omega t)] + \sin\left(\frac{\pi}{2}\alpha\right)\sin(2\omega t)\right\} \end{aligned} \tag{5.13}$$

The average power $P_{C\alpha}$ of the fractional-order capacitor in a period can be given by

$$P_{C\alpha} = \frac{1}{T_S}\int_0^{T_S} p_{C\alpha}(t)\mathrm{d}t = \frac{\omega^\alpha C_\alpha U_{Cm}^2}{2}\cos\left(\frac{\pi}{2}\alpha\right) \tag{5.14}$$

where T_S is the period.

It can be found that when $\alpha < 1$, $P_{C\alpha} > 0$, the fractional-order capacitor consumes power, which means that the real part of the impedance of the fractional-order capacitor is characterized as a resistance, and the fractional-order capacitor is passive. When $\alpha > 1$, $P_{C\alpha} < 0$, the fractional-order capacitor provides power, which indicates that the real part of the impedance of the fractional-order capacitor is regarded as a negative resistance, and the fractional-order capacitor is active. When $\alpha = 1$, $P_{C\alpha} = 0$, the fractional-order capacitor becomes an integer-order capacitor.

Moreover, the reactive power $Q_{C\alpha}$ of the fractional-order capacitor can be derived as

$$Q_{C\alpha} = \frac{\omega^{\alpha} C_{\alpha} U_{Cm}^2}{2} \sin\left(\frac{\pi}{2}\alpha\right) = \frac{1}{2} U_{Cm} I_{Cm} \sin\left(\frac{\pi}{2}\alpha\right) \tag{5.15}$$

where $I_{Cm} = \omega^{\alpha} C_{\alpha} U_{Cm}$ is the amplitude of the current $i_C(t)$ of the fractional-order capacitor.

Based on (5.14) and (5.15), the apparent power $S_{C\alpha}$ of the fractional-order capacitor can be calculated as

$$S_{C\alpha} = \sqrt{P_{C\alpha}^2 + Q_{C\alpha}^2} = \frac{\omega^{\alpha} C_{\alpha} U_{Cm}^2}{2} = \frac{1}{2} U_{Cm} I_{Cm} \tag{5.16}$$

5.2.2 Fractional-Order Inductor

In the same way, by applying Laplace transform to (2.8), the impedance of the fractional-order inductor can be described as $Z_{\beta}(s) = L_{\beta} s^{\beta}$. Replacing s with $j\omega$, the impedance of the fractional-order inductor can be expressed as

$$Z_{\beta}(j\omega) = L_{\beta}(j\omega)^{\beta} = L_{\beta}\omega^{\beta} e^{j\frac{\pi\beta}{2}} = L_{\beta}\omega^{\beta}\left[\cos\left(\frac{\pi}{2}\beta\right) + \mathrm{j}\sin\left(\frac{\pi}{2}\beta\right)\right] \tag{5.17}$$

As can be seen from (5.17), the impedance of the fractional-order inductor has a real part and an imaginary part, which are determined by the pseudo-inductance value L_{β}, the operating angular frequency ω, and the fractional order β. In addition, when $\beta < 1$, the real part of the impedance is positive, and when $\beta > 1$, the real part of the impedance is negative.

Assuming that the current flowing through the fractional-order inductor is sinusoidal, i.e., $i_L(t) = I_{Lm}\sin(\omega t + \theta_2)$, the corresponding phasor can be written as

$$\dot{I}_L = I_{Lm}\angle\theta_2 = \sqrt{2} I_L \angle\theta_2 \tag{5.18}$$

where I_{Lm} and I_L are the amplitude and effective values of the current $i_L(t)$.

Then, the voltage phasor across the fractional-order inductor can be deduced as

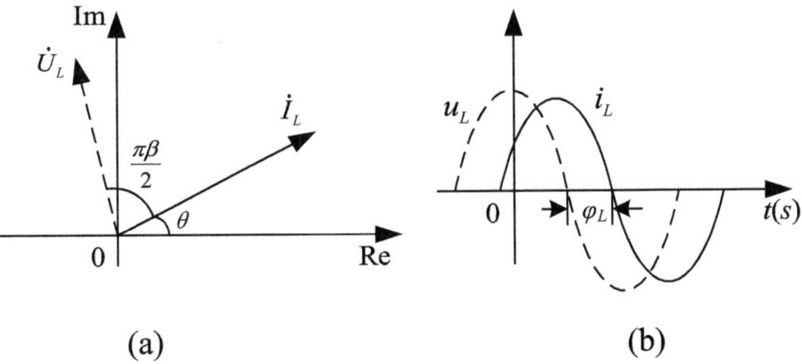

Fig. 5.3 Voltage and current of fractional-order inductor: **a** Phasor diagram. **b** Time-domain waveforms

$$\dot{U}_L = \dot{I}_{Lm}\dot{Z}_\beta = \omega^\beta L_\beta I_{Lm}\angle\left(\theta_2 + \frac{\pi\beta}{2}\right) \tag{5.19}$$

From (5.19), the time-domain expression of the voltage can be written as

$$u_L(t) = \omega^\beta L_\beta I_{Lm}\sin\left(\omega t + \theta_2 + \frac{\pi\beta}{2}\right) \tag{5.20}$$

Figure 5.3 depicts the phasor diagram and time-domain waveforms of the current and voltage of a fractional-order inductor. $\varphi_L = \pi\beta/2$ is the phase difference between voltage and current waveforms of the fractional-order inductor, which only depends on β. Thus, the fractional-order inductor is a constant phase element [2, 4].

Without loss of generality, assuming that the initial phase angle of the current is zero, that is, $\theta_2 = 0$, the current of the fractional-order inductor can be simplified as

$$i_L(t) = I_{Lm}\sin(\omega t) \tag{5.21}$$

Then, the voltage of the fractional-order inductor can be calculated as

$$u_L(t) = \omega^\beta L_\beta I_{Lm}\sin\left(\omega t + \frac{\pi\beta}{2}\right) \tag{5.22}$$

Therefore, the instantaneous power $p_{L\beta}(t)$ of the fraction-order inductor can be obtained as

$$\begin{aligned} p_{L\beta}(t) &= u_L(t)i_L(t) \\ &= \frac{\omega^\alpha L_\beta I_{Lm}^2}{2}\left\{\cos\left(\frac{\pi}{2}\beta\right)[1-\cos(2\omega t)] + \sin\left(\frac{\pi}{2}\beta\right)\sin(2\omega t)\right\} \end{aligned} \tag{5.23}$$

The average power $P_{L\beta}$ of the fractional-order inductor in a period is

$$P_{L\beta} = \frac{1}{T_S}\int_0^{T_S} p_{L\beta}(t)\mathrm{d}t = \frac{\omega^\beta L_\beta I_{Lm}^2}{2}\cos\left(\frac{\pi}{2}\beta\right) \tag{5.24}$$

As can be seen from (5.24), when $\beta < 1$, $P_{L\beta} > 0$, the fractional-order inductor consumes power, which denotes that the real part of the impedance of the fractional-order inductor is characterized as a resistance, and the fractional-order inductor is passive. If $\beta > 1$, $P_{L\beta} < 0$, the fractional-order inductor provides power, which indicates that the real part of the impedance of the fractional-order inductor is a negative resistance, and the fractional-order inductor is active. When $\beta = 1$, $P_{L\beta} = 0$, the inductor is an integer-order inductor.

Besides, according to the definition of the reactive power, the reactive power $Q_{L\beta}$ of the fractional-order inductor can be deduced as

$$Q_{L\beta} = \frac{\omega^\beta L_\beta I_{Lm}^2}{2}\sin\left(\frac{\pi}{2}\beta\right) = \frac{1}{2}U_{Lm}I_{Lm}\sin\left(\frac{\pi}{2}\beta\right) \tag{5.25}$$

where $U_{Lm} = \omega^\beta L_\beta I_{Lm}$ is the amplitude of the voltage $u_L(t)$.

The apparent power $S_{L\beta}$ of the fraction-order inductor can be calculated as [5, 6]

$$S_{L\beta} = \sqrt{P_{L\beta}^2 + Q_{L\beta}^2} = \frac{\omega^\alpha L_\beta I_{Lm}^2}{2} = \frac{1}{2}U_{Lm}I_{Lm} \tag{5.26}$$

5.2.3 *Fractional-Order Mutual Inductor*

Mutual inductance is defined as the inductance caused by the mutual coupling of a circuit and other circuits. Mutual coupling of two coils means that the magnetic field generated by one coil is linking or crossing the other coil, and vice versa [7, 8].

For the fractional-order mutual inductor shown in Fig. 2.3, its differential equations can be rewritten by using the Laplace transform as

$$\begin{cases} U_1(s) = s^{\beta_1}L_{\beta 1}I_1(s) + s^\gamma M_\gamma I_2(s) \\ U_2(s) = s^{\beta_2}L_{\beta 2}I_2(s) + s^\gamma M_\gamma I_1(s) \end{cases} \tag{5.27}$$

Considering $s = j\omega$, (5.27) can be expressed in phasor form as

$$\begin{cases} \dot{U}_1 = (j\omega)^{\beta_1}L_{\beta 1}\dot{I}_1 + (j\omega)^\gamma M_\gamma \dot{I}_2 \\ \dot{U}_2 = (ss\,j\omega)^{\beta_2}L_{\beta 2}\dot{I}_2 + (j\omega)^\gamma M_\gamma \dot{I}_1 \end{cases} \tag{5.28}$$

If $\beta_1 = \beta_2 = \gamma$ and $M_\gamma = M$, the fractional-order mutual inductor can be equivalent to a T-type circuit, as shown in Fig. 5.4.

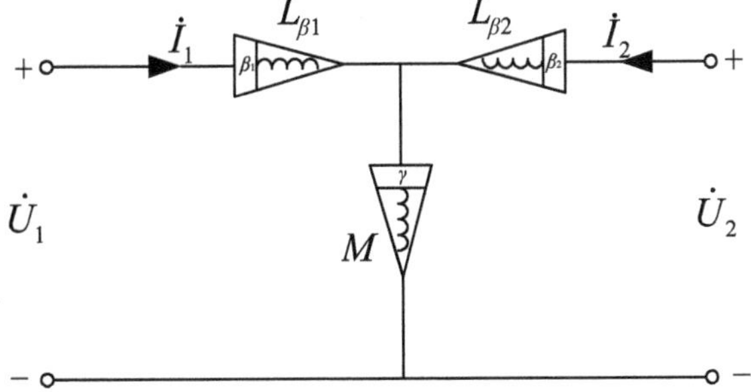

Fig. 5.4 T-type equivalent circuit of fractional-order mutual inductor

5.2.4 *Fractional-Order Circuit*

According to the definition of phasor, KCL and KVL can be expanded into phasor forms, which can be described as

$$\begin{cases} \sum^{(n)} \dot{I}_k = 0 \\ \sum^{(n)} \dot{U}_k = 0 \end{cases} \tag{5.29}$$

The phasor representations of several typical fractional-order circuits are described as follows:

(1) Fractional-order $RL_\beta C_\alpha$ branch circuit

Figure 5.5 shows a simple fractional-order $RL_\beta C_\alpha$ branch circuit, where the associated reference direction is selected for the voltage and current [9–11].

Based on KVL, the circuit equation can be represented in phasor form as

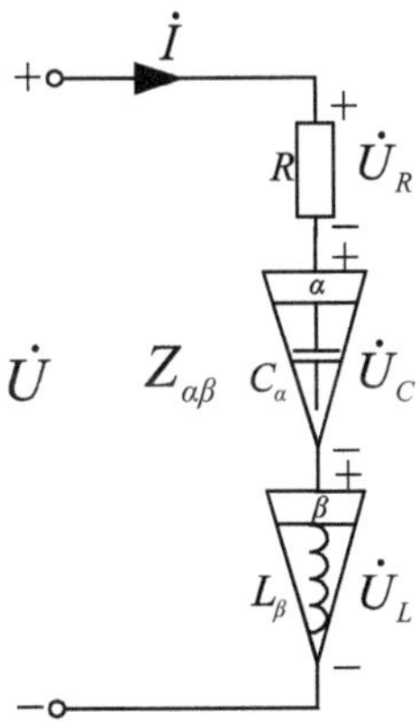

Fig. 5.5 Fractional-order $RL_\beta C_\alpha$ branch circuit

$$\dot{U} = \dot{I} Z_{\alpha\beta} = \dot{I}\left[R + \frac{1}{(j\omega)^{\alpha} C_{\alpha}} + (j\omega)^{\beta} L_{\beta}\right] \tag{5.30}$$

where $Z_{\alpha\beta}$ is the impedance of the fractional-order $RL_{\beta}C_{\alpha}$ branch circuit.

(2) n fractional-order branch circuits in series

The series connection of n fractional-order branch circuits is shown in Fig. 5.6, β_k is the order of the fractional-order inductor $L_{\beta k}$, α_k is the order of the fractional-order capacitor $C_{\alpha k}$, R_k is the resistor, $Z_{\alpha\beta\_k}$ is the impedance of the $R_k L_{\beta k} C_{\alpha k}$ branch circuit, $k = 1, 2, \ldots, n$ [12].

The phasor equations of the circuit can be written as

$$\dot{I}_1 = \dot{I}_2 = \cdots = \dot{I}_k = \cdots = \dot{I}_n = \dot{I} \tag{5.31}$$

$$\dot{U} = \sum_{k=1}^{n} \dot{U}_k = \sum_{k=1}^{n} \dot{I}_k Z_{\alpha\beta\_k} = \sum_{k=1}^{n} \dot{I} Z_{\alpha\beta\_k} \tag{5.32}$$

where $\dot{U}_k$ and $\dot{I}_k$ are phasor forms of the voltage and current of $R_k L_{\beta k} C_{\alpha k}$ branch circuit, respectively.

Therefore, the equivalent impedance $Z_{\alpha\beta}$ of the circuit shown in Fig. 5.6 is equal to the sum of the impedance of each branch circuit, that is

$$Z_{\alpha\beta} = \frac{\dot{U}}{\dot{I}} = \sum_{k=1}^{n} Z_{\alpha\beta\_k} = \sum_{k=1}^{n} R_k + \sum_{k=1}^{n} \frac{1}{(j\omega)^{\alpha_k} C_{\alpha k}} + \sum_{k=1}^{n} (j\omega)^{\beta_k} L_{\beta k} \tag{5.33}$$

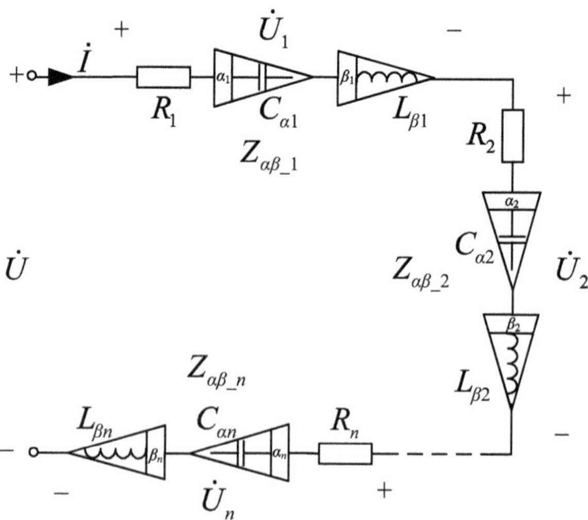

Fig. 5.6 n fractional-order branch circuits in series

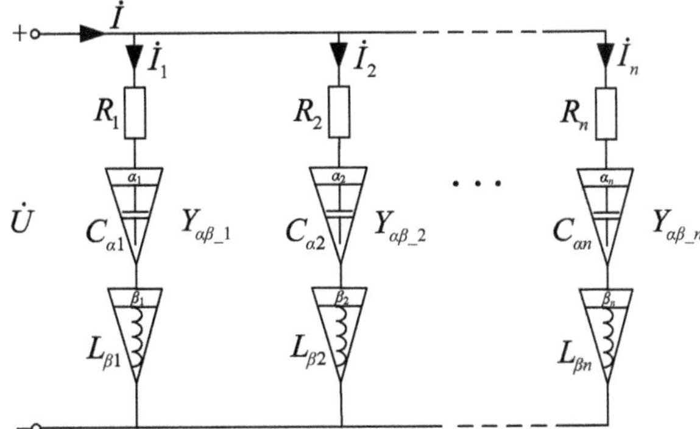

Fig. 5.7 *n* fractional-order branch circuits in parallel

(3) *n* fractional-order branch circuits in parallel

The parallel connection of *n* fractional-order branch circuits is described in Fig. 5.7, where β_k is the order of the fractional-order inductor $L_{\beta k}$, α_k is the order of the fractional-order capacitor $C_{\alpha k}$, R_k is the resistor, $Y_{\alpha\beta\_k}$ is the admittance of the $R_k L_{\beta k} C_{\alpha k}$ branch circuit, and $k = 1, 2, \ldots, n$.

The phasor equations of the circuit are

$$\dot{U}_1 = \dot{U}_2 = \cdots = \dot{U}_n = \dot{U} \tag{5.34}$$

$$\dot{I} = \sum_{k=1}^{n} \dot{I}_k = \sum_{k=1}^{n} \dot{U}_k Y_{\alpha\beta\_k} = \sum_{k=1}^{n} \dot{U} Y_{\alpha\beta\_k} \tag{5.35}$$

Similarly, the equivalent admittance *Y* of the circuit shown in Fig. 5.7 is equal to the sum of the admittance of each branch circuit, it is given by

$$Y_{\alpha\beta} = \frac{\dot{I}}{\dot{U}} = \sum_{k=1}^{n} Y_{\alpha\beta\_k} = \sum_{k=1}^{n} G_k + \sum_{k=1}^{n} (j\omega)^{\alpha_k} C_{\alpha k} + \sum_{k=1}^{n} \frac{1}{(j\omega)^{\beta_k} L_{\beta k}} \tag{5.36}$$

The series–parallel connection of *n* fractional-order branch circuits can be simplified step by step by the principle of series and parallel.

5.3 Fractional-Order Resonant Circuits

In the analysis of fractional-order circuits, the phasor method can be used to analyze and calculate the sinusoidal variables of circuits first, and then convert them into time-domain expressions [13–16].

5.3.1 Resonance in Series Fractional-Order $RL_\beta C_\alpha$ Circuit

The resonance of a sinusoidal steady-state fractional-order circuit is a very special phenomenon. When the circuit resonates, the voltage or current of some branch circuits will be much greater than the voltage or current of the circuit ports [17, 18].

For a certain port or branch of a circuit, if the voltage and current on the port or branch are in phase, it means

$$\mathrm{Im}[Z(j\omega)] = 0 \quad \text{or} \quad \mathrm{Im}[Y(j\omega)] = 0 \tag{5.37}$$

which indicates that this port or this branch is in resonance.

For the series fractional-order $RL_\beta C_\alpha$ circuit shown in Fig. 5.8, the impedance of the circuit can be expressed as

$$\begin{aligned} Z_{\alpha\beta} &= R + \frac{1}{(j\omega)^\alpha C_\alpha} + (j\omega)^\beta L_\beta \\ &= \left[R + \frac{\cos(\frac{\pi\alpha}{2})}{\omega^\alpha C_\alpha} + \omega^\beta L_\beta \cos\left(\frac{\pi\beta}{2}\right)\right] + j\left[\omega^\beta L_\beta \sin\left(\frac{\pi\beta}{2}\right) - \frac{\sin(\frac{\pi\alpha}{2})}{\omega^\alpha C_\alpha}\right] \end{aligned} \tag{5.38}$$

Separating the real and imaginary parts of (5.38), the imaginary part of the impedance $Z_{\alpha\beta}$ is given by

$$\mathrm{Im}(Z_{\alpha\beta}) = X = \omega^\beta L_\beta \sin\left(\frac{\pi\beta}{2}\right) - \frac{1}{\omega^\alpha C_\alpha} \sin\left(\frac{\pi\alpha}{2}\right) \tag{5.39}$$

If $X = 0$, there is

$$\omega^\beta L_\beta \sin\left(\frac{\pi}{2}\beta\right) - \frac{1}{\omega^\alpha C_\alpha} \sin\left(\frac{\pi}{2}\alpha\right) = 0 \tag{5.40}$$

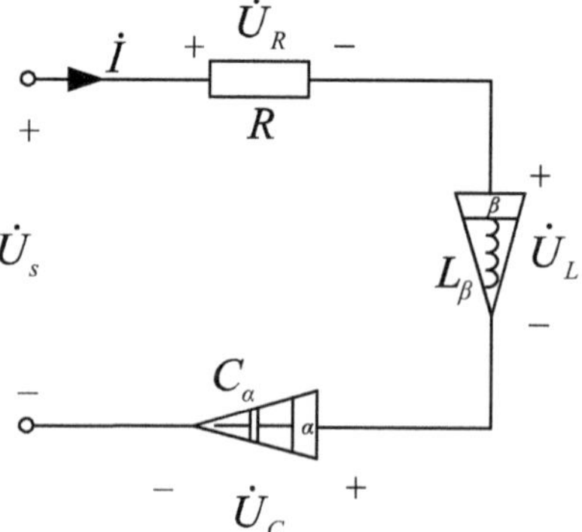

Fig. 5.8 Series fractional-order $RL_\beta C_\alpha$ circuit

the series fractional-order $RL_\beta C_\alpha$ circuit resonates. At this time, the resonance angular frequency can be derived as

$$\omega_r = \left[\frac{\sin\left(\frac{\pi\alpha}{2}\right)}{L_\beta C_\alpha \sin\left(\frac{\pi\beta}{2}\right)}\right]^{\frac{1}{\alpha+\beta}} \tag{5.41}$$

From (5.41), it can be seen that the resonance angular frequency of the series fractional-order $RL_\beta C_\alpha$ circuit is not only determined by L_β and C_α, but also depends on the fractional orders α and β.

When the circuit resonates, the impedance of the series fractional-order $RL_\beta C_\alpha$ circuit is equivalent to a pure resistance $R_{\alpha\beta}$, which is

$$R_{\alpha\beta} = R + \frac{1}{\omega_r^\alpha C_\alpha}\cos\left(\frac{\pi\alpha}{2}\right) + \omega_r^\beta L_\beta \cos\left(\frac{\pi\beta}{2}\right) \tag{5.42}$$

If fractional orders are identical, i.e., $\alpha = \beta$, the equivalent resistance can be further simplified as

$$R'_{\alpha,\beta} = R + 2\sqrt{\frac{L_\beta}{C_\alpha}}\cos\left(\frac{\pi\alpha}{2}\right) \tag{5.43}$$

The resonance current of the circuit can be obtained as

$$\dot{I}_r = \frac{\dot{U}_s}{R'_{\alpha\beta}} = \frac{\dot{U}_s}{R + \frac{1}{\omega_r^\alpha C_\alpha}\cos\left(\frac{\pi\alpha}{2}\right) + \omega_r^\beta L_\beta \cos\left(\frac{\pi\beta}{2}\right)} \tag{5.44}$$

5.3.2 Resonance in Parallel Fractional-Order $RL_\beta C_\alpha$ Circuit

The parallel fractional-order $RL_\beta C_\alpha$ circuit is shown in Fig. 5.9 [19–21], its admittance $Y_{\alpha\beta}$ is

$$\begin{aligned} Y_{\alpha\beta} &= \frac{1}{R} + (j\omega)^\alpha C_\alpha + \frac{1}{(j\omega)^\beta L_\beta} \\ &= \left[\frac{1}{R} + \omega^\alpha C_\alpha \cos\left(\frac{\pi\alpha}{2}\right) + \frac{\cos\left(\frac{\pi\beta}{2}\right)}{\omega^\beta L_\beta}\right] + j\left[\omega^\alpha C_\alpha \sin\left(\frac{\pi\alpha}{2}\right) - \frac{\sin\left(\frac{\pi\beta}{2}\right)}{\omega^\beta L_\beta}\right] \end{aligned} \tag{5.45}$$

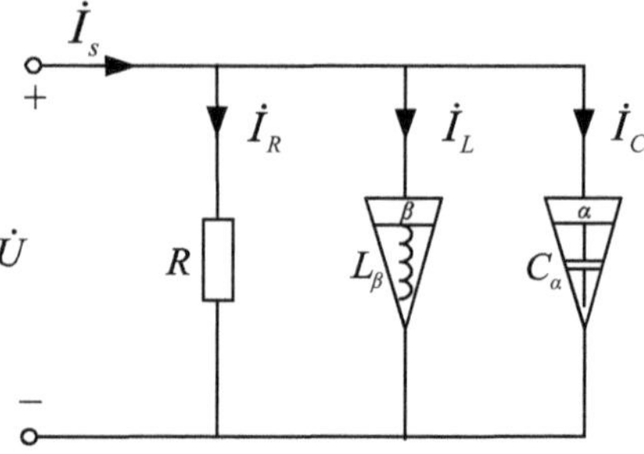

Fig. 5.9 Parallel fractional-order $RL_\beta C_\alpha$ circuit

Separating the real and imaginary parts of (5.45), the imaginary part of the admittance $Y_{\alpha\beta}$ is

$$\mathrm{Im}\left(Y_{\alpha\beta}\right) = X = \omega^\alpha C_\alpha \sin\left(\frac{\pi\alpha}{2}\right) - \frac{\sin\left(\frac{\pi\beta}{2}\right)}{\omega^\beta L_\beta} \tag{5.46}$$

Similarly, if X satisfies

$$\omega^\alpha C_\alpha \sin\left(\frac{\pi\alpha}{2}\right) - \frac{\sin\left(\frac{\pi\beta}{2}\right)}{\omega^\beta L_\beta} = 0 \tag{5.47}$$

the parallel fractional-order $RL_\beta C_\alpha$ circuit resonates. At this time, the resonance angular frequency can be deduced as

$$\omega_r = \left[\frac{\sin\left(\frac{\pi\beta}{2}\right)}{L_\beta C_\alpha \sin\left(\frac{\pi\alpha}{2}\right)}\right]^{\frac{1}{\alpha+\beta}} \tag{5.48}$$

It can be seen from (5.48) that the resonance angular frequency of the parallel fractional-order $RL_\beta C_\alpha$ circuit is not only determined by L_β and C_α, but also depends on the fractional orders α and β.

When the circuit is in resonance, the admittance of the parallel fractional-order $RL_\beta C_\alpha$ circuit is equivalent to a pure conductance $G_{\alpha\beta}$, which is given by

$$G_{\alpha\beta} = \frac{1}{R} + \omega_r^\alpha C_\alpha \cos\left(\frac{\pi\alpha}{2}\right) + \frac{1}{\omega_r^\beta L_\beta}\cos\left(\frac{\pi\beta}{2}\right) \tag{5.49}$$

the port voltage can be derived as

$$\dot{U} = \frac{1}{\frac{1}{R} + \omega_r^\alpha C_\alpha \cos\left(\frac{\pi\alpha}{2}\right) + \frac{1}{\omega_r^\beta L_\beta}\cos\left(\frac{\pi\beta}{2}\right)}\dot{I}_s \tag{5.50}$$

5.4 Fractional-Order Mutual Inductance Coupling Circuit

Figure 5.10 shows a typical example of the practical application of fractional-order mutual inductor. The energy of the primary circuit is transferred to the secondary circuit through the magnetic coupling of two coils. Here, r_1 and r_2 are internal resistors of coils, R_L is a resistance load.

Based on KVL and KCL, the phasor equations of the fractional-order mutual coupling circuit are

$$\begin{cases} \left[(j\omega)^{\beta_1} L_{\beta 1} + \dfrac{1}{(j\omega)^{\alpha_1} C_{\alpha 1}} + r_1\right]\dot{I}_1 + (\mathrm{j}\omega)^{\gamma} M_{\gamma}\dot{I}_2 = \dot{U}_s \\ (j\omega)^{\gamma} M_{\gamma}\dot{I}_1 + \left[(j\omega)^{\beta_2} L_{\beta 2} + \dfrac{1}{(j\omega)^{\alpha_2} C_{\alpha 2}} + r_2 + R_L\right]\dot{I}_2 = 0 \end{cases} \tag{5.51}$$

Then, the currents flowing through the primary and secondary circuits can be derived as

$$\begin{cases} \dot{I}_1 = \dfrac{\dot{U}_s}{Z_{\alpha\beta\_1} - \frac{(j\omega)^{2\gamma} M_{\gamma}^2}{Z_{\alpha\beta\_2}}} \\ \dot{I}_2 = \dfrac{-\frac{(j\omega)^{\gamma} M_{\gamma}}{Z_{\alpha\beta\_1}}\dot{U}_s}{Z_{\alpha\beta\_2} - \frac{(j\omega)^{2\gamma} M_{\gamma}^2}{Z_{\alpha\beta\_1}}} \end{cases} \tag{5.52}$$

where are $Z_{\alpha\beta\_1}$ and $Z_{\alpha\beta\_2}$ are impedances of the primary and secondary circuits, respectively. They are given by

$$\begin{cases} Z_{\alpha\beta\_1} = (\mathrm{j}\omega)^{\beta_1} L_{\beta 1} + \dfrac{1}{(j\omega)^{\alpha_1} C_{\alpha 1}} + r_1 \\ Z_{\alpha\beta\_2} = (j\omega)^{\beta_2} L_{\beta 2} + \dfrac{1}{(j\omega)^{\alpha_2} C_{\alpha 2}} + r_2 + R_L \end{cases} \tag{5.53}$$

When the primary and secondary circuits resonate, there are

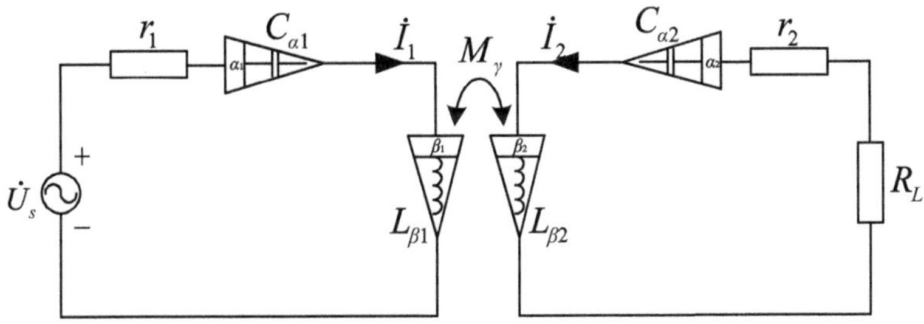

Fig. 5.10 Fractional-order mutual coupling circuit

$$\omega = \omega_{r1} = \omega_{r2} \tag{5.54}$$

where ω is the operating angular frequency of the voltage source $\dot{U}_s$. ω_{r1} and ω_{r2} are natural resonance frequencies of the primary and secondary circuits, respectively. They are defined as

$$\begin{cases} \omega_{r1} = \left[\dfrac{\sin\left(\frac{\pi\alpha_1}{2}\right)}{L_{\beta 1} C_{\alpha 1} \sin\left(\frac{\pi\beta_1}{2}\right)} \right]^{\frac{1}{\alpha_1+\beta_1}} \\ \omega_{r2} = \left[\dfrac{\sin\left(\frac{\pi\alpha_2}{2}\right)}{L_{\beta 2} C_{\alpha 2} \sin\left(\frac{\pi\beta_2}{2}\right)} \right]^{\frac{1}{\alpha_2+\beta_2}} \end{cases} \tag{5.55}$$

In order to simplify the analysis, assuming that the mutual inductance is integer-order, i.e., $\gamma = 1$ and $M_\gamma = M$, the currents of primary and secondary circuits can be obtained as

$$\begin{cases} \dot{I}_1 = \dfrac{\left[r_2 + R_L + \frac{1}{\omega^{\alpha_2} C_{\alpha 2}} \cos\left(\frac{\pi\alpha_2}{2}\right) + \omega^{\beta_2} L_{\beta 2} \cos\left(\frac{\pi\beta_2}{2}\right)\right] \dot{U}_s}{\left\{ \begin{array}{l} \left[r_1 + \dfrac{1}{\omega^{\alpha_1} C_{\alpha 1}} \cos\left(\dfrac{\pi\alpha_1}{2}\right) + \omega^{\beta_1} L_{\beta 1} \cos\left(\dfrac{\pi\beta_1}{2}\right)\right] \cdot \\ \left[r_2 + R_L + \dfrac{1}{\omega^{\alpha_2} C_{\alpha 2}} \cos\left(\dfrac{\pi\alpha_2}{2}\right) + \omega^{\beta_2} L_{\beta 2} \cos\left(\dfrac{\pi\beta_2}{2}\right)\right] + \omega^2 M^2 \end{array} \right\}} \\ \dot{I}_2 = \dfrac{-j\omega M \dot{U}_s}{\left\{ \begin{array}{l} \left[r_1 + \dfrac{1}{\omega^{\alpha_1} C_{\alpha 1}} \cos\left(\dfrac{\pi\alpha_1}{2}\right) + \omega^{\beta_1} L_{\beta 1} \cos\left(\dfrac{\pi\beta_1}{2}\right)\right] \cdot \\ \left[r_2 + R_L + \dfrac{1}{\omega^{\alpha_2} C_{\alpha 2}} \cos\left(\dfrac{\pi\alpha_2}{2}\right) + \omega^{\beta_2} L_{\beta 2} \cos\left(\dfrac{\pi\beta_2}{2}\right)\right] + \omega^2 M^2 \end{array} \right\}} \end{cases} \tag{5.56}$$

Then, the power absorbed by the load is

$$\begin{aligned} P_L &= \left|\dot{I}_2\right|^2 R_L \\ &= \dfrac{\omega^2 M^2 U_s^2 R_L}{\left\{ \begin{array}{l} \left[r_1 + \dfrac{1}{\omega^{\alpha_1} C_{\alpha 1}} \cos\left(\dfrac{\pi\alpha_1}{2}\right) + \omega^{\beta_1} L_{\beta 1} \cos\left(\dfrac{\pi\beta_1}{2}\right)\right] \cdot \\ \left[r_2 + R_L + \dfrac{1}{\omega^{\alpha_2} C_{\alpha 2}} \cos\left(\dfrac{\pi\alpha_2}{2}\right) + \omega^{\beta_2} L_{\beta 2} \cos\left(\dfrac{\pi\beta_2}{2}\right)\right] + \omega^2 M^2 \end{array} \right\}^2} \end{aligned} \tag{5.57}$$

where U_s is the effective value of the voltage source $\dot{U}_s$.

5.5 Fractional-Order Filter Circuits

In engineering, it is usually necessary to design a specific circuit between the input and output ports to select the signal of the required frequency and suppress the noise of the undesired frequency band. This circuit is called a filter. The frequency band that passes through the filter without attenuation is called the passband, and the frequency band that is attenuated by the filter is called the stopband. According to the frequency range of the passband and stopband, the filters can be divided into four types: low-pass, high-pass, band-pass, and band-stop [22].

In order to improve the traditional filter circuit, the fractional-order capacitor and the fractional-order inductor are introduced to construct a fractional-order $RL_\beta C_\alpha$ filter. The corresponding four filter circuits are shown in Fig. 5.11 [23–25].

As shown in Fig. 5.11a, the transmission function of the fractional-order low-pass filter is expressed as

$$H(j\omega) = \frac{\dot{U}_0}{\dot{U}_{in}} = \frac{\frac{1}{(j\omega)^\alpha C_\alpha}}{R + \frac{1}{(j\omega)^\alpha C_\alpha} + (j\omega)^\beta L_\beta} = \frac{\frac{1}{L_\beta C_\alpha}}{(j\omega)^{\alpha+\beta} + (j\omega)^\alpha \left(\frac{R}{L_\beta}\right) + \frac{1}{L_\beta C_\alpha}} \tag{5.58}$$

Figure 5.12 shows the bode diagram of the fractional-order low-pass filter, where $\beta = 1, R = 1\,\Omega, L_\beta = L_1 = 1\,\text{H}, C_\alpha = 1\text{F/s}^{1-\alpha}$. It can be found that only low-frequency signals can pass through the filter without attenuation [26].

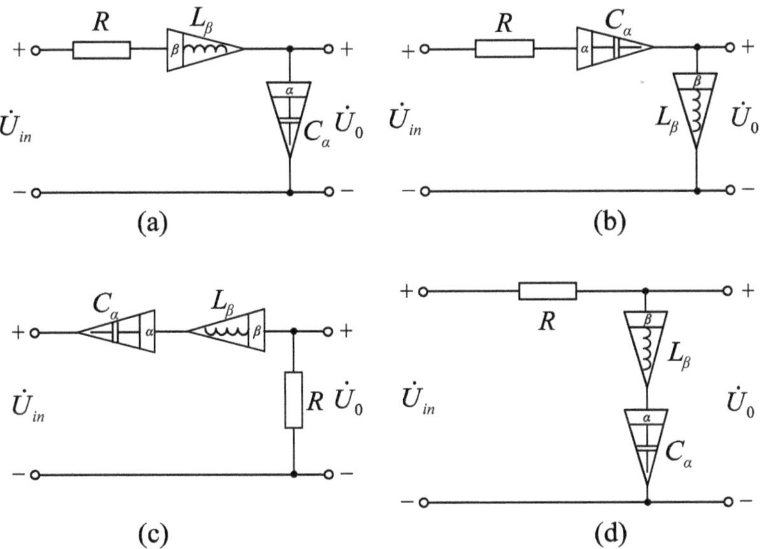

Fig. 5.11 Fractional-order filter circuits: **a** Fractional-order low-pass filter. **b** Fractional-order high-pass filter. **c** Fractional-order band-pass filter. **d** Fractional-order band-reject filter

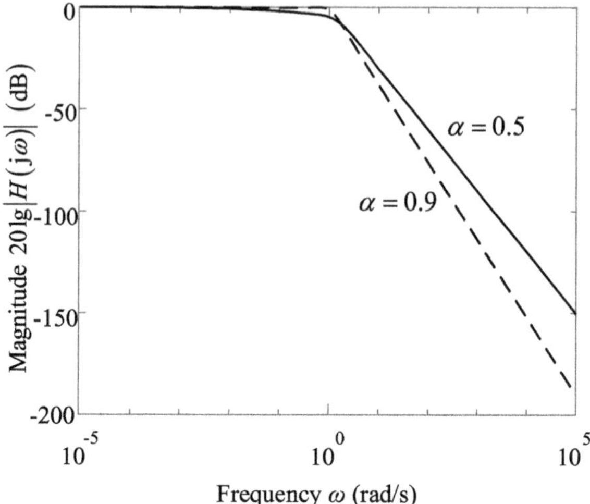

Fig. 5.12 Bode diagram of fractional-order low-pass filter

For the fractional-order high-pass filter shown in Fig. 5.11b, the transfer function is

$$H(j\omega) = \frac{\dot{U}_0}{\dot{U}_{in}} = \frac{(j\omega)^\beta L_\beta}{R + \frac{1}{(j\omega)^\alpha C_\alpha} + (j\omega)^\beta L_\beta} = \frac{(j\omega)^{\alpha+\beta}}{(j\omega)^{\alpha+\beta} + (j\omega)^\alpha \left(\frac{R}{L_\beta}\right) + \frac{1}{L_\beta C_\alpha}} \tag{5.59}$$

The Bode diagram of the fractional-order high-pass filter with different orders α is drawn in Fig. 5.13, where $\beta = 1$, $L_\beta = L_1 = 1\text{H}$, $R = 1\Omega$, and $C_\alpha = 1\text{F}/s^{1-\alpha}$.

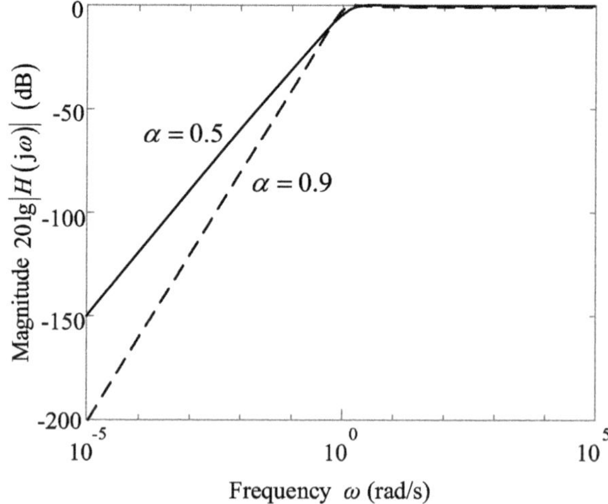

Fig. 5.13 Bode diagram of fractional-order high-pass filter

It can be seen that only high-frequency signals can pass through the filter without attenuation.

For the fractional-order band-pass filter shown in Fig. 5.11c, the transfer function is

$$H(j\omega) = \frac{\dot{U}_0}{\dot{U}_{in}} = \frac{R}{R + \frac{1}{(j\omega)^{\alpha} C_{\alpha}} + (j\omega)^{\beta} L_{\beta}} = \frac{(j\omega)^{\alpha}\left(\frac{R}{L_{\beta}}\right)}{(j\omega)^{\alpha+\beta} + (j\omega)^{\alpha}\left(\frac{R}{L_{\beta}}\right) + \frac{1}{L_{\beta} C_{\alpha}}} \tag{5.60}$$

Figure 5.14 depicts the bode diagram of the fractional-order band-pass filter, where $\beta = 1$, $L_{\beta} = L_1 = 1$ H, $R = 1\ \Omega$, and $C_{\alpha} = 1\text{F}/s^{1-\alpha}$. It can be known that only signals within a certain frequency band can pass through the filter.

Figure 5.11d is the fractional band-stop filter, its transfer function is

$$H(j\omega) = \frac{\dot{U}_0}{\dot{U}_{in}} = \frac{\frac{1}{(j\omega)^{\alpha} C_{\alpha}} + (j\omega)^{\beta} L_{\beta}}{R + \frac{1}{(j\omega)^{\alpha} C_{\alpha}} + (j\omega)^{\beta} L_{\beta}} = \frac{(j\omega)^{\alpha+\beta} + \frac{1}{L_{\beta} C_{\alpha}}}{(j\omega)^{\alpha+\beta} + (j\omega)^{\alpha}\left(\frac{R}{L_{\beta}}\right) + \frac{1}{L_{\beta} C_{\alpha}}} \tag{5.61}$$

From (5.61), the bode diagram of the fractional band-stop filter can be drawn in Fig. 5.15, where the circuit parameters are $\beta = 1$, $L_{\beta} = L_1 = 1$ H, $R = 1\ \Omega$, and $C_{\alpha} = 1\text{F}/s^{1-\alpha}$. It can be seen that all signals except for a certain frequency band can pass the filter without attenuation well.

From the above analysis, it is clear that more flexibility in shaping the filter response can be obtained via a fractional-order filter [27, 28].

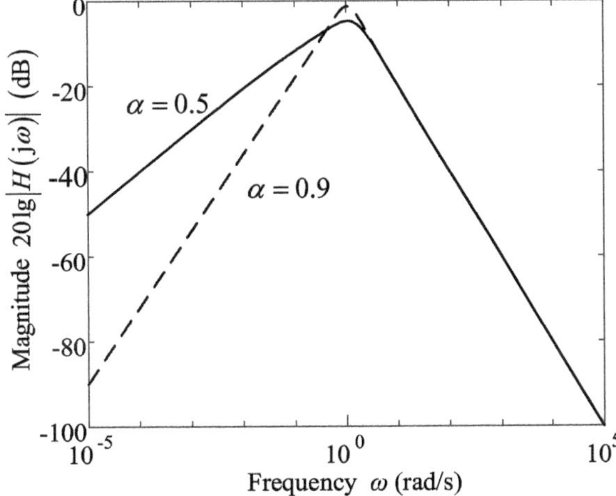

Fig. 5.14 Bode diagram of fractional-order band-pass filter

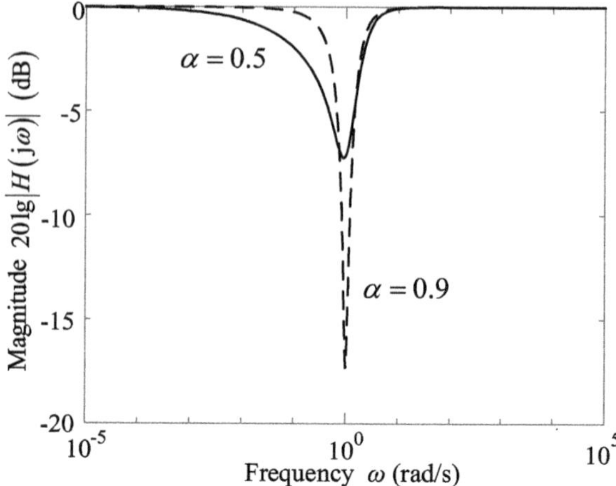

Fig. 5.15 Bode diagram of fractional-order band-stop filter

5.6 Summary

This chapter mainly introduces the sinusoidal steady-state analysis of fractional-order circuits. Based on the definition of the phasor method, the phasor representations of the fractional-order capacitor, fractional-order inductor, and fractional-order mutual inductor are given first, which lay the foundation for the analysis of fractional-order circuits. Then, on this basis, the phasor is introduced into the sinusoidal steady-state analysis of the basic fractional-order circuit, and a special state of the fractional-order circuit called resonance is presented. Moreover, a fractional-order mutual inductance coupling circuit is proposed, and its sinusoidal steady-state response under resonance is elaborated. Finally, several fractional-order filters composed of two fractional-order components of different orders α and β are illustrated, including fractional-order low-pass, high-pass, band-pass, and band-stop filters.

References

1. Sugi M, Hirano Y, Miura YF (1999) Simulation of fractal immittance by analog circuits: an approach to the optimized circuits. IEICE Trans Fundam Electron Commun Comput Sci 82(8):1627–1635
2. Elwakil AS (2010) Fractional-order circuits and systems: an emerging interdisciplinary research area. IEEE Circuits Syst Mag 10(4):40–50
3. Jesus IS, Tenreiro Machado JA, Boaventure Cunha J (2008) Fractional electrical impedances in botanical elements. J Vib Control 14(9–10):1389–1402.
4. Bohannan GW, Hurst SK, Spangler L (2006) Electrical component with fractional order impedance. US Patent 20060267595
5. Hartley TT, Trigeassou JC, Lorenzo CF, Maamri N (2015) Energy storage and loss in fractional-order systems. Circ Dev Syst 10(6):227–235
6. Fouda ME, Elwakil AS, Radwan AG, Allagui A (2016) Power and energy analysis of fractional-order electrical energy storage devices. Energy 111(15):785–792

7. Soltan A, Radwan AG, Soliman AM (2016) Fractional-order mutual inductance: analysis and design. Int J Circuit Theory Appl 44(1):85–97
8. Sarathi VP, Uma G, Umapathy M (2018) Realization of fractional order inductive transducer. IEEE Sens J 18(21):8803–8811
9. Radwan AG, Salama KN (2011) Passive and active elements using fractional LβCα circuit. IEEE Trans Circuits Syst I Regul Pap 58(10):2388–2397
10. Abbisso S, Caponetto R, Diamante O, Fortuna L, Porto D (2001) Non-integer order integration by using neural networks. In: ISCAS 2001. The 2001 IEEE International symposium on circuits and systems (Cat. No. 01CH37196), vol. 2. Sydney, NSW, Australia, pp 688–691
11. Biswas K, Sen S, Dutta PK (2006) Realization of a constant phase element and its performance study in a differentiator circuit. IEEE Trans Circuits Syst II Express Briefs 53(9):802–806
12. Radwan AG (2012) Stability analysis of the fractional-order RLβCα circuit. J Fract Calc Appl 3(1):1–15
13. Yang N, Xu C, Wu C, Jia R, Liu C (2019) Fractional-order cubic nonlinear flux-controlled memristor: theoretical analysis, numerical calculation and circuit simulation. Nonlinear Dyn 97(1):33–44
14. Radwan AG, Elwakil AS, Soliman AM (2008) Fractional-order sinusoidal oscillators: design procedure and practical examples. IEEE Trans Circuits Syst I Regul Pap 55(7):2051–2063
15. Maundy B, Elwakil A, Gift S (2010) On a multivibrator that employs a fractional-order capacitor. Analog Integr Circ Sig Process 62(1):99–103
16. Radwan AG, Salama KN (2012) Fractional-order RC and RL circuits. Circ Syst Signal Process 31(6):1901–1915
17. Radwan AG, Soliman AM, Elwakil AS (2007) Design equations for fractional-order sinusoidal oscillators: practical circuit examples. In: 2007 International conference on microelectronics. Cairo, Egypt, pp 89–92
18. Radwan AG, Soliman AM, Elwakil AS (2008) Design equations for fractional-order sinusoidal oscillators: Four practical circuit examples. Int J Circuit Theory Appl 36(4):473–492
19. Fouda ME, Radwan AG (2013) On the fractional-order memristor model. J Fract Calc Appl 4(1):1–7
20. Walczak J, Jakubowska A (2017) Analysis of the transient state in a parallel circuit of the class RLβCα. Appl Math Comput 35(6):1831–1853
21. Radwan AG (2013) Resonance and quality factor of the RLαCβ fractional circuit. IEEE J Emerg Sel Top Circ Syst 3(3):377–385
22. Radwan AG, Elwakil AS, Soliman AM (2009) On the generalization of second-order filters to the fractional-order domain. J Circ Syst Comput 18(2):361–386
23. Radwan AG, Fouda ME (2013) Optimization of fractional-order RLC filters. Circ Syst Signal Process 32(5):2097–2118
24. Freeborn TJ, Maundy B, Elwakil A (2013) Fractional resonance-based RLβCα filters. Math Probl Eng 2013:1–10
25. Soltan A, Radwan AG, Soliman AM (2012) Fractional order filter with two fractional elements of dependant orders. Microelectron J 43(11):818–827
26. Jurišić D, Emanović E, Lutovac B, Moschytz GS (2018) Noise analysis of fractional-order two-integrator CCII low-pass filter using Pspice. In: 2018 7th Mediterranean conference on embedded computing (MECO). Budva, pp 1–5
27. Ates A, Kavuran G, Alagoz BB, Yeroglu C (2016) Improvement of IIR filter discretization for fractional order filter by discrete stochastic optimization. In: 2016 39th International conference on telecommunications and signal processing (TSP). Vienna, Austria, pp 583–586
28. Kwan HK, Jiang A (2009) FIR, Allpass, and IIR variable fractional delay digital filter design. IEEE Trans Circuits Syst I Regul Pap 56(9):2064–2074

Chapter 6
Sinusoidal Steady-State Analysis of Three-Phase Fractional-Order Circuits

At present, the three-phase power system has been widely used in the power system of all countries in the world, and it is the most extensive production system. The three-phase power system is the basis for the production, transmission, and utilization of electric energy. And the structure of the system has been highly standardized to meet the needs of industrialized production. A complete three-phase power system is mainly composed of a three-phase power source, three-phase loads, and power transmission circuits. The three-phase power system is widely applied in the power system due to the following reasons [1–3]:

1. Under the same conditions, using the three-phase system to transfer the same power, the cross-sectional area of the three-phase transmission line is only half of the single-phase transmission line. The copper and steel materials used in the three-phase transmission line are only 3/4 of the single-phase transmission line, which saves 25% of the copper and steel materials.
2. In the same capacity, the electrical equipment of a three-phase system is cheaper than a single-phase system, such as motors and transformers.
3. When the three-phase system already has a group of three-phase loads, it can be connected into a three-phase four-wire type. This connection method can easily connect single-phase loads and makes the three-phase system as flexible as the single-phase system.

In a traditional three-phase system, the basic components that make up the circuit are integer-order components. However, many studies have shown that a fractional-order system has completely different characteristics and advantages compared to an integer-order system. Therefore, the analysis in this chapter is based on a sinusoidal steady-state three-phase circuit with fractional-order components, which means that all load impedances in the circuit contain fractional-order components. Here, this unique three-phase circuit is called the three-phase fractional-order circuit. When analyzing the three-phase fractional-order circuits, the characteristics of conventional three-phase circuits and the inherent properties of the fractional-order component can simplify the problem.

B. Zhang and X. Shu, *Fractional-Order Electrical Circuit Theory*, CPSS Power Electronics Series, https://doi.org/10.1007/978-981-16-2822-1_6

According to the connection type of the power source and the fractional-order load, the three-phase fractional-order circuits can be divided into a symmetrical three-phase fractional-order circuit and an asymmetrical three-phase fractional-order circuit. In a symmetrical three-phase fractional-order circuit, the neutral-point voltage of the power source and the load are both zero. Therefore, a wire can be used to connect the neutral point between the power source and the load in a symmetrical circuit. Due to the symmetry, a three-phase fractional-order circuit can be transformed into a single-phase fractional-order circuit. As for an asymmetric three-phase fractional-order circuit, it is essentially a general sinusoidal steady-state circuit. This means that conventional methods can be used to complete the analysis of asymmetric circuits. Of course, we can also use bitmaps and vector diagrams to analyze asymmetric three-phase fractional-order circuits, which can sometimes achieve a multiplier effort.

6.1 Preliminaries

This section introduces some basic concepts of sinusoidal steady-state three-phase fractional-order circuits. It mainly includes the symmetrical three-phase power source, the connection type of a three-phase circuit, and the phasor of circuit and line quantity. Learning this basic knowledge will help us analyze three-phase fractional-order circuits in later chapters.

6.1.1 Connection Types

The symmetrical three-phase power source is composed of three sinusoidal voltage sources with the same amplitude and angular frequency, and their initial phases lag 120° in turn. These three power sources are called phase *A*, phase *B*, and phase *C*, respectively. The expressions of their instantaneous voltages can be described as

$$\begin{cases} u_A(t) = U_m \sin(\omega t + \phi) = \sqrt{2}U \sin(\omega t + \phi) \\ u_B(t) = U_m \sin(\omega t + \phi - 120^\circ) = \sqrt{2}U \sin(\omega t + \phi - 120^\circ) \\ u_C(t) = U_m \sin(\omega t + \phi + 120^\circ) = \sqrt{2}U \sin(\omega t + \phi + 120^\circ) \end{cases} \tag{6.1}$$

and their phasors can be expressed as

$$\begin{cases} \dot{U}_A = \sqrt{2}U\angle 0^\circ \\ \dot{U}_B = \sqrt{2}U\angle - 120^\circ \\ \dot{U}_C = \sqrt{2}U\angle 120^\circ \end{cases} \tag{6.2}$$

where the voltage of phase A is $u_A(t)$, it is usually used as the sinusoidal reference phasor. U_m and U are the amplitude and effective value of the sinusoidal voltage, respectively.

In actual production, the generator can provide a set of symmetrical three-phase voltages in general. The power frequency of the three-phase system in China is $f = 50$ Hz, and the effective value of the household voltage is 220 V, while in Europe and America, they are 60 Hz and 110 V, respectively [4]. The waveform and phasor diagram of each phase are displayed in Fig. 6.1a, b.

Generally, the symmetrical three-phase voltage should satisfy

$$u_A(t) + u_B(t) + u_C(t) = 0 \tag{6.3}$$

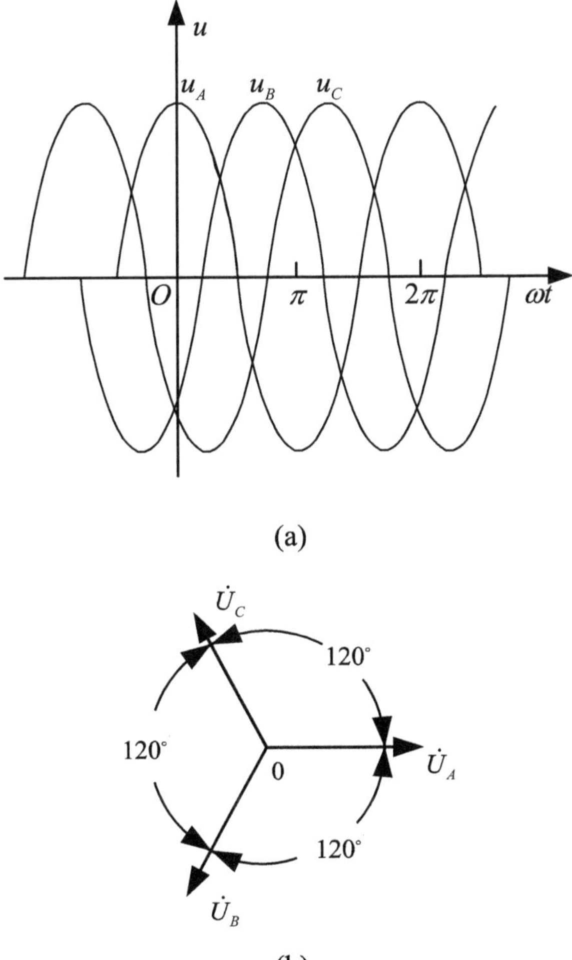

Fig. 6.1 Voltage waveforms and phasor diagram of the symmetrical three-phase voltage: **a** waveform diagram, **b** phasor diagram

The phase sequence A, B, and C of those above-mentioned sinusoidal steady-state three-phase voltage is called the positive sequence. On the contrary, if the initial phases of A, B and C lead by 120° in turn, this phase sequence is called the negative-sequence or the reverse-sequence. If the initial phase difference of A, B and C is zero, the phase sequence is zero-sequence. Since the positive sequence has been widely used in power systems, unless otherwise specified in this book, it refers to the positive sequence.

In order to conveniently describe the three-phase circuits in engineering, the symbol "a" is usually used to represent the unit phasor operator. It is a unit vector with a length of 1 and an argument of 120°. If the unit vector is multiplied by any vector, the vector needs to be rotated 120° counterclockwise, but the length remains unchanged.

The unit phasor operator can be described as

$$\begin{cases} a = -\frac{1}{2} + j\frac{\sqrt{3}}{2} = 1\angle - 240^\circ \\ a^2 = \left(-\frac{1}{2} + j\frac{\sqrt{3}}{2}\right)^2 = 1\angle - 120^\circ \\ a^3 = \left(-\frac{1}{2} + j\frac{\sqrt{3}}{2}\right)^3 = 1\angle 360^\circ = 1 \\ a + a^2 + a^3 = 1 + a + a^2 = 0 \end{cases} \tag{6.4}$$

Therefore, we can use the phasor operator to describe the sinusoidal three-phase voltage again. It can be expressed by

$$\begin{cases} \dot{U}_A = \sqrt{2}U\angle 0^\circ \\ \dot{U}_B = a^2\dot{U}_A \\ \dot{U}_C = a\dot{U}_A \end{cases} \tag{6.5}$$

In a sinusoidal steady-state three-phase fractional-order circuit, the power source has two connection types: star (Y) and delta (Δ).

The Y connection method connects the back ends (or head ends) of the three power supplies to form a neutral-point N, and the other three ends are led out and connected to three transmission lines. Figure 6.2 shows the Y-connection mode of a three-phase power source, which is usually called the Y power source. It can be seen that the terminal line is drawn from the positive terminals A, B and C of the three-phase power source, and the neutral line is drawn from the neutral point N [5].

Δ connection method connects the first and last of the three power supplies in series to form a closed triangle (closed-loop). The three vertices of the triangle are led out and connected to three power transmission lines. Figure 6.3 shows the Δ connection of the three-phase power source, which is usually called a Δ power source. It can be seen from Fig. 6.3 that the Δ power source cannot lead out a neutral line. Since

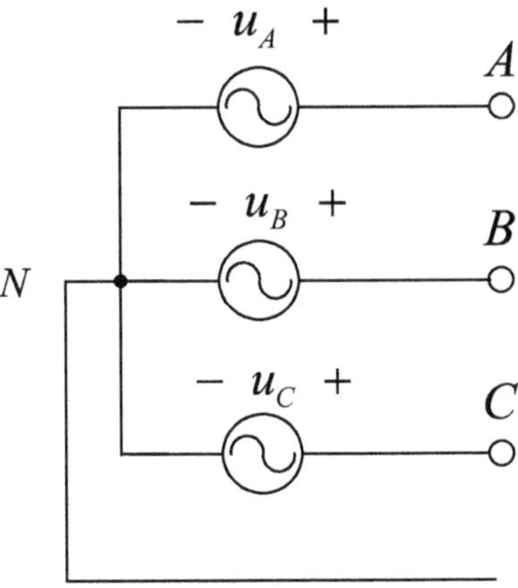

Fig. 6.2 Y connection of a three-phase power source

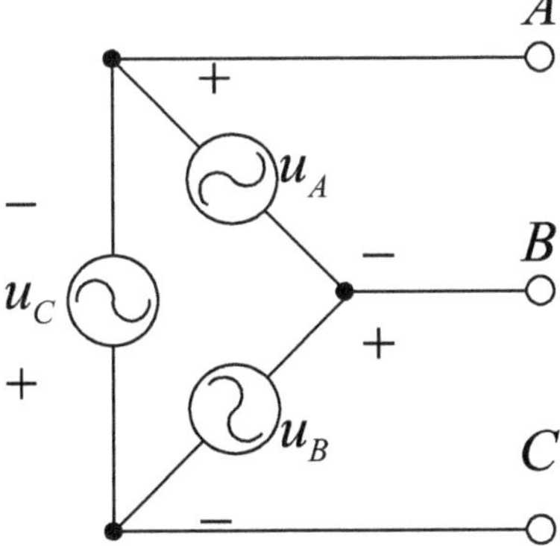

Fig. 6.3 $\triangle$ connection of a three-phase power source

the sinusoidal three-phase power source generally adopts the Y connection method in engineering, the connection of a three-phase power source is the Y connection unless otherwise specified in this book.

Like a three-phase power source, three fractional-order loads with the same impedance or admittance can form a group of symmetrical three-phase fractional-order loads. As shown in Fig. 6.4a, b, three fractional-order loads can be connected in Y or $\triangle$ type. For the Y connection, the fractional-order loads have a neutral-point N'. A wire can be connected to the neutral-point N of power source to form a three-phase four-wire system. This wire is usually called a neutral line in engineering. And this kind of three-phase four-wire connection is called a Y_0 connection [6].

A three-phase fractional-order circuit is composed of the three-phase power source and load. Therefore, the three-phase fractional-order circuit has five connection modes according to the different connections of the power source and load, which are Y/Y, $Y/\triangle$, $\triangle/Y$, $\triangle/\triangle$, and Y_0/Y_0. The Y_0/Y_0 connection means that a neutral line

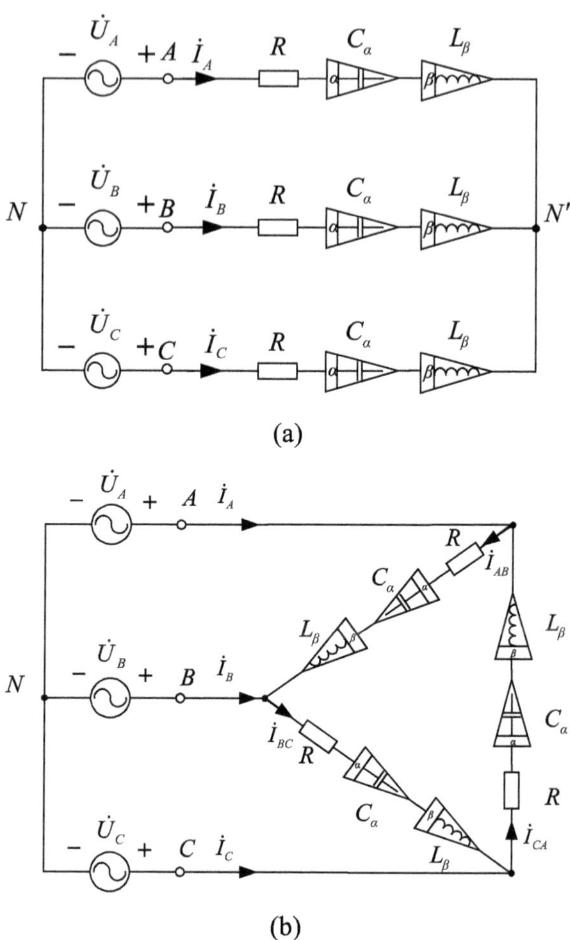

Fig. 6.4 Connection modes of fractional-order three-phase loads: **a** Y connection. **b** $\triangle$ connection

connects the neutral point between the power source and load. It is called the three-phase four-wire system, and the other connection methods belong to three-phase three-wire systems.

6.1.2 Relationship Between Phase and Line Voltages

The power source and fractional-order load of a symmetrical three-phase fractional-order circuit are composed of three component groups. Therefore, no matter the power source and the load are connected in Y or $\triangle$ type, the voltage and current on them are collectively referred to as phase voltage and phase current. There are three transmission lines connecting the power source and the fractional-order load in a symmetrical three-phase fractional-order circuit. The current and voltage on the transmission line are collectively referred to as line voltage and line current.

If there is impedance on the transmission line, the voltage drop caused by the line current on the transmission line is called impedance drop. For a three-phase four-wire system, if there is a current flowing through the neutral wire, it is usually called the neutral current. And the voltage between the neutral points of the load and the power supply is called the neutral-point voltage.

In a three-phase system, the relationship between line quantity and phase quantity is related to the connection modes. For a symmetrical three-phase Y power source, assuming that the line voltages are $\dot{U}_{AB}$, $\dot{U}_{BC}$ and $\dot{U}_{CA}$, respectively. Similarly, the phase voltages can be set to $\dot{U}_A$, $\dot{U}_B$ and $\dot{U}_C$ (or $\dot{U}_{AN}$, $\dot{U}_{BN}$ and $\dot{U}_{CN}$). As shown in Fig. 6.2, it can be obtained as

$$\begin{cases} \dot{U}_{AB} = \dot{U}_A - \dot{U}_B = (1 - a^2)\dot{U}_A = \sqrt{3}\dot{U}_A \angle 30^\circ \\ \dot{U}_{BC} = \dot{U}_B - \dot{U}_C = (1 - a^2)\dot{U}_B = \sqrt{3}\dot{U}_B \angle 30^\circ \\ \dot{U}_{CA} = \dot{U}_C - \dot{U}_A = (1 - a^2)\dot{U}_C = \sqrt{3}\dot{U}_C \angle 30^\circ \end{cases} \tag{6.6}$$

From (6.6), it can be concluded that $\dot{U}_{AB} + \dot{U}_{BC} + \dot{U}_{CA} = 0$. Hence, there are only two equations that are independent in (6.6). A particular voltage phasor diagram can represent the relationship between the line voltages and the phase voltages of the symmetrical three-phase Y power source. As shown in Fig. 6.5, the phasors represented by the above three formulas in (6.6) are described, from which we can find that $\dot{U}_{AB} = \dot{U}_{AN} + (-\dot{U}_{BN})$. The voltage triangle can describe the relationship between the line voltage and phase voltage of a symmetrical three-phase Y connection. It can be clearly seen that when the phase voltages are symmetrical, the line voltages must also be symmetrical. Moreover, according to the nature of the isosceles triangle, it can be known that the amplitude of line voltage is $\sqrt{3}$ times of the phase voltage, and its phase leads the corresponding phase voltage by 30°, which can be described

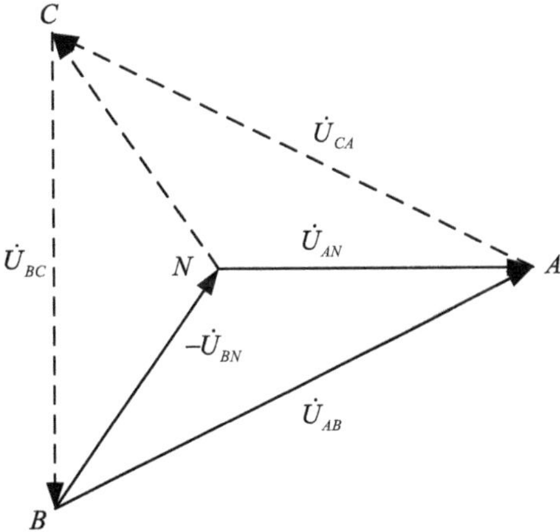

Fig. 6.5 Phasor diagram between line voltages and phase voltages

by

$$\begin{cases} \dot{U}_{AB} = \sqrt{3}\dot{U}_A \angle 30^\circ \\ \dot{U}_{BC} = \sqrt{3}\dot{U}_B \angle 30^\circ \\ \dot{U}_{CA} = \sqrt{3}\dot{U}_C \angle 30^\circ \end{cases} \tag{6.7}$$

In the actual calculation, $\dot{U}_{BC}$ and $\dot{U}_{CA}$ can be directly written in turn when the $\dot{U}_{AB}$ is known.

For a symmetrical three-phase $\triangle$ power source, as shown in Fig. 6.3. It is evident that the line voltage is equal to the phase voltage, which can be expressed as

$$\begin{cases} \dot{U}_{AB} = \dot{U}_A \\ \dot{U}_{BC} = \dot{U}_B \\ \dot{U}_{CA} = \dot{U}_C \end{cases} \tag{6.8}$$

The above relationship between line voltage and phase voltage also applies to symmetrical Y or $\triangle$ loads. When the line voltages of the symmetrical three-phase system are set as $\dot{U}_{AB}$,$\dot{U}_{BC}$ and $\dot{U}_{CA}$, and the phase voltages are $\dot{U}_A$, $\dot{U}_B$ and $\dot{U}_C$ (or $\dot{U}_{AN}$, $\dot{U}_{BN}$ and $\dot{U}_{CN}$), respectively, the relationship between line voltages and phase voltages in a symmetrical sinusoidal steady-state three-phase fractional-order system can be described as (6.9) and (6.10).

For the Y connection, it can be derived as

$$\begin{cases} \dot{U}_{AB} = \sqrt{3}\dot{U}_A \angle 30^\circ \\ \dot{U}_{BC} = \sqrt{3}\dot{U}_B \angle 30^\circ \\ \dot{U}_{CA} = \sqrt{3}\dot{U}_C \angle 30^\circ \end{cases} \tag{6.9}$$

For the $\triangle$ connection, it can be written as

$$\begin{cases} \dot{U}_{AB} = \dot{U}_A \\ \dot{U}_{BC} = \dot{U}_B \\ \dot{U}_{CA} = \dot{U}_C \end{cases} \tag{6.10}$$

6.1.3 Relationship Between Phase and Line Currents

The relationship between line current and phase current in a symmetrical three-phase fractional-order circuit is also related to the connection modes of fractional-order circuits. For the symmetrical three-phase Y-connected power source, the line current is equal to the phase current. As can be seen from Fig. 6.4b, the symmetrical three-phase power source is Y-connected. Here, the symmetrical phase currents can be assumed as $\dot{I}_{AB}$, $\dot{I}_{BC}(=a^2\dot{I}_{AB})$ and $\dot{I}_{CA}(=a\dot{I}_{AB})$, and the line currents are $\dot{I}_A$, $\dot{I}_B$ and $\dot{I}_C$, respectively. According to KCL, the phase currents and line currents can be derived as

$$\begin{cases} \dot{I}_A = \dot{I}_{AB} - \dot{I}_{CA} = \left(1-a^2\right)\dot{I}_{AB} = \sqrt{3}\dot{I}_{AB}\angle -30^\circ \\ \dot{I}_B = \dot{I}_{BC} - \dot{I}_{AB} = \left(1-a^2\right)\dot{I}_{BC} = \sqrt{3}\dot{I}_{BC}\angle -30^\circ \\ \dot{I}_C = \dot{I}_{CA} - \dot{I}_{BC} = \left(1-a^2\right)\dot{I}_{CA} = \sqrt{3}\dot{I}_{CA}\angle -30^\circ \end{cases} \tag{6.11}$$

From (6.11), it can be seen that $\dot{I}_A + \dot{I}_B + \dot{I}_C = 0$. Therefore, only two equations are independent in (6.11).

The relationship between the line current and the phase current can also be described by the phasor diagram shown in Fig. 6.6, in which the current triangle can illustrate the relationship between the line current and the phase currents of the Δ-connection load. When the phase currents are symmetrical, the line currents must also be symmetrical. The amplitude of the line current is $\sqrt{3}$ times of the phase current, and the phase lags the corresponding phase current by 30°. In actual calculations, as long as $\dot{I}_A$ has been calculated, the other two currents $\dot{I}_B = a^2\dot{I}_A$ and $\dot{I}_C = a\dot{I}_A$ can be directly written out.

The above relationship between line currents and phase currents also applies to symmetrical Y or Δ-connected three-phase power source. When the line currents of the symmetrical three-phase system are assumed as $\dot{I}_A$, $\dot{I}_B$ and $\dot{I}_C$, and the phase currents are $\dot{I}_{AB}$, $\dot{I}_{BC}$ and $\dot{I}_{CA}$, the relationship between line currents and phase currents in a symmetrical three-phase fractional-order system can be described as (6.12) and (6.13).

For the Y connection, it can be expressed as

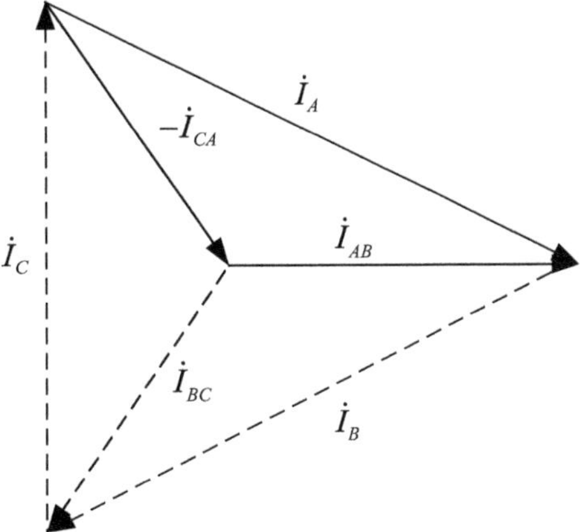

Fig. 6.6 Phasor diagram of circuit with Δ-connection loads

$$\begin{cases} \dot{I}_A = \dot{I}_{AB} \\ \dot{I}_B = \dot{I}_{BC} \\ \dot{I}_C = \dot{I}_{CA} \end{cases} \tag{6.12}$$

For the $\triangle$ connection, it can be written as

$$\begin{cases} \dot{I}_A = \dot{I}_{AB} - \dot{I}_{CA} = (1 - a^2)\dot{I}_{AB} = \sqrt{3}\dot{I}_{AB}\angle - 30^\circ \\ \dot{I}_B = \dot{I}_{BC} - \dot{I}_{AB} = (1 - a^2)\dot{I}_{BC} = \sqrt{3}\dot{I}_{BC}\angle - 30^\circ \\ \dot{I}_C = \dot{I}_{CA} - \dot{I}_{BC} = (1 - a^2)\dot{I}_{CA} = \sqrt{3}\dot{I}_{CA}\angle - 30^\circ \end{cases} \tag{6.13}$$

Finally, it must be noted that all discussions on the relationship between phase values should be conducted in the specified order and reference direction. In the actual analysis, it should be ensured that the relationship can be expressed in a simple and orderly manner. Generally, the reference phasor cannot be arbitrarily selected in a specific circuit. Otherwise, the expression of the problem will become disorderly [7, 8].

6.2 Three-Phase Fractional-Order Circuits

6.2.1 Symmetrical Three-Phase Fractional-Order Circuits

This section introduces the analysis of a symmetrical sinusoidal steady-state three-phase fractional-order circuit. The key is how to combine the characteristics of the symmetrical three-phase circuit with the features of fractional-order components. In this section, the loads are all fractional-order, which are represented by $Z_{\alpha\beta}$. α is the order of the fractional-order capacitor, and β is the order of the fractional-order inductor.

The symmetrical three-phase fractional-order circuit is a special sinusoidal steady-state circuit. The conventional phasor method is entirely suitable for the analysis of the balanced three-phase fractional-order circuit. And in actual analysis, the characteristics of the traditional three-phase circuit can help simplify the symmetrical three-phase fractional-order circuit.

As shown in Fig. 6.7, a symmetrical *Y*/*Y* three-phase four-wire fractional-order circuit is taken as an example for analysis. Among them, $Z_{\alpha\beta}$ represents the impedance of the fractional-order load, and $Z_{\alpha\beta\_N}$ is the fractional-order impedance on the neutral line. For N and N', they are the neutral points of the three-phase power source and fractional-order load, respectively. The impedances in the circuit are

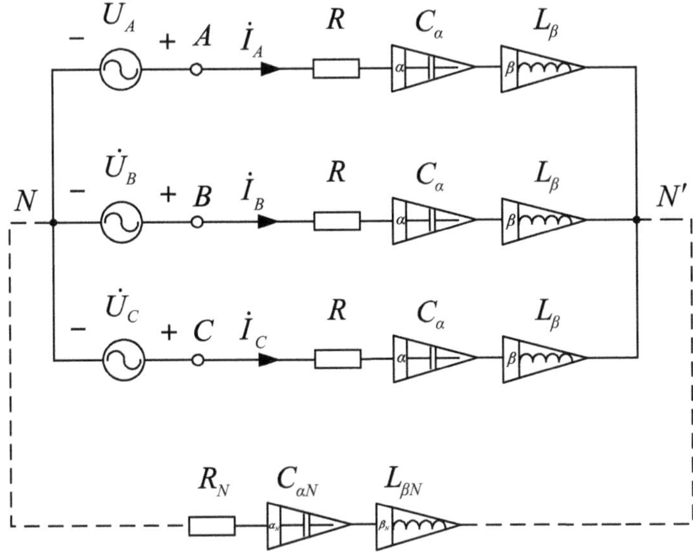

Fig. 6.7 Symmetrical three-phase four-wire fractional-order circuit

$$\begin{cases} Z_{\alpha\beta} = R + \dfrac{1}{(j\omega)^{\alpha} C_{\alpha}} + (j\omega)^{\beta} L_{\beta} \\ Z_{\alpha\beta\_N} = R + \dfrac{1}{(j\omega)^{\alpha_N} C_{\alpha N}} + (j\omega)^{\beta_N} L_{\beta N} \end{cases} \tag{6.14}$$

For the circuit shown in Fig. 6.7, taking N as the reference node and assuming that $\dot{U}_A = \sqrt{2}U\angle 0°$, the voltage between the neutral point N and N' can be derived by (6.15).

$$\left(\frac{1}{Z_{\alpha\beta\_N}} + \frac{3}{Z_{\alpha\beta}}\right)\dot{U}_{NN'} = \frac{1}{Z_{\alpha\beta}}\left(\dot{U}_A + \dot{U}_B + \dot{U}_C\right) \tag{6.15}$$

Considering that $\dot{U}_A + \dot{U}_B + \dot{U}_C = 0$, the voltage $\dot{U}_{NN'} = 0$ can be concluded from (6.15). On account of the Y/Y connection, the line current is equal to the corresponding phase current. According to circuit theory, the current of phase A can be written as

$$\begin{aligned} \dot{I}_A &= \frac{\dot{U}_A - \dot{U}_{NN'}}{Z_{\alpha\beta}} = \frac{\dot{U}_A}{Z_{\alpha\beta}} = \frac{\dot{U}_A}{R + \frac{1}{(j\omega)^{\alpha} C_{\alpha}} + (j\omega)^{\beta} L_{\beta}} \\ &= \frac{\sqrt{2}U\angle 0°}{R + \frac{1}{\omega^{\alpha} C_{\alpha}} \cos\frac{\pi\alpha}{2} + \omega^{\beta} L_{\beta} \cos\frac{\pi\beta}{2} + j\left(\omega^{\beta} L_{\beta} \sin\frac{\pi\beta}{2} - \frac{1}{\omega^{\alpha} C_{\alpha}} \sin\frac{\pi\alpha}{2}\right)} \\ &= I_m\angle - \varphi = \sqrt{2}I\angle - \varphi \end{aligned} \tag{6.16}$$

where I is the effective value of phase current $\dot{I}_A$, φ is the phase angle of $\dot{I}_A$, and $I_m = \sqrt{2}\,I$ is the amplitude of $\dot{I}_A$.

Similarly, the current of phase B and C can be derived as

$$\begin{aligned}\dot{I}_B &= \frac{\dot{U}_B - \dot{U}_{NN'}}{Z_{\alpha\beta}} = \frac{\dot{U}_B}{Z_{\alpha\beta}} = \frac{\dot{U}_B}{R + \frac{1}{(j\omega)^{\alpha} C_{\alpha}} + (j\omega)^{\beta} L_{\beta}} \\ &= \frac{\sqrt{2}U\angle -120^{\circ}}{R + \frac{1}{\omega^{\alpha} C_{\alpha}}\cos\frac{\pi\alpha}{2} + \omega^{\beta} L_{\beta}\cos\frac{\pi\beta}{2} + j\left(\omega^{\beta} L_{\beta}\sin\frac{\pi\beta}{2} - \frac{1}{\omega^{\alpha} C_{\alpha}}\sin\frac{\pi\alpha}{2}\right)} \\ &= \sqrt{2}I\angle(-\varphi - 120^{\circ})\end{aligned} \tag{6.17}$$

$$\begin{aligned}\dot{I}_C &= \frac{\dot{U}_C - \dot{U}_{NN'}}{Z_{\alpha\beta}} = \frac{\dot{U}_C}{R + \frac{1}{(j\omega)^{\alpha} C_{\alpha}} + (j\omega)^{\beta} L_{\beta}} \\ &= \frac{\sqrt{2}U\angle 120^{\circ}}{R + \frac{1}{\omega^{\alpha} C_{\alpha}}\cos\frac{\pi\alpha}{2} + \omega^{\beta} L_{\beta}\cos\frac{\pi\beta}{2} + j\left(\omega^{\beta} L_{\beta}\sin\frac{\pi\beta}{2} - \frac{1}{\omega^{\alpha} C_{\alpha}}\sin\frac{\pi\alpha}{2}\right)} \\ &= \sqrt{2}I\angle(-\varphi + 120^{\circ})\end{aligned} \tag{6.18}$$

where the effective value I and phase φ of each current can be expressed as

$$\begin{cases} I = \dfrac{U}{\sqrt{\left(R + \frac{1}{\omega^{\alpha} C_{\alpha}}\cos\frac{\pi\alpha}{2} + \omega^{\beta} L_{\beta}\cos\frac{\pi\beta}{2}\right)^2 + \left(\omega^{\beta} L_{\beta}\sin\frac{\pi\beta}{2} - \frac{1}{\omega^{\alpha} C_{\alpha}}\sin\frac{\pi\alpha}{2}\right)^2}} \\ \varphi = \arctan\left(\dfrac{\omega^{\beta} L_{\beta}\sin\frac{\pi\beta}{2} - \frac{1}{\omega^{\alpha} C_{\alpha}}\sin\frac{\pi\alpha}{2}}{R + \frac{1}{\omega^{\alpha} C_{\alpha}}\cos\frac{\pi\alpha}{2} + \omega^{\beta} L_{\beta}\cos\frac{\pi\beta}{2}}\right) \end{cases} \tag{6.19}$$

It can be seen from (6.19) that the effective value and phase of phase currents are related to the orders of the fractional-order components in a three-phase fractional-order circuit, which is different from the traditional sinusoidal steady-state three-phase circuit. The phase current could be changed by adjusting the orders α and β of the fractional-order components without changing the circuit structure. On the other hand, the impedance characteristics of three-phase fractional-order circuits also vary with the orders of the fractional-order components [9, 10].

When $\frac{1}{\omega^{\alpha} C_{\alpha}}\cos\frac{\pi\alpha}{2} + \omega^{\beta} L_{\beta}\cos\frac{\pi\beta}{2} > 0$, the fractional-order loads absorb active power. When $\frac{1}{\omega^{\alpha} C_{\alpha}}\cos\frac{\pi\alpha}{2} + \omega^{\beta} L_{\beta}\cos\frac{\pi\beta}{2} < 0$, the fractional-order loads emit active energy.

When $\omega^{\beta} L_{\beta}\sin\frac{\pi\beta}{2} - \frac{1}{\omega^{\alpha} C_{\alpha}}\sin\frac{\pi\alpha}{2} > 0$, the sinusoidal steady-state three-phase fractional-order circuit is inductive. When $\omega^{\beta} L_{\beta}\sin\frac{\pi\beta}{2} - \frac{1}{\omega^{\alpha} C_{\alpha}}\sin\frac{\pi\alpha}{2} < 0$, the sinusoidal steady-state three-phase fractional-order circuit is capacitive.

Here, some specific examples can be used to illustrate the characteristics of a three-phase fractional-order circuit in detail.

1. When the fractional order $\beta = 1$ and α changes, Fig. 6.8 shows that the current of phase A changes with order α.

For the three-phase fractional-order circuit where the load contains an integer-order inductor and a fractional-order capacitor, the amplitude, and phase of the phase current will vary with the fractional order α. As shown in Fig. 6.8, the amplitude of the current first increases as α increases, and when the amplitude reaches its maximum value, it slowly decreases and eventually stabilizes. It can be found that when α reaches a certain value, the amplitude is not sensitive to the fractional order. The same is true for the phase of the current. As α increases, the entire circuit appears capacitive. After the maximum value appears, the phase begins to decrease. When α approaches 1, the entire circuit becomes inductive. At the same time, the phase is not sensitive to the variations of the order.

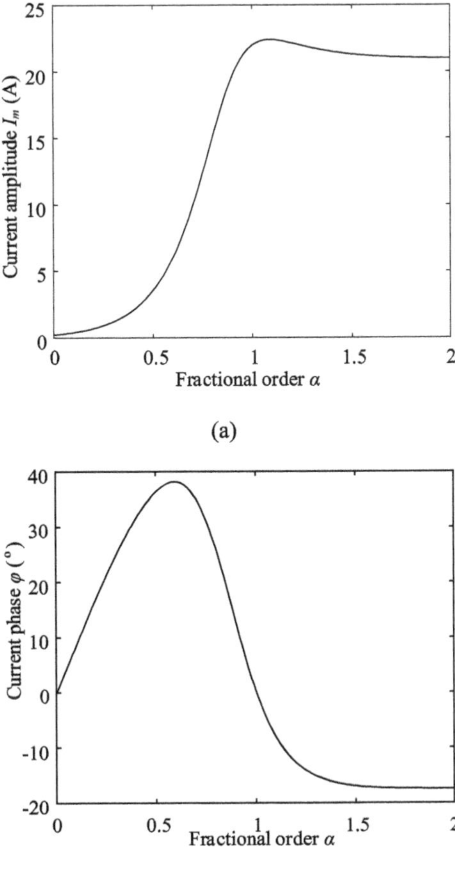

Fig. 6.8 The variation of current with fractional order α: **a** The amplitude versus fractional order α. **b** The phase versus fractional order α (Circuit parameters are $U = 220$ V, $f = 50$ Hz, $R = 10\ \Omega$, $C_\alpha = 0.001$ F/s$^{1-\alpha}$, $L_\beta = L_1 = 0.01$ H)

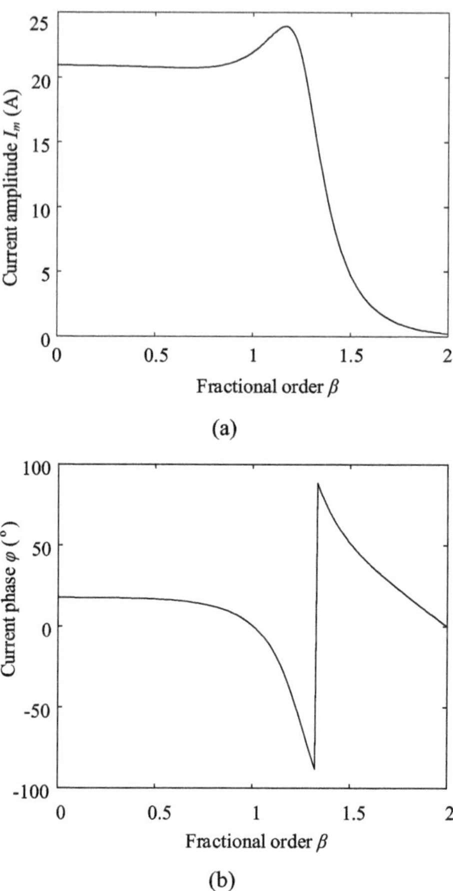

Fig. 6.9 The variation of current with fractional order β: **a** The current amplitude versus fractional order β. **b** The current phase versus fractional order β (Circuit parameters are $U = 220$ V, $f = 50$ Hz, $R = 10\ \Omega$, $C_\alpha = C_1 = 0.001$ F, $L_\beta = 0.01$ H/s$^{1-\beta}$)

2. When the fractional order $\alpha = 1$ and β changes, Fig. 6.9 shows that the current of phase A changes with the fractional order β.

For the three-phase fractional-order circuit whose load is composed of an integer-order capacitor and a fractional-order inductor, the variations of the amplitude and phase of phase current with the fractional order β is shown in Fig. 6.9. It can be found that when the fractional order β is close to 1, the amplitude and phase of the current are not sensitive to changes in β. As shown in Fig. 6.9a, the current amplitude starts to drop sharply after it reaches its maximum value. For the phase of the current, it appears a big jump when β is a particular value, and the entire circuit begins to appear inductive.

3. When the order α and β change simultaneously, Figs. 6.10 and Fig. 6.11 show the relationship between the current of phase A and the fractional orders α and β, respectively.

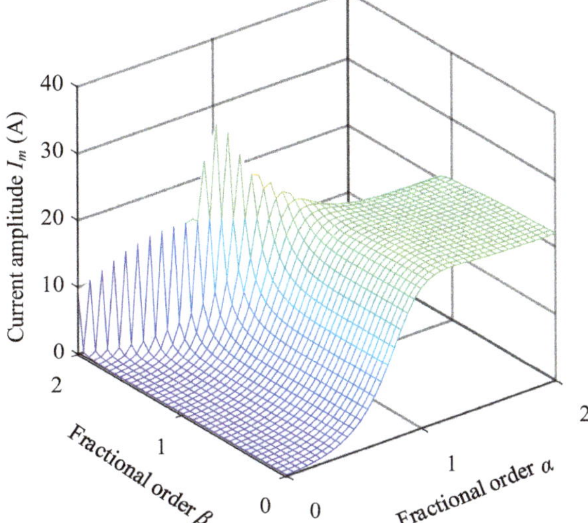

Fig. 6.10 The amplitude of phase current versus the fractional orders α and β (Circuit parameters are $U = 220$ V, $f = 50$ Hz, $R = 10\ \Omega$, $C_\alpha = 0.001\ \text{F/s}^{1-\alpha}$, $L_\beta = 0.01\ \text{H/s}^{1-\beta}$)

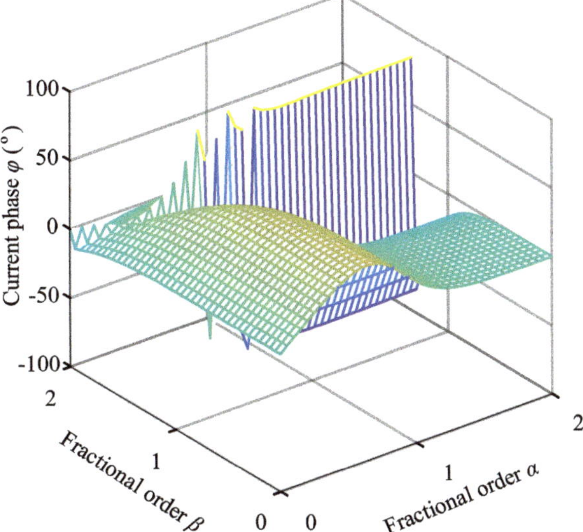

Fig. 6.11 The phase of current versus the fractional orders α and β (Circuit parameters are $U = 220$ V, $f = 50$ Hz, $R = 10\ \Omega$, $C_\alpha = 0.001\ \text{F/s}^{1-\alpha}$, $L_\beta = 0.01\ \text{H/s}^{1-\beta}$)

Figure 6.10 shows the relationship between the amplitude of phase current and the fractional orders α and β in the three-phase fractional-order load, where the capacitor and inductor are both fractional-order components. As shown in Fig. 6.10, in the square area enclosed by $0 < \alpha < 1$, $1 < \beta < 2$, the current amplitude is very sensitive to fractional order, and there are many current spikes. The main reason for this phenomenon is the negative resistance characteristic of the fractional-order capacitor, which can be equivalent to the series of a negative resistor and an integer-order capacitor when the order is greater than 1. This negative resistance characteristic will make the equivalent impedance of the entire circuit become zero. Therefore, it leads to many current spikes when the fractional orders α and β change simultaneously. The actual sinusoidal steady-state three-phase fractional-order circuit should be avoided to work in these peak regions.

4. When the fractional orders α and β are constant, Figs. 6.12 and Fig. 6.13 show curves of the current of phase A with the power frequency.

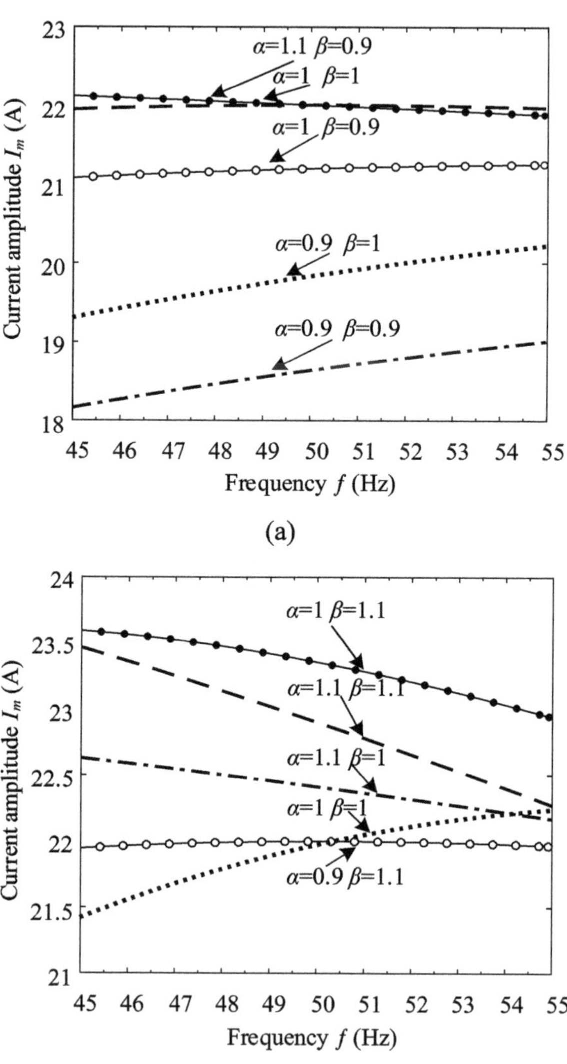

Fig. 6.12 The amplitude of current versus the frequency of the power supply: **a** $0 < \alpha \leq 1$ and $0 < \beta \leq 1$, **b** $1 \leq \alpha \leq 2$ and $1 \leq \beta \leq 2$ (Circuit parameters are $U = 220$ V, $R = 10\ \Omega$, $C_\alpha = 0.001\ \text{F/s}^{1-\alpha}$, $L_\beta = 0.01\text{H/s}^{1-\beta}$)

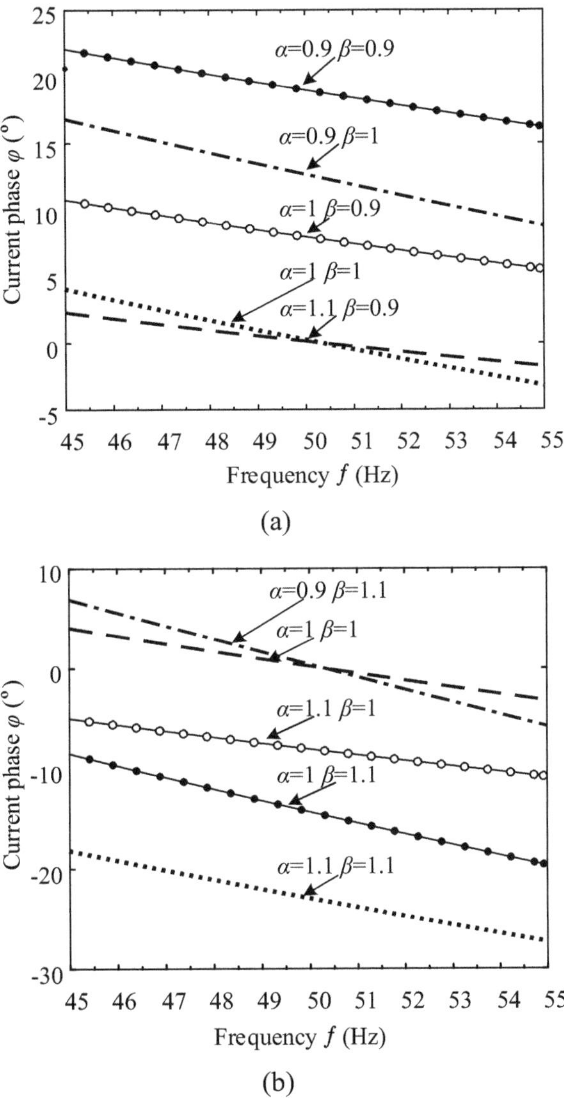

Fig. 6.13 The phase of current versus the frequency of the power supply: **a** $0 \le \alpha \le 2$, $0 \le \beta \le 1$, **b** $0 \le \alpha \le 2$, $1 \le \beta \le 2$ (Circuit parameters are $U = 220$ V, $R = 10\ \Omega$, $C_\alpha = 0.001$ F/s$^{1-\alpha}$, $L_\beta = 0.01$ H/s$^{1-\beta}$)

Figures 6.12 and 6.13 discusses the three-phase fractional-order circuit in which the capacitor and inductor in the load are both fractional-order components. As can be seen from Figs. 6.12 and 6.13, when the frequency of the power source changes, the stability of the entire circuit can be maintained by adjusting the fractional order. By selecting the appropriate fractional order, the frequency anti-interference of the whole circuit can be improved.

Through the above analysis, it can be seen that $\dot{U}_{NN'} = 0$ is a necessary and sufficient condition for each phase to be independent. Therefore, the symmetrical

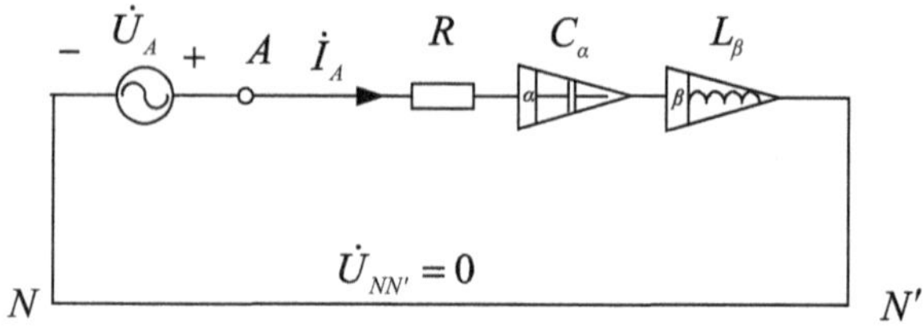

Fig. 6.14 Single-phase equivalent circuit

Y/*Y* circuit can be divided into three independent single-phase circuits. When one phase has been calculated, the other two phases can be directly written according to the symmetry. In other words, the symmetrical three-phase fractional-order circuit connected by *Y*/*Y* can be simplified to a single-phase circuit for analysis. Figure 6.14 shows the single-phase equivalent circuit of a three-phase fractional-order circuit. (Taking the phase *A* as an example).

In a symmetrical three-phase fractional-order circuit connected by *Y*/*Y*, the voltage $\dot{U}_{NN'}$ is always equal to zero whether $Z_{\alpha\beta\_N}$ is 0 or ∞. And by using KCL, the neutral-point current $\dot{I}_N$ can be obtained by

$$\dot{I}_N = \dot{I}_A + \dot{I}_B + \dot{I}_C = \sqrt{2}I\angle\varphi + \sqrt{2}I\angle(\varphi - 120^\circ) + \sqrt{2}I\angle(\varphi + 120^\circ) = 0 \tag{6.20}$$

From (6.20), it can be seen that the neutral-point current $\dot{I}_N$ is equal to zero. Therefore, the neutral points *N* and *N'* can be connected with a wire. In simple terms, the symmetrical three-phase fractional-order circuit connected by *Y*/*Y* does not need the neutral line in theory, and it can be removed directly.

6.2.2 Complex Symmetrical Three-Phase Fractional-Order Circuits

A complex symmetrical three-phase fractional-order circuit generally has many sets of loads and power supplies. These loads and power supplies might be connected in Δ or *Y* mode in the circuit. As shown in Fig. 6.15, this is a typical complex symmetrical three-phase fractional-order circuit.

The complex symmetrical three-phase fractional-order circuit can be simplified into a *Y*/*Y* connection circuit for analysis. It needs to transform the power supplies and loads into a *Y*-connection, and then use the single-phase equivalent circuit for analysis. The specific steps for calculating the symmetrical three-phase fractional-order circuit are as follows:

1. Use the relationship between line and phasor values to convert the symmetrical three-phase Δ-connection power source into a *Y*-connection. For a positive sinusoidal steady-state three-phase fractional-order system, the conversion method is

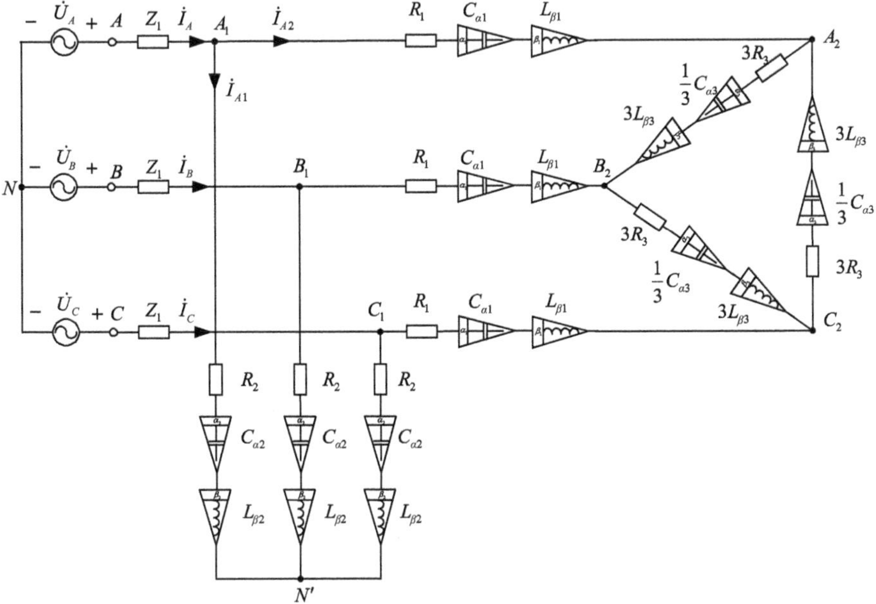

Fig. 6.15 Complex symmetrical three-phase fractional-order circuit

$$
\begin{cases}
\dot{U}_A = \dot{U}_{AB}\dfrac{1}{\sqrt{3}}\angle - 30^\circ \\
\dot{U}_B = a^2\dot{U}_A \\
\dot{U}_C = a\dot{U}_A
\end{cases}
\tag{6.21}
$$

2. The Δ/Y equivalent transformation formula can convert a symmetrical three-phase fractional-order Δ-connection load into a Y-connection. It can be described as

$$
Z_Y = \frac{1}{3}Z_\Delta \tag{6.22}
$$

3. Connect the neutral points N and N', a single-phase equivalent circuit can be obtained as shown in Fig. 6.16. According to the analysis method of asymmetrical three-phase system, the current $\dot{I}_A$, $\dot{I}_{A1}$ and $\dot{I}_{A2}$ can be calculated.

Z_l represents the line impedance of the sinusoidal steady-state three-phase fractional-order circuit, and $Z_{\alpha\beta\_1}$, $Z_{\alpha\beta\_2}$, and $Z_{\alpha\beta\_3}$ are the impedance of loads. These impedances can be written as

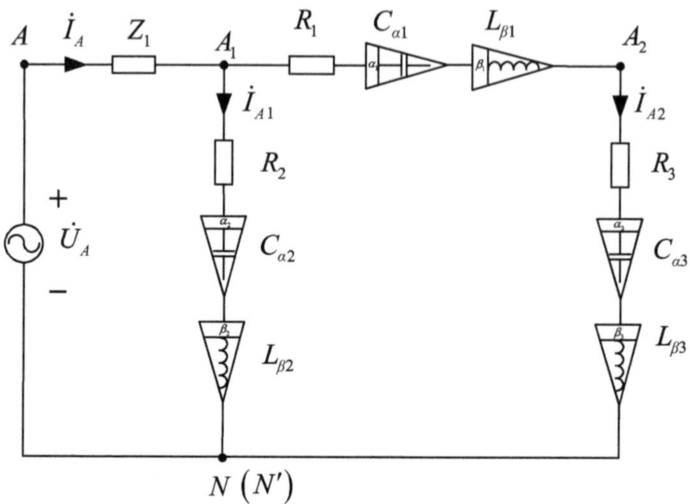

Fig. 6.16 Single-phase equivalent circuit

$$\begin{cases} Z_l = R + \dfrac{1}{j\omega C} + j\omega L \\ Z_{\alpha\beta\_1} = R_1 + \dfrac{1}{(j\omega)^{\alpha_1} C_{\alpha 1}} + (j\omega)^{\beta_1} L_{\beta 1} \\ Z_{\alpha\beta\_2} = R_2 + \dfrac{1}{(j\omega)^{\alpha_2} C_{\alpha 2}} + (j\omega)^{\beta_2} L_{\beta 2} \\ 3Z_{\alpha\beta\_3} = 3R_3 + \dfrac{3}{(j\omega)^{\alpha_3} C_{\alpha 3}} + (j\omega)^{\beta_3} 3L_{\beta 3} \end{cases} \tag{6.23}$$

Assuming that $\dot{U}_A = \sqrt{2}U\angle 0^\circ$, the parallel impedance can be calculated as

$$\begin{aligned} Z_{\alpha\beta} &= Z_{\alpha\beta\_1} \| (Z_{\alpha\beta\_2} + Z_{\alpha\beta\_3}) = \frac{Z_{\alpha\beta\_1}(Z_{\alpha\beta\_2} + Z_{\alpha\beta\_3})}{Z_{\alpha\beta\_1} + (Z_{\alpha\beta\_2} + Z_{\alpha\beta\_3})} \\ &= \frac{(Z_{\alpha\beta\_2} + Z_{\alpha\beta\_3})}{Z_{\alpha\beta\_1} + (Z_{\alpha\beta\_2} + Z_{\alpha\beta\_3})} Z_{\alpha\beta\_1} \\ &= K \begin{bmatrix} R_1 + \dfrac{1}{\omega^{\alpha_1} C_{\alpha 1}} \cos\dfrac{\pi\alpha_1}{2} + \omega^{\beta_1} L_{\beta 1} \cos\dfrac{\pi\beta_1}{2} \\ +j\left(\omega^{\beta_1} L_{\beta 1} \sin\dfrac{\pi\beta_1}{2} - \dfrac{1}{\omega^{\alpha_1} C_{\alpha 1}} \sin\dfrac{\pi\alpha_1}{2}\right) \end{bmatrix} \end{aligned} \tag{6.24}$$

The coefficient K and phase φ_K are

$$\begin{cases} K = |K|\angle\varphi_K = \dfrac{M_{12} + jN_{12}}{M_{13} + jN_{13}} = \sqrt{\dfrac{M_{12}^2 + N_{12}^2}{M_{13}^2 + N_{13}^2}}\angle\varphi_K \\ \varphi_K = \arctan\dfrac{\mathrm{Im}\left(\frac{M_{12}+jN_{12}}{M_{13}+jN_{13}}\right)}{\mathrm{Re}\left(\frac{M_{12}+jN_{12}}{M_{13}+jN_{13}}\right)} \end{cases} \tag{6.25}$$

where

$$\begin{cases} M_{12} = \sum\limits_{i=1}^{2}\left(R_i + \dfrac{1}{\omega^{\alpha_i}C_{\alpha i}}\cos\dfrac{\pi\alpha_i}{2} + \omega^{\beta_i}L_{\beta i}\cos\dfrac{\pi\beta_i}{2}\right) \\ N_{12} = \sum\limits_{i=1}^{2}\left(\omega^{\beta_i}L_{\beta i}\sin\dfrac{\pi\beta_i}{2} - \dfrac{1}{\omega^{\alpha_i}C_{\alpha i}}\sin\dfrac{\pi\alpha_i}{2}\right) \\ M_{13} = \sum\limits_{i=1}^{3}\left(R_i + \dfrac{1}{\omega^{\alpha_i}C_{\alpha i}}\cos\dfrac{\pi\alpha_i}{2} + \omega^{\beta_i}L_{\beta i}\cos\dfrac{\pi\beta_i}{2}\right) \\ N_{13} = \sum\limits_{i=1}^{3}\left(\omega^{\beta_i}L_{\beta i}\sin\dfrac{\pi\beta_i}{2} - \dfrac{1}{\omega^{\alpha_i}C_{\alpha i}}\sin\dfrac{\pi\alpha_i}{2}\right) \end{cases} \tag{6.26}$$

The current of phase A can be obtained as

$$\begin{aligned} \dot{I}_A &= \frac{\sqrt{2}U\angle 0^\circ}{Z_l + Z_{\alpha\beta}} = \frac{\dot{U}_A}{Z_l + Z_{\alpha\beta\_1}\|(Z_{\alpha\beta\_2} + Z_{\alpha\beta\_3})} \\ &= \frac{\sqrt{2}U\angle 0^\circ}{R + \frac{1}{j\omega C} + j\omega L + K\cdot Z_{\alpha\beta\_1}} \\ &= \sqrt{2}I\angle-\varphi \end{aligned} \tag{6.27}$$

where

$$\begin{cases} I = \dfrac{U}{\left|R + \frac{1}{j\omega C} + j\omega L + \frac{M_{12}+jN_{12}}{M_{13}+jN_{13}}Z_{\alpha\beta\_1}\right|} \\ \varphi = \arctan\dfrac{\mathrm{Im}\left(R + \frac{1}{j\omega C} + j\omega L + \frac{M_{12}+jN_{12}}{M_{13}+jN_{13}}Z_{\alpha\beta\_1}\right)}{\mathrm{Re}\left(R + \frac{1}{j\omega C} + j\omega L + \frac{M_{12}+jN_{12}}{M_{13}+jN_{13}}Z_{\alpha\beta\_1}\right)} \end{cases} \tag{6.28}$$

Similarly, the current of phase B can be derived as

$$\dot{I}_{A1} = \frac{Z_{\alpha\beta\_1}}{Z_{\alpha\beta\_1}\|(Z_{\alpha\beta\_2} + Z_{\alpha\beta\_3})}\dot{I}_A$$

$$
= \frac{\sum_{i=1}^{3}\left[\begin{array}{l} R_i + \frac{1}{\omega^{\alpha_i} C_{\alpha i}} \cos\frac{\pi\alpha_i}{2} + \omega^{\beta_i} L_{\beta i} \cos\frac{\pi\beta_i}{2} \\ +j\left(\omega^{\beta_i} L_{\beta i} \sin\frac{\pi\beta_i}{2} - \frac{1}{\omega^{\alpha_i} C_{\alpha i}} \sin\frac{\pi\alpha_i}{2}\right) \end{array}\right]}{\sum_{i=1}^{2}\left[\begin{array}{l} R_i + \frac{1}{\omega^{\alpha_i} C_{\alpha i}} \cos\frac{\pi\alpha_i}{2} + \omega^{\beta_i} L_{\beta i} \cos\frac{\pi\beta_i}{2} \\ +j\left(\omega^{\beta_i} L_{\beta i} \sin\frac{\pi\beta_i}{2} - \frac{1}{\omega^{\alpha_i} C_{\alpha i}} \sin\frac{\pi\alpha_i}{2}\right) \end{array}\right]} \dot{I}_A
$$

$$
= \frac{\dot{I}_A}{K} = \frac{\sqrt{2}I}{|K|} \angle(-\varphi - \varphi_K) \tag{6.29}
$$

In the same way, the current of phase C can be calculated as

$$
\dot{I}_{A2} = \frac{\left(Z_{\alpha\beta\_2} + \frac{1}{3} Z_{\alpha\beta\_3}\right)}{Z_{\alpha\beta\_1} \| \left(Z_{\alpha\beta\_2} + \frac{1}{3} Z_{\alpha\beta\_3}\right)} \dot{I}_A
$$

$$
= \frac{\sum_{i=1}^{2}\left[\begin{array}{l} R_i + \frac{1}{\omega^{\alpha_i} C_{\alpha i}} \cos\frac{\pi\alpha_i}{2} + \omega^{\beta_i} L_{\beta i} \cos\frac{\pi\beta_i}{2} \\ +j\left(\omega^{\beta_i} L_{\beta i} \sin\frac{\pi\beta_i}{2} - \frac{1}{\omega^{\alpha_i} C_{\alpha i}} \sin\frac{\pi\alpha_i}{2}\right) \end{array}\right]}{K\left[\begin{array}{l} R_1 + \frac{1}{\omega^{\alpha_1} C_{\alpha 1}} \cos\frac{\pi\alpha_1}{2} + \omega^{\beta_1} L_{\beta 1} \cos\frac{\pi\beta_1}{2} \\ +j\left(\omega^{\beta_1} L_{\beta 1} \sin\frac{\pi\beta_1}{2} - \frac{1}{\omega^{\alpha_1} C_{\alpha 1}} \sin\frac{\pi\alpha_1}{2}\right) \end{array}\right]} \dot{I}_A
$$

$$
= \left(1 - \frac{1}{K}\right) \dot{I}_A = \frac{|K-1| I}{|K|} \angle(-\varphi + \varphi_{K-1} - \varphi_K) \tag{6.30}
$$

It can be seen from the above analysis that a sinusoidal steady-state three-phase fractional-order circuit has better flexibility due to the fractional orders. The amplitude and phase of each current in the circuit can be changed by adjusting the fractional orders of the capacitor and inductor. Similarly, by adjusting the fractional orders, the three-phase fractional-order circuit can be characterized as capacitive or inductive.

4. Calculate the phase current and phase voltage of each equivalent Y-type load based on the equivalent circuit. The voltage and current of the original circuit can be derived from the relationship between the line and the phasor values. From here on, any uncertain symmetrical fractional-order load will be treated as a Y connection.

The analysis and calculation of the complex symmetrical sinusoidal steady-state three-phase fractional-order circuit can finally be simplified into a Y/Y circuit, we will no longer analyze how the current in the three-phase fractional-order circuit is affected by other factors.

6.2.3 Asymmetric Three-Phase Fractional-Order Circuits

In a three-phase fractional-order circuit, as long as there is a part of asymmetry, it is called an asymmetrical three-phase fractional-order circuit. Asymmetrical sinusoidal steady-state three-phase fractional-order circuit is a complex sinusoidal steady-state circuit. It can be analyzed by the solution method of the traditional sinusoidal steady-state circuit. However, the sinusoidal steady-state three-phase fractional-order circuit has more abundant characteristics than the traditional three-phase circuit. In a symmetrical three-phase fractional-order circuit, the neutral-point voltage $\dot{U}_{NN'}$ is always equal to zero, but it is not true in the asymmetric three-phase fractional-order circuit. For an asymmetrical circuit, the neutral-point voltage $\dot{U}_{NN'}$ needs to be calculated first. Then, the voltage and current of each phase can be calculated based on the neutral point voltage of the load. Therefore, according to whether the neutral-point voltage of the load is zero, there are two kinds of asymmetric three-phase fractional-order circuits. The first is that the neutral-point voltage is equal to zero, and the other is that the neutral-point voltage is not equal to zero.

The asymmetric sinusoidal steady-state three-phase fractional-order circuit is shown in Fig. 6.17. Here, the fractional-order loads are symmetrically Y-connected, all of which are $Z_{\alpha\beta\_1}$. In addition, there are two fractional-order loads $Z_{\alpha\beta\_2}$ and

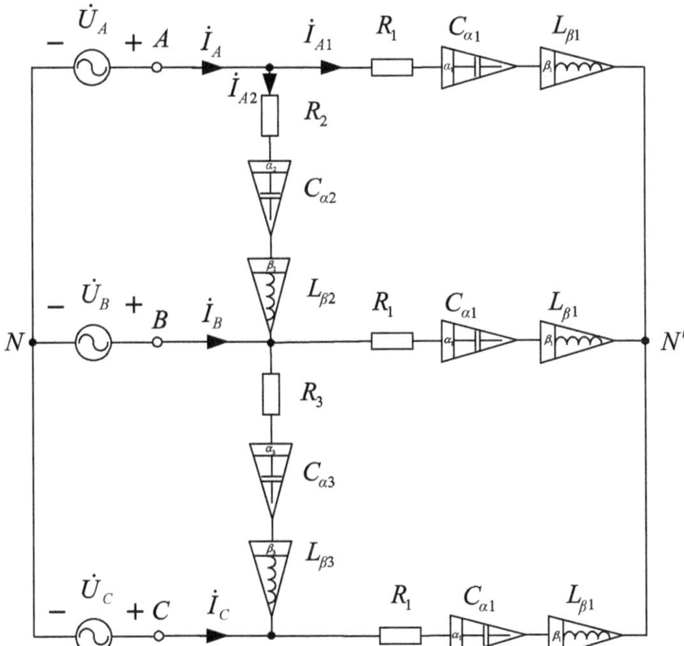

Fig. 6.17 The first-type asymmetric three-phase fractional-order circuit

$Z_{\alpha\beta\_3}$ connected across AB and BC. This kind of circuit belongs to the first-type asymmetric sinusoidal steady-state three-phase fractional-order circuit [11, 12].

As shown in Fig. 6.17, the neutral-point voltage of the load can be calculated by the node-voltage method, that is

$$\left(\frac{1}{Z_{\alpha\beta\_1}}+\frac{1}{Z_{\alpha\beta\_1}}+\frac{1}{Z_{\alpha\beta\_1}}\right)\dot{U}_{NN'}=\frac{1}{Z_{\alpha\beta\_1}}\left(\dot{U}_A+\dot{U}_B+\dot{U}_C\right) \tag{6.31}$$

Then, since $\dot{U}_A+\dot{U}_B+\dot{U}_C=0$, we can know from (6.31) that $\dot{U}_{NN'}=0$. Therefore, for the first-type asymmetric three-phase fractional-order circuit, although it is an asymmetrical three-phase fractional-order circuit, the neutral-point voltage of the load is still equal to zero.

To calculate the current in the asymmetrical three-phase fractional-order circuit, we assume that the voltage phasor is $\dot{U}_A=\sqrt{2}U\angle 0°$, and then $\dot{U}_{AB}=\sqrt{3}\dot{U}_A\angle 30°$ can be derived by the relationship between the line and phase value.

According to KCL, the current of phase A can be calculated as

$$\begin{aligned}\dot{I}_A&=\dot{I}_{A1}+\dot{I}_{A2}=\frac{\dot{U}_A}{Z_{\alpha\beta\_1}}+\frac{\dot{U}_{AB}}{Z_{\alpha\beta\_2}}\\&=\frac{\sqrt{2}U\angle 0°}{R_1+\frac{1}{(j\omega)^{\alpha_1}C_{\alpha 1}}+(j\omega)^{\beta_1}L_{\beta 1}}+\frac{\sqrt{6}U\angle 30°}{R_2+\frac{1}{(j\omega)^{\alpha_2}C_{\alpha 2}}+(j\omega)^{\beta_2}L_{\beta 2}}\\&=\frac{M_A\sqrt{2}U\angle 0°+N_A\sqrt{6}U\angle 30°}{\prod_{i=1}^{2}\left[\begin{array}{l}R_i+\frac{1}{\omega^{\alpha_i}C_{\alpha i}}\cos\frac{\pi\alpha_i}{2}+\omega^{\beta_i}L_{\beta i}\cos\frac{\pi\beta_i}{2}\\+j\left(\omega^{\beta_i}L_{\beta i}\sin\frac{\pi\beta_i}{2}-\frac{1}{\omega^{\alpha_i}C_{\alpha i}}\sin\frac{\pi\alpha_i}{2}\right)\end{array}\right]}\end{aligned} \tag{6.32}$$

where

$$\begin{cases}M_A=\left[\begin{array}{l}R_2+\frac{1}{\omega^{\alpha_2}C_{\alpha 2}}\cos\frac{\pi\alpha_2}{2}+\omega^{\beta_2}L_{\beta 2}\cos\frac{\pi\beta_2}{2}\\+j\left(\omega^{\beta_2}L_{\beta 2}\sin\frac{\pi\beta_2}{2}-\frac{1}{\omega^{\alpha_2}C_{\alpha 2}}\sin\frac{\pi\alpha_2}{2}\right)\end{array}\right]\\N_A=\left[\begin{array}{l}R_1+\frac{1}{\omega^{\alpha_1}C_{\alpha 1}}\cos\frac{\pi\alpha_1}{2}+\omega^{\beta_1}L_{\beta 1}\cos\frac{\pi\beta_1}{2}\\+j\left(\omega^{\beta_1}L_{\beta 1}\sin\frac{\pi\beta_1}{2}-\frac{1}{\omega^{\alpha_1}C_{\alpha 1}}\sin\frac{\pi\alpha_1}{2}\right)\end{array}\right]\end{cases} \tag{6.33}$$

In the same way, the current of phase B can be derived as

$$\dot{I}_B=\frac{\dot{U}_B}{Z_{\alpha\beta\_1}}+\frac{\dot{U}_{BA}}{Z_{\alpha\beta\_2}}+\frac{\dot{U}_{BC}}{Z_{\alpha\beta\_3}}$$

$$= \frac{M_B\sqrt{2}U\angle-120^\circ + N_B\sqrt{6}U\angle-150^\circ + R\sqrt{6}U\angle-90^\circ}{\prod_{i=1}^{3}\begin{bmatrix} R_i + \frac{1}{\omega^{\alpha_i}C_{\alpha i}}\cos\frac{\pi\alpha_i}{2} + \omega^{\beta_i}L_{\beta i}\cos\frac{\pi\beta_i}{2} \\ +j\left(\omega^{\beta_i}L_{\beta i}\sin\frac{\pi\beta_i}{2} - \frac{1}{\omega^{\alpha_i}C_{\alpha i}}\sin\frac{\pi\alpha_i}{2}\right)\end{bmatrix}} \tag{6.34}$$

where

$$\begin{cases} M_B = \prod_{i=2}^{3}\begin{bmatrix} R_i + \frac{1}{\omega^{\alpha_i}C_{\alpha i}}\cos\frac{\pi\alpha_i}{2} + \omega^{\beta_i}L_{\beta i}\cos\frac{\pi\beta_i}{2} \\ +j\left(\omega^{\beta_i}L_{\beta i}\sin\frac{\pi\beta_i}{2} - \frac{1}{\omega^{\alpha_i}C_{\alpha i}}\sin\frac{\pi\alpha_i}{2}\right)\end{bmatrix} \\ N_B = \prod_{i=1,i\neq 2}^{3}\begin{bmatrix} R_i + \frac{1}{\omega^{\alpha_i}C_{\alpha i}}\cos\frac{\pi\alpha_i}{2} + \omega^{\beta_i}L_{\beta i}\cos\frac{\pi\beta_i}{2} \\ +j\left(\omega^{\beta_i}L_{\beta i}\sin\frac{\pi\beta_i}{2} - \frac{1}{\omega^{\alpha_i}C_{\alpha i}}\sin\frac{\pi\alpha_i}{2}\right)\end{bmatrix} \\ R = \prod_{i=1}^{2}\begin{bmatrix} R_i + \frac{1}{\omega^{\alpha_i}C_{\alpha i}}\cos\frac{\pi\alpha_i}{2} + \omega^{\beta_i}L_{\beta i}\cos\frac{\pi\beta_i}{2} \\ +j\left(\omega^{\beta_i}L_{\beta i}\sin\frac{\pi\beta_i}{2} - \frac{1}{\omega^{\alpha_i}C_{\alpha i}}\sin\frac{\pi\alpha_i}{2}\right)\end{bmatrix} \end{cases} \tag{6.35}$$

Similarly, the current of phase C is

$$\begin{aligned} \dot{I}_C &= \frac{\dot{U}_C}{Z_{\alpha\beta\_1}} + \frac{\dot{U}_{CB}}{Z_{\alpha\beta\_3}} \\ &= \frac{M_C\sqrt{2}U\angle 120^\circ + N_C\sqrt{6}U\angle 150^\circ}{\prod_{i=1,i\neq 2}^{3}\begin{bmatrix} R_i + \frac{1}{\omega^{\alpha_i}C_{\alpha i}}\cos\frac{\pi\alpha_i}{2} + \omega^{\beta_i}L_{\beta i}\cos\frac{\pi\beta_i}{2} \\ +j\left(\omega^{\beta_i}L_{\beta i}\sin\frac{\pi\beta_i}{2} - \frac{1}{\omega^{\alpha_i}C_{\alpha i}}\sin\frac{\pi\alpha_i}{2}\right)\end{bmatrix}} \end{aligned} \tag{6.36}$$

where

$$\begin{cases} M_C = \begin{bmatrix} R_3 + \frac{1}{\omega^{\alpha_3}C_{\alpha 3}}\cos\frac{\pi\alpha_3}{2} + \omega^{\beta_3}L_{\beta 3}\cos\frac{\pi\beta_3}{2} \\ +j\left(\omega^{\beta_3}L_{\beta 3}\sin\frac{\pi\beta_3}{2} - \frac{1}{\omega^{\alpha_3}C_{\alpha 3}}\sin\frac{\pi\alpha_3}{2}\right)\end{bmatrix} \\ N_C = \begin{bmatrix} R_1 + \frac{1}{\omega^{\alpha_1}C_{\alpha 1}}\cos\frac{\pi\alpha_1}{2} + \omega^{\beta_1}L_{\beta 1}\cos\frac{\pi\beta_1}{2} \\ +j\left(\omega^{\beta_1}L_{\beta 1}\sin\frac{\pi\beta_1}{2} - \frac{1}{\omega^{\alpha_1}C_{\alpha 1}}\sin\frac{\pi\alpha_1}{2}\right)\end{bmatrix} \end{cases} \tag{6.37}$$

It can be seen that for this kind of asymmetrical three-phase fractional-order circuit with zero neutral-point voltage, the load can be divided into symmetrical and asymmetrical parts for analysis. The load of the symmetrical part can be analyzed by the characteristics of the symmetrical three-phase fractional-order circuit and the load of the asymmetric part can be calculated by the method of a general sinusoidal steady-state circuit.

Figure 6.18 shows another type of asymmetric sinusoidal steady-state three-phase fractional-order circuit, in which the neutral-point voltage of load is not equal to zero. Here, the neutral potential of the three-phase power source is generally considered to be equal to zero, i.e., $\dot{U}_N = 0$. And the neutral potential of the load is not equal to zero, i.e., $\dot{U}_{NN'} \neq 0$. Therefore, the first key is to obtain the neutral-point voltage of the load $\dot{U}_{NN'}$, and then the current of each phase can be calculated in turn [13].

According to Fig. 6.18, the node voltage can be given by

$$\left(\frac{1}{Z_{\alpha\beta\_1}} + \frac{1}{Z_{\alpha\beta\_2}} + \frac{1}{Z_{\alpha\beta\_3}}\right)\dot{U}_{NN'} = \frac{\dot{U}_A}{Z_{\alpha\beta\_1}} + \frac{\dot{U}_B}{Z_{\alpha\beta\_2}} + \frac{\dot{U}_C}{Z_{\alpha\beta\_3}} \tag{6.38}$$

Then, the neutral-point voltage of the load can be obtained as

$$\begin{aligned}\dot{U}_{NN'} &= \frac{\dot{U}_A Z_{\alpha\beta\_2} Z_{\alpha\beta\_3} + \dot{U}_B Z_{\alpha\beta\_1} Z_{\alpha\beta\_3} + \dot{U}_C Z_{\alpha\beta\_1} Z_{\alpha\beta\_2}}{Z_{\alpha\beta\_1} Z_{\alpha\beta\_2} + Z_{\alpha\beta\_2} Z_{\alpha\beta\_3} + Z_{\alpha\beta\_1} Z_{\alpha\beta\_3}} \\ &= \frac{\dot{U}_A X_1 + \dot{U}_B X_2 + \dot{U}_C X_3}{X_1 + X_2 + X_3}\end{aligned} \tag{6.39}$$

where

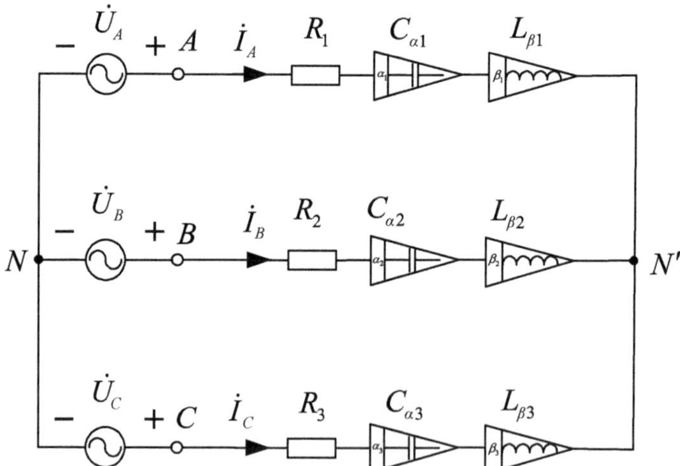

Fig. 6.18 The second-type asymmetric three-phase fractional-order circuit

$$\begin{cases} X_1 = \prod_{i=2}^{3} \left[\begin{array}{l} R_i + \dfrac{1}{\omega^{\alpha_i} C_{\alpha i}} \cos\dfrac{\pi\alpha_i}{2} + \omega^{\beta_i} L_{\beta i} \cos\dfrac{\pi\beta_i}{2} \\ +j\left(\omega^{\beta_i} L_{\beta i} \sin\dfrac{\pi\beta_i}{2} - \dfrac{1}{\omega^{\alpha_i} C_{\alpha i}} \sin\dfrac{\pi\alpha_i}{2}\right) \end{array} \right] \\ X_2 = \prod_{i=1, i\neq 2}^{3} \left[\begin{array}{l} R_i + \dfrac{1}{\omega^{\alpha_i} C_{\alpha i}} \cos\dfrac{\pi\alpha_i}{2} + \omega^{\beta_i} L_{\beta i} \cos\dfrac{\pi\beta_i}{2} \\ +j\left(\omega^{\beta_i} L_{\beta i} \sin\dfrac{\pi\beta_i}{2} - \dfrac{1}{\omega^{\alpha_i} C_{\alpha i}} \sin\dfrac{\pi\alpha_i}{2}\right) \end{array} \right] \\ X_3 = \prod_{i=1}^{2} \left[\begin{array}{l} R_i + \dfrac{1}{\omega^{\alpha_i} C_{\alpha i}} \cos\dfrac{\pi\alpha_i}{2} + \omega^{\beta_i} L_{\beta i} \cos\dfrac{\pi\beta_i}{2} \\ +j\left(\omega^{\beta_i} L_{\beta i} \sin\dfrac{\pi\beta_i}{2} - \dfrac{1}{\omega^{\alpha_i} C_{\alpha i}} \sin\dfrac{\pi\alpha_i}{2}\right) \end{array} \right] \end{cases} \tag{6.40}$$

It can be found from (6.39) that when the load of the three-phase circuit is asymmetrical, the neutral-point voltage of the load is not always equal to zero, which may cause the voltage on the three-phase load to change sharply. However, the neutral-point voltage $\dot{U}_{NN'}$ can be effectively changed by adjusting the fractional order of the capacitor or inductor. Therefore, adjusting fractional orders can ensure that the three-phase fractional-order circuit maintains the neutral point voltage equal to zero under an asymmetrical load.

At this time, the current of phase A is

$$\begin{aligned} \dot{I}_A &= \frac{\dot{U}_A - \dot{U}_{NN'}}{Z_{\alpha\beta\_1}} = \frac{(\dot{U}_A - \dot{U}_B) Z_{\alpha\beta\_3} + (\dot{U}_A - \dot{U}_C) Z_{\alpha\beta\_2}}{Z_{\alpha\beta\_1} Z_{\alpha\beta\_2} + Z_{\alpha\beta\_2} Z_{\alpha\beta\_3} + Z_{\alpha\beta\_1} Z_{\alpha\beta\_3}} \\ &= \frac{P_A \dot{U}_{AB} + Q_A \dot{U}_{AC}}{X_1 + X_2 + X_3} \end{aligned} \tag{6.41}$$

where

$$\begin{cases} P_A = \left[\begin{array}{l} R_3 + \dfrac{1}{\omega^{\alpha_3} C_{\alpha 3}} \cos\dfrac{\pi\alpha_3}{2} + \omega^{\beta_3} L_{\beta 3} \cos\dfrac{\pi\beta_3}{2} \\ +j\left(\omega^{\beta_3} L_{\beta 3} \sin\dfrac{\pi\beta_3}{2} - \dfrac{1}{\omega^{\alpha_3} C_{\alpha 3}} \sin\dfrac{\pi\alpha_3}{2}\right) \end{array} \right] \\ Q_A = \left[\begin{array}{l} R_2 + \dfrac{1}{\omega^{\alpha_2} C_{\alpha 2}} \cos\dfrac{\pi\alpha_2}{2} + \omega^{\beta_2} L_{\beta 2} \cos\dfrac{\pi\beta_2}{2} \\ +j\left(\omega^{\beta_2} L_{\beta 2} \sin\dfrac{\pi\beta_2}{2} - \dfrac{1}{\omega^{\alpha_2} C_{\alpha 2}} \sin\dfrac{\pi\alpha_2}{2}\right) \end{array} \right] \end{cases} \tag{6.42}$$

In the same way, the currents of phase B and phase C can be deduced as

$$\begin{cases} \dot{I}_B = \dfrac{\dot{U}_B - \dot{U}_{NN'}}{Z_{\alpha\beta\_2}} = \dfrac{(\dot{U}_B - \dot{U}_A)Z_{\alpha\beta\_3} + (\dot{U}_B - \dot{U}_C)Z_{\alpha\beta\_1}}{Z_{\alpha\beta\_1}Z_{\alpha\beta\_2} + Z_{\alpha\beta\_2}Z_{\alpha\beta\_3} + Z_{\alpha\beta\_1}Z_{\alpha\beta\_3}} \\ \quad = \dfrac{P_B\dot{U}_{BA} + Q_B\dot{U}_{BC}}{X_1 + X_2 + X_3} \\ \dot{I}_C = \dfrac{\dot{U}_C - \dot{U}_{NN'}}{Z_{\alpha\beta\_3}} = \dfrac{(\dot{U}_C - \dot{U}_A)Z_{\alpha\beta\_2} + (\dot{U}_C - \dot{U}_B)Z_{\alpha\beta\_1}}{Z_{\alpha\beta\_1}Z_{\alpha\beta\_2} + Z_{\alpha\beta\_2}Z_{\alpha\beta\_3} + Z_{\alpha\beta\_1}Z_{\alpha\beta\_3}} \\ \quad = \dfrac{P_C\dot{U}_{CA} + Q_C\dot{U}_{CB}}{X_1 + X_2 + X_3} \end{cases} \tag{6.43}$$

where

$$\begin{cases} P_B = \begin{bmatrix} R_3 + \dfrac{1}{\omega^{\alpha_3}C_{\alpha3}}\cos\dfrac{\pi\alpha_3}{2} + \omega^{\beta_3}L_{\beta3}\cos\dfrac{\pi\beta_3}{2} + \\ j\left(\omega^{\beta_3}L_{\beta3}\sin\dfrac{\pi\beta_3}{2} - \dfrac{1}{\omega^{\alpha_3}C_{\alpha3}}\sin\dfrac{\pi\alpha_3}{2}\right) \end{bmatrix} \\ Q_B = \begin{bmatrix} R_1 + \dfrac{1}{\omega^{\alpha_1}C_{\alpha1}}\cos\dfrac{\pi\alpha_1}{2} + \omega^{\beta_1}L_{\beta1}\cos\dfrac{\pi\beta_1}{2} \\ +j\left(\omega^{\beta_1}L_{\beta1}\sin\dfrac{\pi\beta_1}{2} - \dfrac{1}{\omega^{\alpha_1}C_{\alpha1}}\sin\dfrac{\pi\alpha_1}{2}\right) \end{bmatrix} \end{cases} \tag{6.44}$$

$$\begin{cases} P_C = \begin{bmatrix} R_2 + \dfrac{1}{\omega^{\alpha_2}C_{\alpha2}}\cos\dfrac{\pi\alpha_2}{2} + \omega^{\beta_2}L_{\beta2}\cos\dfrac{\pi\beta_2}{2} \\ +j\left(\omega^{\beta_2}L_{\beta2}\sin\dfrac{\pi\beta_2}{2} - \dfrac{1}{\omega^{\alpha_2}C_{\alpha2}}\sin\dfrac{\pi\alpha_2}{2}\right) \end{bmatrix} \\ Q_C = \begin{bmatrix} R_1 + \dfrac{1}{\omega^{\alpha_1}C_{\alpha1}}\cos\dfrac{\pi\alpha_1}{2} + \omega^{\beta_1}L_{\beta1}\cos\dfrac{\pi\beta_1}{2} \\ +j\left(\omega^{\beta_1}L_{\beta1}\sin\dfrac{\pi\beta_1}{2} - \dfrac{1}{\omega^{\alpha_1}C_{\alpha1}}\sin\dfrac{\pi\alpha_1}{2}\right) \end{bmatrix} \end{cases} \tag{6.45}$$

From the above analysis, it can be known that for an asymmetrical sinusoidal steady-state three-phase fractional-order circuit, it is first important to obtain the neutral-point voltage of the load, and then combine it with the relevant characteristics of the three-phase circuit and the sinusoidal steady-state circuit to obtain the line and phase quantities of the asymmetrical three-phase fractional-order circuit.

6.3 Power Measurement of Three-Phase Fractional-Order Circuits

The power of a sinusoidal steady-state three-phase fractional-order circuit mainly refers to the energy consumed by the three-phase fractional-order load, including active power, reactive power, and apparent power. Since the neutral-point voltage of load in the asymmetric sinusoidal steady-state three-phase fractional-order circuit is not zero, this section only discusses the power of the symmetrical three-phase fractional-order circuit and its measurement.

Figure 6.4a depicts the symmetrical *Y*/*Y* sinusoidal steady-state three-phase fractional-order circuit. Due to symmetry, the neutral-point voltage shown in Fig. 6.4a is equal to zero. Assuming that the impedance angle of each load is φ, and the reference voltage is $\dot{U}_A = \sqrt{2}U\angle 0°$, the current of phase *A* can be calculated as

$$\begin{aligned}\dot{I}_A &= \frac{\dot{U}_A - \dot{U}_{NN'}}{Z_{\alpha\beta}} = \frac{\dot{U}_A}{R + \frac{1}{(j\omega)^\alpha C_\alpha} + (j\omega)^\beta L_\beta}\\ &= \frac{\sqrt{2}U\angle 0°}{R + \frac{1}{\omega^\alpha C_\alpha}\cos\frac{\pi\alpha}{2} + \omega^\beta L_\beta \cos\frac{\pi\beta}{2} + j\left(\omega^\beta L_\beta \sin\frac{\pi\beta}{2} - \frac{1}{\omega^\alpha C_\alpha}\sin\frac{\pi\alpha}{2}\right)}\\ &= \sqrt{2}I\angle -\varphi\end{aligned} \tag{6.46}$$

where

$$\begin{cases} I = \dfrac{U}{\sqrt{\left(R + \frac{1}{\omega^\alpha C_\alpha}\cos\frac{\pi\alpha}{2} + \omega^\beta L_\beta\cos\frac{\pi\beta}{2}\right)^2 + \left(\omega^\beta L_\beta\sin\frac{\pi\beta}{2} - \frac{1}{\omega^\alpha C_\alpha}\sin\frac{\pi\alpha}{2}\right)^2}} \\ \varphi = \arctan\left(\dfrac{\omega^\beta L_\beta\sin\frac{\pi\beta}{2} - \frac{1}{\omega^\alpha C_\alpha}\sin\frac{\pi\alpha}{2}}{R + \frac{1}{\omega^\alpha C_\alpha}\cos\frac{\pi\alpha}{2} + \omega^\beta L_\beta\cos\frac{\pi\beta}{2}}\right)\end{cases} \tag{6.47}$$

According to the symmetry of the three-phase fractional-order circuit, the expressions of the voltage and current $\dot{U}_B$, $\dot{U}_C$, $\dot{I}_B$ and $\dot{I}_C$ can be written sequentially. Then, the active power *P*, reactive power *Q*, and apparent power *S* can be obtained.

The active power can be calculated as

$$\begin{aligned}P &= \mathrm{Re}\left[\dot{U}_A\dot{I}_A^* + \dot{U}_B\dot{I}_B^* + \dot{U}_C\dot{I}_C^*\right] = U_AI_A\cos\varphi + U_BI_B\cos\varphi + U_CI_C\cos\varphi\\ &= 3UI\cos\varphi\\ &= \frac{3U^2\cos\left[\arctan\left(\frac{\omega^\beta L_\beta\sin\frac{\pi\beta}{2} - \frac{1}{\omega^\alpha C_\alpha}\sin\frac{\pi\alpha}{2}}{R + \frac{1}{\omega^\alpha C_\alpha}\cos\frac{\pi\alpha}{2} + \omega^\beta L_\beta\cos\frac{\pi\beta}{2}}\right)\right]}{\sqrt{\left(R + \frac{1}{\omega^\alpha C_\alpha}\cos\frac{\pi\alpha}{2} + \omega^\beta L_\beta\cos\frac{\pi\beta}{2}\right)^2 + \left(\omega^\beta L_\beta\sin\frac{\pi\beta}{2} - \frac{1}{\omega^\alpha C_\alpha}\sin\frac{\pi\alpha}{2}\right)^2}}\end{aligned} \tag{6.48}$$

The reactive power can be obtained as

$$\begin{aligned}Q &= \mathrm{Im}\left[\dot{U}_A\dot{I}_A^* + \dot{U}_B\dot{I}_B^* + \dot{U}_C\dot{I}_C^*\right]\\ &= U_AI_A\sin\varphi + U_BI_B\sin\varphi + U_CI_C\sin\varphi\end{aligned}$$

$$= 3UI\sin\varphi$$

$$= \frac{3U^2\sin\left[\arctan\left(\frac{\omega^\beta L_\beta\sin\frac{\pi\beta}{2}-\frac{1}{\omega^\alpha C_\alpha}\sin\frac{\pi\alpha}{2}}{R+\frac{1}{\omega^\alpha C_\alpha}\cos\frac{\pi\alpha}{2}+\omega^\beta L_\beta\cos\frac{\pi\beta}{2}}\right)\right]}{\sqrt{\left(R+\frac{1}{\omega^\alpha C_\alpha}\cos\frac{\pi\alpha}{2}+\omega^\beta L_\beta\cos\frac{\pi\beta}{2}\right)^2+\left(\omega^\beta L_\beta\sin\frac{\pi\beta}{2}-\frac{1}{\omega^\alpha C_\alpha}\sin\frac{\pi\alpha}{2}\right)^2}} \tag{6.49}$$

Then, the apparent power can be derived as

$$S=\sqrt{P^2+Q^2}=\sqrt{(3UI\cos\varphi)^2+(3UI\sin\varphi)^2}=3UI$$
$$=\frac{3U^2}{\sqrt{\left(R+\frac{1}{\omega^\alpha C_\alpha}\cos\frac{\pi\alpha}{2}+\omega^\beta L_\beta\cos\frac{\pi\beta}{2}\right)^2+\left(\omega^\beta L_\beta\sin\frac{\pi\beta}{2}-\frac{1}{\omega^\alpha C_\alpha}\sin\frac{\pi\alpha}{2}\right)^2}} \tag{6.50}$$

Figure 6.4b shows the symmetrical Y/Δ sinusoidal steady-state three-phase fractional-order circuit. Similarly, suppose that the impedance angle of each load is φ, and the reference voltage is $\dot{U}_A=\sqrt{2}U\angle 0°$, the line voltages $\dot{U}_{AB}$ and $\dot{U}_{CA}$ can be derived by (6.2) and (6.6) as

$$\begin{cases}\dot{U}_{AB}=\sqrt{3}\dot{U}_A\angle 30°=\sqrt{6}U\angle 30°\\ \dot{U}_{CA}=\sqrt{3}\dot{U}_C\angle 30°=\sqrt{6}U\angle 150°\end{cases} \tag{6.51}$$

and the current of phase A can be deduced by KCL as

$$\dot{I}_A=\dot{I}_{AB}-\dot{I}_{CA}=\frac{\dot{U}_{AB}}{Z_{\alpha\beta}}-\frac{\dot{U}_{CA}}{Z_{\alpha\beta}}=\frac{\dot{U}_{AB}-\dot{U}_{CA}}{R+\frac{1}{(j\omega)^\alpha C_\alpha}+(j\omega)^\beta L_\beta}$$
$$=\frac{3\sqrt{2}U\angle 0°}{R+\frac{1}{\omega^\alpha C_\alpha}\cos\frac{\pi\alpha}{2}+\omega^\beta L_\beta\cos\frac{\pi\beta}{2}+j\left(\omega^\beta L_\beta\sin\frac{\pi\beta}{2}-\frac{1}{\omega^\alpha C_\alpha}\sin\frac{\pi\alpha}{2}\right)}$$
$$=\sqrt{2}I\angle-\varphi \tag{6.52}$$

where

$$\begin{cases}I=\frac{3U}{\sqrt{\left(R+\frac{1}{\omega^\alpha C_\alpha}\cos\frac{\pi\alpha}{2}+\omega^\beta L_\beta\cos\frac{\pi\beta}{2}\right)^2+\left(\omega^\beta L_\beta\sin\frac{\pi\beta}{2}-\frac{1}{\omega^\alpha C_\alpha}\sin\frac{\pi\alpha}{2}\right)^2}}\\ \varphi=\arctan\left(\frac{\omega^\beta L_\beta\sin\frac{\pi\beta}{2}-\frac{1}{\omega^\alpha C_\alpha}\sin\frac{\pi\alpha}{2}}{R+\frac{1}{\omega^\alpha C_\alpha}\cos\frac{\pi\alpha}{2}+\omega^\beta L_\beta\cos\frac{\pi\beta}{2}}\right)\end{cases} \tag{6.53}$$

Then, the line current $\dot{I}_{AB}$ can be calculated by (6.11) as

$$\dot{I}_{AB} = \frac{\dot{I}_A}{\sqrt{3}}\angle 30^\circ = \frac{\sqrt{2}I}{\sqrt{3}}\angle 30^\circ - \varphi \tag{6.54}$$

Thus, the active power of the Y/Δ symmetrical sinusoidal steady-state three-phase fractional-order circuit can be calculated as

$$\begin{aligned} P &= \mathrm{Re}\left[\dot{U}_{AB}\dot{I}_{AB}^* + \dot{U}_{BC}\dot{I}_{BC}^* + \dot{U}_{CA}\dot{I}_{CA}^*\right] \\ &= U_{AB}I_{AB}\cos\varphi + U_{BC}I_{BC}\cos\varphi + U_{CA}I_{CA}\cos\varphi = 3UI\cos\varphi \\ &= \frac{9U^2\cos\left[\arctan\left(\frac{\omega^\beta L_\beta \sin\frac{\pi\beta}{2} - \frac{1}{\omega^\alpha C_\alpha}\sin\frac{\pi\alpha}{2}}{R + \frac{1}{\omega^\alpha C_\alpha}\cos\frac{\pi\alpha}{2} + \omega^\beta L_\beta\cos\frac{\pi\beta}{2}}\right)\right]}{\sqrt{\left(R + \frac{1}{\omega^\alpha C_\alpha}\cos\frac{\pi\alpha}{2} + \omega^\beta L_\beta\cos\frac{\pi\beta}{2}\right)^2 + \left(\omega^\beta L_\beta\sin\frac{\pi\beta}{2} - \frac{1}{\omega^\alpha C_\alpha}\sin\frac{\pi\alpha}{2}\right)^2}} \end{aligned} \tag{6.55}$$

In the same way, the reactive power can be given by

$$Q = \frac{9U^2\sin\left[\arctan\left(\frac{\omega^\beta L_\beta \sin\frac{\pi\beta}{2} - \frac{1}{\omega^\alpha C_\alpha}\sin\frac{\pi\alpha}{2}}{R + \frac{1}{\omega^\alpha C_\alpha}\cos\frac{\pi\alpha}{2} + \omega^\beta L_\beta\cos\frac{\pi\beta}{2}}\right)\right]}{\sqrt{\left(R + \frac{1}{\omega^\alpha C_\alpha}\cos\frac{\pi\alpha}{2} + \omega^\beta L_\beta\cos\frac{\pi\beta}{2}\right)^2 + \left(\omega^\beta L_\beta\sin\frac{\pi\beta}{2} - \frac{1}{\omega^\alpha C_\alpha}\sin\frac{\pi\alpha}{2}\right)^2}} \tag{6.56}$$

Then, the apparent power S can be derived as

$$S = \frac{9U^2}{\sqrt{\left(R + \frac{1}{\omega^\alpha C_\alpha}\cos\frac{\pi\alpha}{2} + \omega^\beta L_\beta\cos\frac{\pi\beta}{2}\right)^2 + \left(\omega^\beta L_\beta\sin\frac{\pi\beta}{2} - \frac{1}{\omega^\alpha C_\alpha}\sin\frac{\pi\alpha}{2}\right)^2}} \tag{6.57}$$

It can be seen from the above analysis that the active power P, reactive power Q, and apparent power S of the symmetrical three-phase fractional-order circuit are three times the single-phase power, and the power can be changed by adjusting the fractional orders of the capacitor and inductor.

To illustrate the power characteristics of a symmetrical three-phase fractional-order circuit, some specific examples are given as follows:

1. When the fractional order $\beta = 1$ and α changes, the active power P, reactive power Q, and apparent power S of a single-phase circuit vary with order α, as shown in Fig. 6.19.

As presented in Fig. 6.19, the active power and reactive power change slowly at the beginning with the increase of fractional order α, and when α reaches a certain

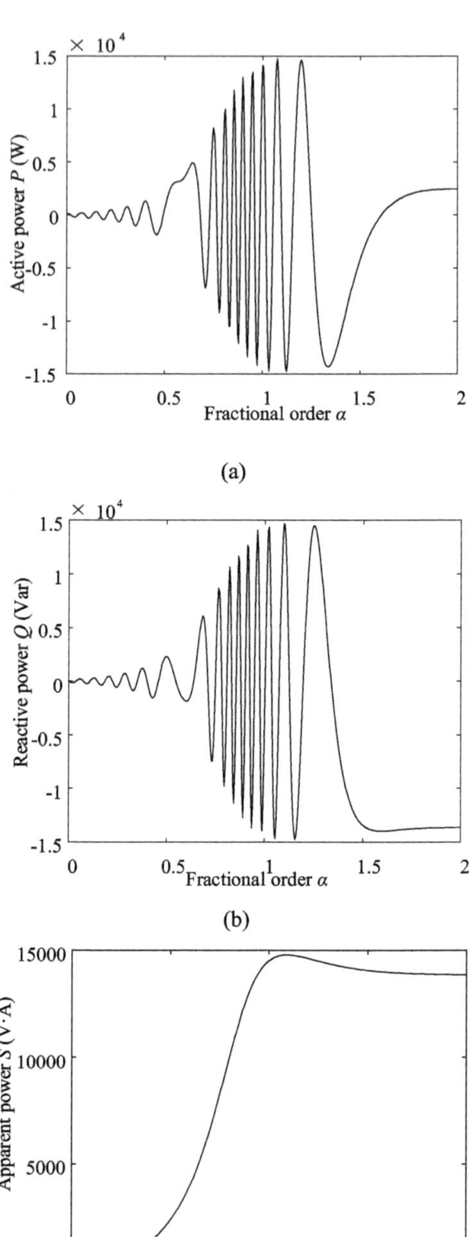

Fig. 6.19 Power versus fractional order α: **a** Active power P versus α. **b** Reactive power Q versus α. **c** Apparent power S versus α (Circuit parameters are $U = 220$ V, $f = 50$ Hz, $R = 10\ \Omega$, $C_\alpha = 0.001$ F/s$^{1-\alpha}$, $L_\beta = L_1 = 0.01$ H)

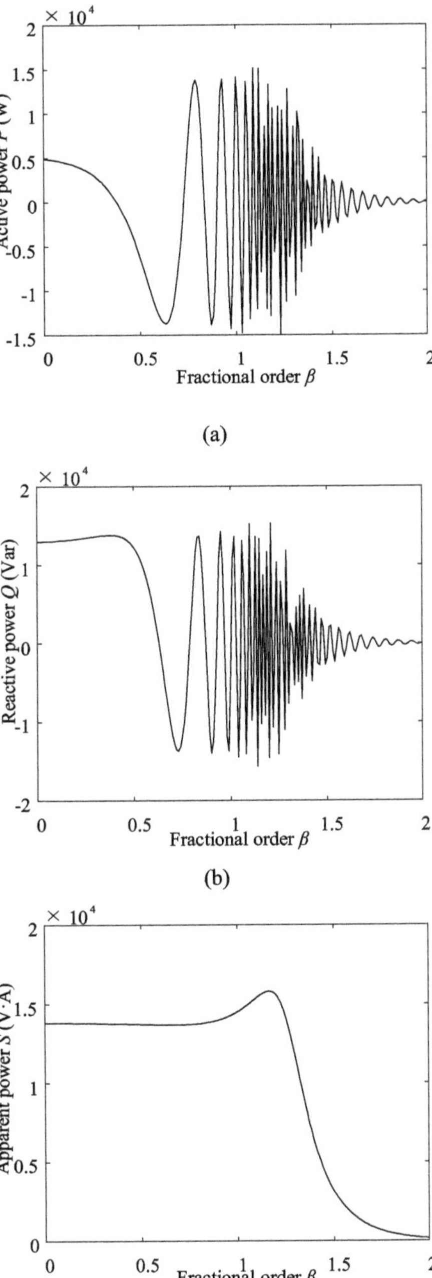

Fig. 6.20 Power versus fractional order β: **a** Active power P versus β. **b** Reactive power Q versus β. **c** Apparent power S versus β (Circuit parameters are $U = 220$ V, $f = 50$ Hz, $R = 10\Omega$, $C_\alpha = C_1 = 0.001$ F, $L_\beta = 0.01$ H/s$^{1-\beta}$)

value, they oscillate violently. Finally, when α is greater than a specific value, the active power and reactive power tend to be constant. The apparent power increases continuously to the maximum value as α increases, and then it starts to decrease slowly with the increase of α. It can be found that the apparent power is ultimately insensitive to the variations of order.

2. When the fractional order $\alpha = 1$ and β changes, the active power P, reactive power Q and apparent power S of a single-phase circuit can be described in Fig. 6.20. It can be found that as β increases, the active power first decreases slowly, then begins to oscillate violently, and finally stabilizes gradually. The reactive power remains constant at first, then begins to oscillate violently, and finally becomes stable when β is greater than a specific value. For the apparent power, as β increases, it remains constant at the beginning. When β exceeds a certain value, the apparent power increases to the maximum value, and then rapidly decrease.

3. When orders α and β change simultaneously, Fig. 6.21 shows the variation of active power P, reactive power Q, and apparent power S versus orders α and β.

As shown in Fig. 6.21, there are many peaks in the square area enclosed by fractional orders $0 < \alpha < 2, 0 < \beta < 2$, the main reason for this phenomenon is that the negative resistance characteristic of fractional-order components. This negative resistance characteristic makes the equivalent impedance of the entire circuit equal to zero under certain orders, resulting in infinite current, which should be avoided by choosing appropriate orders in the actual sinusoidal steady-state three-phase fractional-order circuit.

There are three measurement methods for the active power measurement of the symmetrical sinusoidal steady-state three-phase fractional-order circuit as follows:

1. One wattmeter method uses only one wattmeter to measure the power of a certain phase in the symmetrical three-phase fractional-order circuit and then multiplies the reading values by three times to calculate the power of the three-phase circuit. It should be noted that this measurement method is only suitable for symmetrical three-phase fractional-order circuits.

For a symmetrical Y/Y sinusoidal steady-state three-phase fractional-order circuit, a wattmeter can be used to measure the active power directly. As shown in Fig. 6.22, the potential coil of the wattmeter is connected to the phase voltage, and the current coil is connected to the phase current of the phase current.

Assuming that the voltage phasor is $\dot{U}_A = \sqrt{2}U\angle 0^\circ$, the reading value of the wattmeter P is

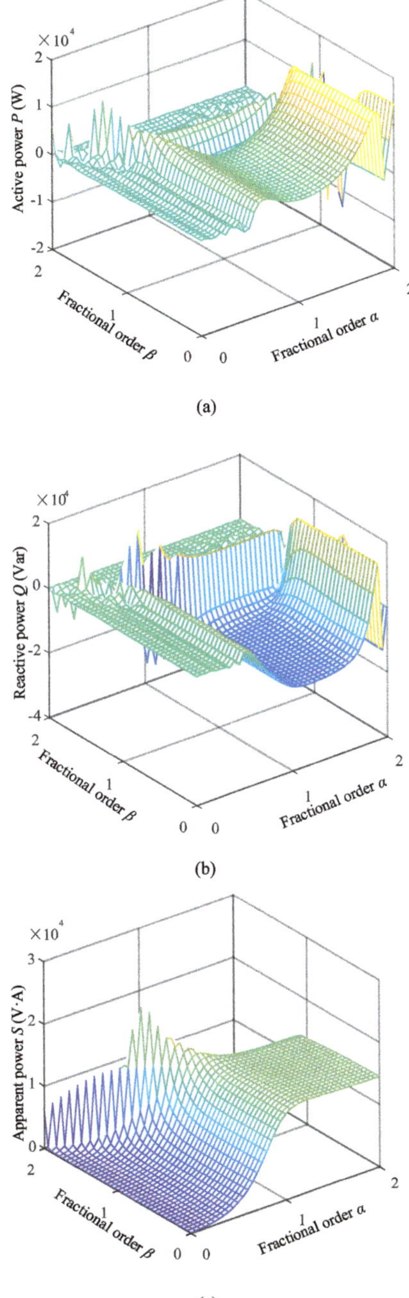

Fig. 6.21 Power versus fractional orders α and β: **a** Active power P versus α and β. **b** Reactive power Q versus α and β. **c** Apparent power S versus α and β (Circuit parameters are $U = 220$ V, $f = 50$ Hz, $R = 10\ \Omega$, $C_\alpha = 0.001$ F/s$^{1-\alpha}$, $L_\beta = 0.01$ H/s$^{1-\beta}$)

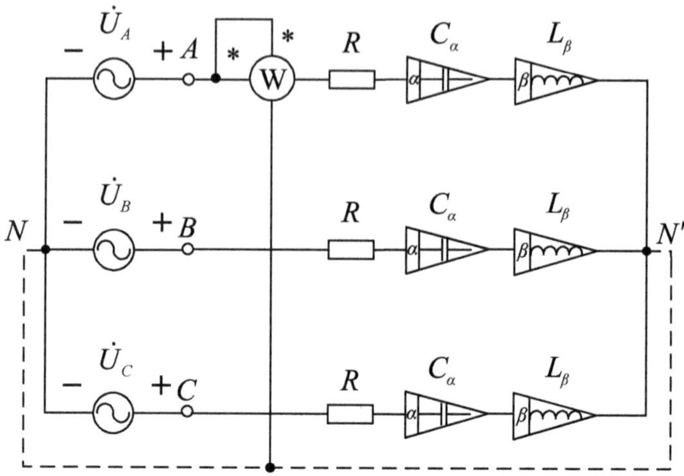

Fig. 6.22 One wattmeter method of the symmetrical *Y*/*Y* three-phase fractional-order circuit

$$P = \mathrm{Re}\left[\dot{U}_A \dot{I}_A^*\right] = \frac{U^2 \cos\left[\arctan\left(\frac{\omega^\beta L_\beta \sin\frac{\pi\beta}{2} - \frac{1}{\omega^\alpha C_\alpha}\sin\frac{\pi\alpha}{2}}{R + \frac{1}{\omega^\alpha C_\alpha}\cos\frac{\pi\alpha}{2} + \omega^\beta L_\beta \cos\frac{\pi\beta}{2}}\right)\right]}{\sqrt{\left(R + \frac{1}{\omega^\alpha C_\alpha}\cos\frac{\pi\alpha}{2} + \omega^\beta L_\beta \cos\frac{\pi\beta}{2}\right)^2 + \left(\omega^\beta L_\beta \sin\frac{\pi\beta}{2} - \frac{1}{\omega^\alpha C_\alpha}\sin\frac{\pi\alpha}{2}\right)^2}} \tag{6.58}$$

The total power of the symmetrical *Y*/*Y* three-phase fractional-order circuit is

$$P_{\mathrm{sum}} = \frac{3U^2 \cos\left[\arctan\left(\frac{\omega^\beta L_\beta \sin\frac{\pi\beta}{2} - \frac{1}{\omega^\alpha C_\alpha}\sin\frac{\pi\alpha}{2}}{R + \frac{1}{\omega^\alpha C_\alpha}\cos\frac{\pi\alpha}{2} + \omega^\beta L_\beta \cos\frac{\pi\beta}{2}}\right)\right]}{\sqrt{\left(R + \frac{1}{\omega^\alpha C_\alpha}\cos\frac{\pi\alpha}{2} + \omega^\beta L_\beta \cos\frac{\pi\beta}{2}\right)^2 + \left(\omega^\beta L_\beta \sin\frac{\pi\beta}{2} - \frac{1}{\omega^\alpha C_\alpha}\sin\frac{\pi\alpha}{2}\right)^2}} \tag{6.59}$$

For the symmetrical *Y*/Δ three-phase fractional-order circuit, it is usually necessary to insert a set of *Y*-connected loads to artificially generate a neutral point. As shown in Fig. 6.23, the current coil of the wattmeter is connected to the line current, and the potential coil is connected to the phase voltage. In this case, the active power of the load can be measured directly by the wattmeter.

Supposing that voltage phasor is $\dot{U}_A = \sqrt{2}U\angle 0°$, according to (6.52), the reading value of the wattmeter *P* can be calculated as

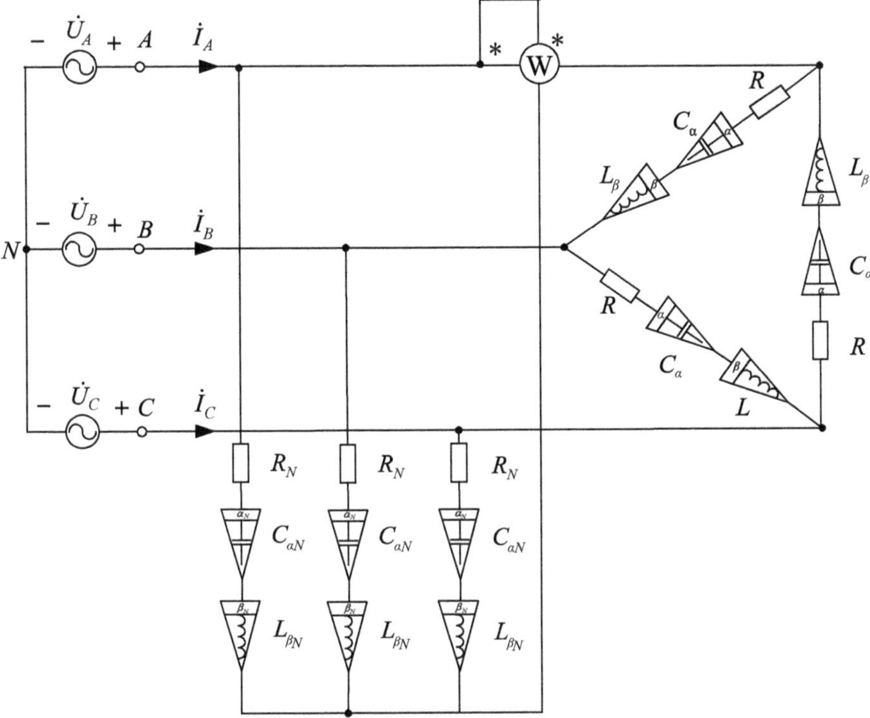

Fig. 6.23 One wattmeter method of a symmetrical Y/Δ three-phase fractional-order circuit

$$P = \mathrm{Re}\big[\dot{U}_A \dot{I}_A^*\big] = \frac{3U^2 \cos\left[\arctan\left(\frac{\omega^\beta L_\beta \sin\frac{\pi\beta}{2} - \frac{1}{\omega^\alpha C_\alpha}\sin\frac{\pi\alpha}{2}}{R + \frac{1}{\omega^\alpha C_\alpha}\cos\frac{\pi\alpha}{2} + \omega^\beta L_\beta \cos\frac{\pi\beta}{2}}\right)\right]}{\sqrt{\left(R + \frac{1}{\omega^\alpha C_\alpha}\cos\frac{\pi\alpha}{2} + \omega^\beta L_\beta \cos\frac{\pi\beta}{2}\right)^2 + \left(\omega^\beta L_\beta \sin\frac{\pi\beta}{2} - \frac{1}{\omega^\alpha C_\alpha}\sin\frac{\pi\alpha}{2}\right)^2}} \tag{6.60}$$

Then, the total power of the symmetrical Y/Δ three-phase fractional-order circuit is

$$P_{\mathrm{sum}} = 3P = \frac{9U^2 \cos\left[\arctan\left(\frac{\omega^\beta L_\beta \sin\frac{\pi\beta}{2} - \frac{1}{\omega^\alpha C_\alpha}\sin\frac{\pi\alpha}{2}}{R + \frac{1}{\omega^\alpha C_\alpha}\cos\frac{\pi\alpha}{2} + \omega^\beta L_\beta \cos\frac{\pi\beta}{2}}\right)\right]}{\sqrt{\left(R + \frac{1}{\omega^\alpha C_\alpha}\cos\frac{\pi\alpha}{2} + \omega^\beta L_\beta \cos\frac{\pi\beta}{2}\right)^2 + \omega^\beta L_\beta \sin\frac{\pi\beta}{2} - \frac{1}{\omega^\alpha C_\alpha}\sin\frac{\pi\alpha}{2}}} \tag{6.61}$$

2. Two wattmeter method is commonly used to measure the active power of the symmetric and three-phase three-wire asymmetrical three-phase fractional-order circuits. The prerequisite for this method to measure the power of a three-phase fractional-order circuit is that the sum of the line currents is zero, that is

$$\dot{I}_A + \dot{I}_B + \dot{I}_C = 0 \tag{6.62}$$

In order to measure the active power of a symmetric three-phase fractional-order circuit, supposing that phase voltage of phase A is $\dot{U}_A = \sqrt{2}U\angle 0°$, the reading values of two wattmeters are P_1, P_2, respectively. Then the total active power can be given by

$$\begin{aligned} P_{\text{sum}} &= \text{Re}\left[\dot{U}_A\dot{I}_A^* + \dot{U}_B\dot{I}_B^* + \dot{U}_C\dot{I}_C^*\right] = \text{Re}\left[\left(\dot{U}_A - \dot{U}_B\right)\dot{I}_A^* + \left(\dot{U}_C - \dot{U}_B\right)\dot{I}_B^*\right] \\ &= \text{Re}\left[\dot{U}_{AB}\dot{I}_A^* + \dot{U}_{CB}\dot{I}_B^*\right] = \text{Re}\left[\dot{U}_{AB}\dot{I}_A^*\right] + \text{Re}\left[\dot{U}_{CB}\dot{I}_B^*\right] \\ &= P_1 + P_2 \end{aligned} \tag{6.63}$$

The reading value of wattmeter 1 is

$$\begin{aligned} P_1 &= \text{Re}\left[\dot{U}_{AB}\dot{I}_A^*\right] \\ &= \frac{\sqrt{3}U^2\cos\left[30° - \arctan\left(\frac{\omega^\beta L_\beta \sin\frac{\pi\beta}{2} - \frac{1}{\omega^\alpha C_\alpha}\sin\frac{\pi\alpha}{2}}{R + \frac{1}{\omega^\alpha C_\alpha}\cos\frac{\pi\alpha}{2} + \omega^\beta L_\beta\cos\frac{\pi\beta}{2}}\right)\right]}{\sqrt{\left(R + \frac{1}{\omega^\alpha C_\alpha}\cos\frac{\pi\alpha}{2} + \omega^\beta L_\beta\cos\frac{\pi\beta}{2}\right)^2 + \left(\omega^\beta L_\beta\sin\frac{\pi\beta}{2} - \frac{1}{\omega^\alpha C_\alpha}\sin\frac{\pi\alpha}{2}\right)^2}} \end{aligned} \tag{6.64}$$

The reading value of wattmeter 2 is

$$\begin{aligned} P_2 &= \text{Re}\left[\dot{U}_{CB}\dot{I}_B^*\right] \\ &= \frac{\sqrt{3}U^2\cos\left[150° + \arctan\left(\frac{\omega^\beta L_\beta \sin\frac{\pi\beta}{2} - \frac{1}{\omega^\alpha C_\alpha}\sin\frac{\pi\alpha}{2}}{R + \frac{1}{\omega^\alpha C_\alpha}\cos\frac{\pi\alpha}{2} + \omega^\beta L_\beta\cos\frac{\pi\beta}{2}}\right)\right]}{\sqrt{\left(R + \frac{1}{\omega^\alpha C_\alpha}\cos\frac{\pi\alpha}{2} + \omega^\beta L_\beta\cos\frac{\pi\beta}{2}\right)^2 + \left(\omega^\beta L_\beta\sin\frac{\pi\beta}{2} - \frac{1}{\omega^\alpha C_\alpha}\sin\frac{\pi\alpha}{2}\right)^2}} \end{aligned} \tag{6.65}$$

From (6.64) and (6.65), it can be seen that as long as the current coil of a wattmeter is connected to the line current of A, the potential coil is connected to the line voltage of AB, while the current coil of the other wattmeter is connected to the line current of C, the potential coil is connected to the line voltage of CB, then the sum of the reading values of the two wattmeters is the active power of the three-phase fractional-order circuit, the connection of the two wattmeters is shown in Fig. 6.24.

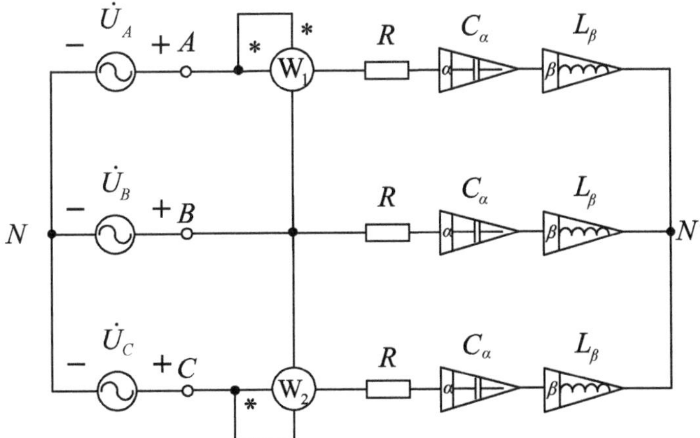

Fig. 6.24 Two wattmeter method

It should be emphasized that when the two wattmeter method is used to measure the active power of the sinusoidal steady-state three-phase fractional-order circuit, only the sum of the reading values of two wattmeters is meaningful. It is meaningless to read only one wattmeter.

3. Three wattmeter method is used to measure the active power of each phase and add them together. This measurement method can measure the active power of any three-phase fractional-order circuits. However, due to the large number of the wattmeter, this method is rarely used in practice applications.

6.4 Summary

The sinusoidal steady-state three-phase fractional-order circuit is a new type of three-phase circuit, which has the properties of fractional-order components and traditional three-phase circuits at the same time. The most prominent feature is that the voltages, currents, and powers of the sinusoidal steady-state three-phase fractional-order circuit can be changed only by adjusting the fractional orders without changing the circuit structure. This chapter first introduces the basic concepts of three-phase fractional-order circuits and then analyzes symmetrical and asymmetrical three-phase fractional-order circuits. Next, the power calculation of the three-phase fractional-order circuit is given. Finally, the power measurement method of the sinusoidal steady-state three-phase fractional-order circuit is stated. Through the above introduction and analysis, beginners can have a preliminary understanding of the three-phase fractional-order circuit. It is hoped that the knowledge in this chapter can provide a theoretical basis for the application of fractional-order components in three-phase circuits.

References

1. Elwakil AS (2010) Fractional-order circuits and systems: An emerging interdisciplinary research area. IEEE Circuits Syst Mag 10:40–50
2. Jesus IS, Tenreiro Machado JA, Boaventure CJ et al (2008) Fractional electrical impedances in botanical components. J Vibr Control 14:1389–1402
3. Hartley TT, Trigeassou JC, Lorenzo CF, Maamri N et al (2015) Energy storage and loss in fractional-order systems. IET Circ Devices Syst 10:227–235
4. Otsuka S (2009) Electric circuit and method for designing electric circuit. U.S. Patent 20090108893
5. Hayt WH Durbin SM (2012) Engineering circuit analysis. McGraw-Hill
6. Bohannan GW, Hurst SK, Spangler L (2006) Electrical component with fractional order impedance. U.S. Patent 20060267595
7. Monje C, Chen Y, Vinagre B et al (2010) Fractional order systems and control—Fundamentals and applications. Springer, London, pp 15–17
8. Westerlund S, Ekstam L (1994) Capacitor theory. IEEE Trans Dielectr Electr Insul 1(5):826–839
9. Charef A, Sun HH, Tsao YY, Onaral B (1992) Fractal system as represented by singularity function. IEEE Trans Autom Control 37(9):1465–1470
10. Özyetkin M, Tan N (2010) Integer order approximation of fractional order systems. In: SIU 2010-IEEE 18th signal processing and communications applications conference, London
11. Radwan AG (2012) Stability analysis of the fractional-order RLC circuit. J Fract Calculus Appl 3(1):1–15
12. Zadeh LA (1962) From circuit theory to system theory. Proc IRE 50(5):856–865
13. Gwynn BT, de Callafon R (2020) Demodulation of three-phase AC power transients in the presence of harmonic distortion. Energies 2(2):1–12

Chapter 7
Non-sinusoidal Periodic Steady-State Analysis of Fractional-Order Circuits

When the fractional-order circuit has nonlinear components or the voltage generated by the generator is non-sinusoidal, stable non-sinusoidal periodic voltage and current will appear in the circuit. The circuit with non-sinusoidal periodic voltage and current in a steady state is defined as the non-sinusoidal periodic steady-state fractional-order circuit, also known as the non-sinusoidal fractional-order circuit.

In the traditional non-sinusoidal periodic steady-state circuit, the basic components that make up the circuit are all integer-order components. However, many studies have shown that the fractional-order circuit has richer characteristics and advantages compared with the integer-order circuit. Therefore, the analysis in this chapter mainly focuses on the non-sinusoidal periodic steady-state circuit with fractional-order components.

In this chapter, only the non-sinusoidal fractional-order circuit composed of linear time-invariant components is discussed. The analysis method is to decompose the non-sinusoidal periodic voltage and current into a series of sinusoidal functions by Fourier series expansion. The frequencies of these sinusoidal functions are integer times of the fundamental frequency. In engineering, they are usually called higher harmonics. Then, the fractional-order circuit can be analyzed separately under a series of harmonics, and the voltage and current can be derived by the superposition theorem [1–3].

7.1 Preliminaries

7.1.1 Harmonic Impedance

In this section, the basic knowledge of harmonic decomposition and harmonic impedance in non-sinusoidal periodic circuits is introduced, which lays the foundation for harmonic analysis in fractional-order circuits.

B. Zhang and X. Shu, *Fractional-Order Electrical Circuit Theory*, CPSS Power Electronics Series, https://doi.org/10.1007/978-981-16-2822-1_7

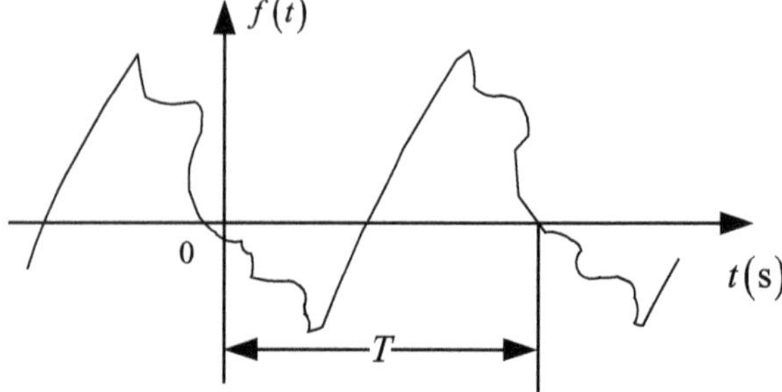

Fig. 7.1 Non-sinusoidal periodic function

The function $f(t)$ that satisfies $f(t) = f(t + kT)$ ($k = 1, 2, ..., n$, T is the period) is defined as the periodic function. If the waveform is non-sinusoidal, it is called the non-sinusoidal periodic function. An example of the non-sinusoidal periodic function is shown in Fig. 7.1.

If $f(t)$ is piecewise continuous and has a finite number of discontinuity points of the first kind and a finite number of maxima and minima in the given interval. This non-sinusoidal periodic function $f(t)$ is said to satisfy the Dirichlet conditions. In general, the non-sinusoidal periodic voltage and current in the fractional-order circuit satisfy the Dirichlet conditions. Any non-sinusoidal periodic function that satisfies Dirichlet conditions can be decomposed into a series of harmonic components by Fourier series expansion.

If the non-sinusoidal periodic function $f(t)$ satisfies the Dirichlet conditions, there is

$$f(t) = \frac{A_0}{2} + \sum_{k=1}^{\infty} B_k \sin k\omega t + \sum_{k=1}^{\infty} C_k \cos k\omega t \tag{7.1}$$

where the coefficients are

$$\begin{cases} A_0 = \frac{2}{T}\int_{-T/2}^{T/2} f(t)\mathrm{d}t = \frac{1}{\pi}\int_{-\pi}^{\pi} f(\omega t)\mathrm{d}(\omega t) \\ B_k = \frac{2}{T}\int_{-T/2}^{T/2} f(t)\sin k\omega t\mathrm{d}t = \frac{1}{\pi}\int_{-\pi}^{\pi} f(t)\sin k\omega t\mathrm{d}(\omega t) \\ C_k = \frac{2}{T}\int_{-T/2}^{T/2} f(t)\cos k\omega t\mathrm{d}t = \frac{1}{\pi}\int_{-\pi}^{\pi} f(t)\cos k\omega t\mathrm{d}(\omega t) \end{cases} \tag{7.2}$$

Figure 7.2 shows a specific example of a half-wave rectification waveform to illustrate the Fourier series expansion. The half-wave rectification waveform can be defined as

$$f(t) = \begin{cases} I_m \sin \omega t, & kT < t < (2k+1)T/2 \\ 0, & (2k+1)T/2 < t < (k+1)T \end{cases} \tag{7.3}$$

According to the Fourier series expansion, the function $f(t)$ can be expressed in the form of (7.1), where the coefficients are

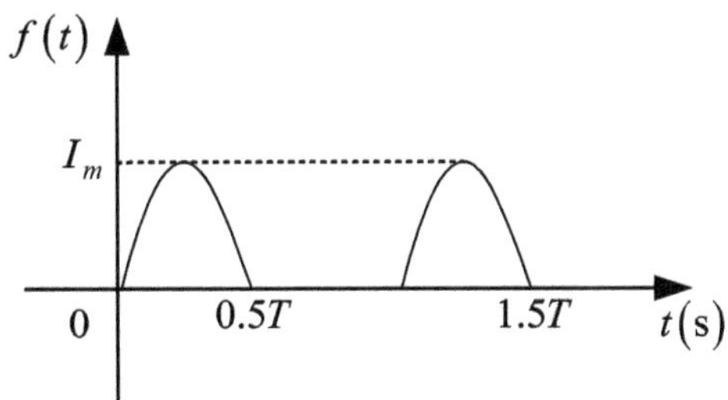

Fig. 7.2 Waveform of the half-wave rectification

$$\begin{cases} A_0 = \frac{1}{\pi}\int_0^{\pi} I_m \sin\omega t \mathrm{d}(\omega t) = \frac{2}{\pi} I_m \\ B_1 = \frac{1}{\pi}\int_0^{\pi} \sin\omega t \mathrm{d}(\omega t) = \frac{1}{2} I_m, \ B_k = 0 \\ C_k = \frac{1}{\pi}\int_0^{\pi} I_m \sin\omega t \cos k\omega t \mathrm{d}(\omega t) \\ \quad = \frac{I_m \sin^2[(k+1)\pi/2]}{(k+1)\pi} - \frac{I_m \sin^2[(k-1)\pi/2]}{(k-1)\pi} \end{cases} (k > 1) \tag{7.4}$$

When k is an even number, the coefficient C_k is $-2I_m/\pi(k+1)(k-1)$. But if k is an odd number, C_k is equal to 0. Then, the Fourier expansion of $f(t)$ in (7.3) can be written as

$$f(t) = \frac{1}{\pi} I_m + \frac{1}{2} I_m \sin\omega t - \frac{2}{3\pi} I_m \cos 2\omega t - \frac{2I_m}{15\pi}\cos 4\omega t + \ldots. \tag{7.5}$$

In the analysis of non-sinusoidal periodic steady-state fractional-order circuits, it should be noted that the impedances of each fractional-order component are different at different frequencies. The impedance of a fractional-order component at the harmonic frequency is called the harmonic impedance.

The capacitive reactance $X_{C\alpha}$ of a fractional-order capacitor is inversely proportional to the frequency. For the k-th harmonic, the capacitive reactance of the fractional-order capacitor is $1/k^{\alpha}$ times that at the fundamental frequency.

The capacitive reactance at the fundamental frequency is

$$X_{C\alpha(1)} = \frac{1}{(j\omega)^{\alpha} C_{\alpha}} = \frac{1}{\omega^{\alpha} C_{\alpha}}\cos\frac{\pi\alpha}{2} - j\frac{1}{\omega^{\alpha} C_{\alpha}}\sin\frac{\pi\alpha}{2} \tag{7.6}$$

Consequently, the capacitive reactance at the k-th harmonic is

$$\begin{aligned} X_{C\alpha(k)} &= \frac{1}{(jk\omega)^{\alpha} C_{\alpha}} = \frac{1}{k^{\alpha}}\frac{1}{(j\omega)^{\alpha} C_{\alpha}} = \frac{1}{k^{\alpha}} X_{C\alpha(1)} \\ &= \frac{1}{k^{\alpha}}\left(\frac{1}{\omega^{\alpha} C_{\alpha}}\cos\frac{\pi\alpha}{2} - j\frac{1}{\omega^{\alpha} C_{\alpha}}\sin\frac{\pi\alpha}{2}\right) \end{aligned} \tag{7.7}$$

For DC, its frequency is regarded as zero, so $X_{C\alpha}(0) = \infty$. Similarly, the inductive reactance $X_{L\beta}$ of the fractional-order inductor is proportional to the frequency. For the k-th harmonic, the inductive reactance of the fractional-order inductor is k^{β} times that at the fundamental frequency.

The inductive reactance at the fundamental frequency is

$$X_{L\beta(1)} = (j\omega)^{\beta} L_{\beta} = \omega^{\beta} L_{\beta} \cos\frac{\pi\beta}{2} + j\omega^{\beta} L_{\beta} \sin\frac{\pi\beta}{2} \tag{7.8}$$

Accordingly, the inductive reactance at the k-th harmonic is given by

$$\begin{aligned} X_{L\beta(k)} &= (jk\omega)^{\beta} L_{\beta} = k^{\beta}(j\omega)^{\beta} L_{\beta} = k^{\beta} X_{L\beta(1)} \\ &= k^{\beta}\left(\omega^{\beta} L_{\beta} \cos\frac{\pi\beta}{2} + j\omega^{\beta} L_{\beta} \sin\frac{\pi\beta}{2}\right) \end{aligned} \tag{7.9}$$

As the frequency approaches zero, the inductive reactance will decrease to zero, causing a short circuit. Thus, for DC, its frequency is regarded as zero, $X_{L\beta(k)} = 0$.

For the ideal resistor, its resistance is constant, that is

$$R_{(k)} = R_{(1)} = R_{(0)} \tag{7.10}$$

Due to the skin effect and the eddy current effect, the resistance increases with increasing frequency. If necessary, a correction coefficient greater than 1 can be introduced, which is related to the frequency. However, for low-frequency circuits, these effects are generally not considered. Unless otherwise specified in this chapter, the resistance is constant.

Therefore, the impedance of the fractional-order series $RL_{\beta}C_{\alpha}$ circuit under the action of the fundamental wave is

$$\begin{aligned} Z_{\alpha\beta(1)} &= R_{(1)} + X_{L\beta(1)} + X_{C\alpha(1)} = R + (j\omega)^{\beta} L_{\beta} + \frac{1}{(j\omega)^{\alpha} C_{\alpha}} \\ &= R + \frac{1}{\omega^{\alpha} C_{\alpha}} \cos\frac{\pi\alpha}{2} + \omega^{\beta} L_{\beta} \cos\frac{\pi\beta}{2} + j\left(\omega^{\beta} L_{\beta} \sin\frac{\pi\beta}{2} - \frac{1}{\omega^{\alpha} C_{\alpha}} \sin\frac{\pi\alpha}{2}\right) \end{aligned} \tag{7.11}$$

The impedances at the k-th harmonic is

$$\begin{aligned} Z_{\alpha\beta(k)} &= R_{(k)} + X_{L\beta(k)} + X_{C\alpha(k)} = R + (jk\omega)^{\beta} L_{\beta} + \frac{1}{(jk\omega)^{\alpha} C_{\alpha}} \\ &= R + \frac{1}{k^{\alpha}} \frac{1}{\omega^{\alpha} C_{\alpha}} \cos\frac{\pi\alpha}{2} + k^{\beta}\omega^{\beta} L_{\beta} \cos\frac{\pi\beta}{2} \\ &\quad + j\left(k^{\beta}\omega^{\beta} L_{\beta} \sin\frac{\pi\beta}{2} - \frac{1}{k^{\alpha}} \frac{1}{\omega^{\alpha} C_{\alpha}} \sin\frac{\pi\alpha}{2}\right) \end{aligned} \tag{7.12}$$

As the DC, the impedances are

$$Z_{\alpha\beta(0)} = R_{(0)} + X_{L\beta(0)} + X_{C\alpha(0)} = \infty \tag{7.13}$$

where $Z_{\alpha\beta(1)}$, $Z_{\alpha\beta(k)}$, and $Z_{\alpha\beta(0)}$ are impedances at the fundamental frequency, k-th harmonic and DC, respectively.

7.1.2 Effective Values

According to the definition of effective value, the effective value of the non-sinusoidal periodic electrical quantity is defined as

$$F = \sqrt{\frac{1}{T}\int_0^T f^2(t)\mathrm{d}t} \tag{7.14}$$

From the above analysis, it can be derived as

$$\begin{aligned} f(t) &= F_{(0)} + \sum_{k=1}^{\infty} B_k \sin k\omega t + \sum_{k=1}^{\infty} C_k \cos k\omega t \\ &= F_{(0)} + \sum_{k=1}^{\infty} F_{(k)} \sin(k\omega t + \varphi_k) \end{aligned} \tag{7.15}$$

$$\begin{aligned} f^2(t) &= F_{(0)}^2 + \sum_{k=1}^{\infty} F_{(k)}^2 \sin^2(k\omega t + \varphi_k) + 2F_0 \sum_{k=1}^{\infty} F_{(k)} \sin(k\omega t + \varphi_k) \\ &= \sum_{\substack{k=1 \\ j=1 \\ k\neq j}}^{\infty} F_{(k)} \sin(k\omega t + \varphi_k) F_{(j)} \sin\left(j\omega t + \varphi_j\right) \end{aligned} \tag{7.16}$$

Based on the orthogonality of trigonometric functions, it can be calculated as

$$\frac{1}{T}\int_0^T f^2(t)\mathrm{d}t = F_{(0)}^2 + \frac{F_{(1)}^2}{2} + \frac{F_{(2)}^2}{2} + \cdots + \frac{F_{(n)}^2}{2} + \cdots = \sum_{k=1}^{\infty} F_{(k)}^2 \tag{7.17}$$

$$F = \sqrt{F_{(0)}^2 + F_{(1)}^2 + F_{(2)}^2 + \cdots + F_{(n)}^2 + \cdots} = \sqrt{\sum_{k=1}^{\infty} F_{(k)}^2} \tag{7.18}$$

Therefore, the effective value of the non-sinusoidal periodic electrical quantity can be calculated by the definition that the effective value is the root mean square value, and it can also be obtained by (7.18). The average value of the non-sinusoidal periodic electrical quantity is its average value in a cycle, also known as its DC component, which is expressed as

$$F_0 = \frac{1}{T}\int_0^T f(t)\mathrm{d}t \tag{7.19}$$

By averaging the absolute value of the non-sinusoidal periodic electrical quantity in a cycle, the average absolute value can be written as

$$F_{\mathrm{a}} = \frac{1}{T}\int_0^T |f(t)|\,\mathrm{d}t \tag{7.20}$$

7.1.3 Power

There are three types of power in non-sinusoidal periodic circuits: active power, reactive power, and apparent power. This section will introduce the power calculation of non-sinusoidal fractional-order circuits [4, 5].

According to the definition of active power, there is

$$\begin{aligned} P &= \frac{1}{T}\int_0^T u(t)i(t)\mathrm{d}t \\ &= \frac{1}{T}\int_0^T \left[\sum_{k=1}^{\infty} U_{(k)}\sin\bigl(k\omega t + \varphi_{u(k)}\bigr) + U_0\right]\left[\sum_{k=1}^{\infty} I_{(k)}\sin\bigl(k\omega t + \varphi_{i(k)}\bigr) + I_0\right]\mathrm{d}t \end{aligned} \tag{7.21}$$

Based on the orthogonality of trigonometric functions, it can be derived as

$$\begin{aligned} P &= U_{(0)}I_{(0)} + \sum_{k=1}^{\infty} U_{(k)}I_{(k)}\cos\bigl(\varphi_{u(k)} - \varphi_{i(k)}\bigr) \\ &= U_{(0)}I_{(0)} + \sum_{k=1}^{\infty} U_{(k)}I_{(k)}\cos\varphi_k = \sum_{k=0}^{\infty} P_{(k)} \end{aligned} \tag{7.22}$$

Rewriting (7.22) in the phasor form, it can be written as

$$P = U_{(0)}I_{(0)} + \sum_{k=1}^{\infty}\mathrm{Re}\bigl[\dot{U}_{(k)}\dot{I}_{(k)}\bigr] = U_{(0)}I_{(0)} + \mathrm{Re}\left[\sum_{k=1}^{\infty}\dot{U}_{(k)}\dot{I}_{(k)}\right] = \mathrm{Re}\left[\sum_{k=0}^{\infty}\dot{U}_{(k)}\dot{I}_{(k)}\right] \tag{7.23}$$

where $\dot{U}_{(k)}$ and $\dot{I}_{(k)}$ are the voltage and current phasors of the k-th harmonic.

Therefore, several conclusions can be derived as follows:

(1) The active power of a non-sinusoidal periodic steady-state fractional-order circuit is the algebraic sum of the active power of each harmonic.
(2) No active power is generated between voltages and currents of different frequencies, and the active power is regarded as zero.

The apparent power S of a non-sinusoidal periodic steady-state fractional-order circuit is defined as the product of the effective values of the branch voltage and current, and the reactive power Q is consumed by a pure inductor or pure capacitor. They can be expressed as

$$\begin{cases} S = UI \\ Q = \sqrt{S^2 - P^2} \end{cases} \tag{7.24}$$

In addition to the active power generated by voltages and currents of the same frequency, power is also exchanged between voltages and currents of different frequencies. Therefore, the reactive power of the non-sinusoidal periodic steady-state circuit is greater than the algebraic sum of the reactive power of each frequency, it can be expressed as

$$\begin{cases} Q = \sqrt{S^2 - P^2} > \sum\limits_{k=1}^{\infty} U_k I_k \sin\varphi_k \\ Q \geq \sum\limits_{k=1}^{\infty} Q_k \end{cases} \tag{7.25}$$

Therefore, in a non-sinusoidal periodic steady-state fractional circuit, if the voltage is sinusoidal and the current is non-sinusoidal, more reactive power will be generated when the same active power is delivered.

7.2 Fractional-Order $RL_\beta C_\alpha$ Circuits

The equivalent sinusoidal wave is an approximate concept. In engineering applications, the non-sinusoidal periodic electric quantity is often approximated by a sinusoidal wave. Therefore, the analysis method of sinusoidal steady-state circuits can be applied to non-sinusoidal steady-state fractional-order circuits. The rules for achieving sinusoidal equivalence are as follows:

(1) The effective value of the equivalent sinusoidal wave is equal to that of the non-sinusoidal periodic wave.
(2) The period of the equivalent sinusoidal wave is set to be equal to that of the fundamental component of the non-sinusoidal periodic quantity.

(3) The phase angle difference between the equivalent voltage and current sinusoidal wave should be set to meet the same condition as the active power of non-sinusoidal and sinusoidal circuits.

In the previous section, some basic knowledge of non-sinusoidal periodic fractional-order circuits was introduced. This section will introduce the analysis method of the non-sinusoidal periodic fractional-order circuit [6–9].

The analytical process of non-sinusoidal periodic fractional-order circuits mainly involves two steps. The first is to perform the Fourier series decomposition on the voltage of the power source, and the second is to calculate the response of the fractional-order circuit at each harmonic frequency.

Taking the series fractional-order $RL_\beta C_\alpha$ circuit shown in Fig. 4.35 as an example, if the voltage source $u_s(t)$ is a non-sinusoidal periodic function, then its Fourier series can be written as

$$\begin{aligned} u_s(t) &= U_{(0)} + \sqrt{2}U_{(1)}\sin(\omega t + \varphi_1) + \sqrt{2}U_{(2)}\sin(2\omega t + \varphi_2) + \cdots \\ &= U_{(0)} + \sum_{k=1}^{\infty}\sqrt{2}U_{(k)}\sin(k\omega t + \varphi_k) \end{aligned} \tag{7.26}$$

Then, the current of the fractional-order circuit under the individual action of each harmonic can be analyzed in turn as follows:

(1) For the DC component, the fractional-order impedance $Z_{\alpha\beta(0)}$ is

$$Z_{\alpha\beta(0)} = R_{(0)} + X_{L\beta(0)} + X_{C\alpha(0)} = \infty \tag{7.27}$$

It can be seen that the entire circuit is equivalent to an open-circuit state, and the current and active power are zero, that is

$$\begin{cases} I_{(0)} = 0 \\ P_{(0)} = 0 \end{cases} \tag{7.28}$$

(2) The fractional-order impedance at the fundamental frequency is

$$\begin{aligned} Z_{\alpha\beta(1)} &= R_{(1)} + X_{L\beta(1)} + X_{C\alpha(1)} = R + (j\omega)^\beta L_\beta + \frac{1}{(j\omega)^\alpha C_\alpha} \\ &= R + \frac{1}{\omega^\alpha C_\alpha}\cos\frac{\pi\alpha}{2} + \omega^\beta L_\beta\cos\frac{\pi\beta}{2} \\ &\quad + j\left(\omega^\beta L_\beta\sin\frac{\pi\beta}{2} - \frac{1}{\omega^\alpha C_\alpha}\sin\frac{\pi\alpha}{2}\right) \end{aligned} \tag{7.29}$$

Assuming that $\dot{U}_{(1)} = \sqrt{2}U_{(1)}\angle 0°$, the current can be calculated as

$$\begin{aligned}\dot{I}_{(1)} &= \frac{\dot{U}_{(1)}}{Z_{\alpha\beta(1)}} = \frac{\dot{U}_{(1)}}{R + \frac{1}{(j\omega)^\alpha C_\alpha} + (j\omega)^\beta L_\beta} \\ &= \frac{\sqrt{2}U_{(1)}\angle 0^\circ}{R + \frac{1}{\omega^\alpha C_\alpha}\cos\frac{\pi\alpha}{2} + \omega^\beta L_\beta \cos\frac{\pi\beta}{2} + j\left(\omega^\beta L_\beta \sin\frac{\pi\beta}{2} - \frac{1}{\omega^\alpha C_\alpha}\sin\frac{\pi\alpha}{2}\right)} \\ &= I_{(1)}\angle - \varphi_1 \end{aligned} \tag{7.30}$$

where

$$\begin{cases} I_{(1)} = \dfrac{\sqrt{2}U_{(1)}}{\sqrt{\begin{array}{l}\left(R + \frac{1}{\omega^\alpha C_\alpha}\cos\frac{\pi\alpha}{2} + \omega^\beta L_\beta \cos\frac{\pi\beta}{2}\right)^2 \\ +\left(\omega^\beta L_\beta \sin\frac{\pi\beta}{2} - \frac{1}{\omega^\alpha C_\alpha}\sin\frac{\pi\alpha}{2}\right)^2\end{array}}} \\ \varphi_1 = \arctan\left(\dfrac{\omega^\beta L_\beta \sin\frac{\pi\beta}{2} - \frac{1}{\omega^\alpha C_\alpha}\sin\frac{\pi\alpha}{2}}{R + \frac{1}{\omega^\alpha C_\alpha}\cos\frac{\pi\alpha}{2} + \omega^\beta L_\beta \cos\frac{\pi\beta}{2}}\right) \end{cases} \tag{7.31}$$

Then, the active power can be obtained as

$$\begin{aligned} P_1 &= \mathrm{Re}\left[\dot{U}_{(1)}\dot{I}^*_{(1)}\right] = \frac{1}{2}I^2_{(1)}\left(R + \frac{1}{\omega^\alpha C_\alpha}\cos\frac{\pi\alpha}{2} + \omega^\beta L_\beta \cos\frac{\pi\beta}{2}\right) \\ &= \frac{U^2_{(1)}\left(R + \frac{1}{\omega^\alpha C_\alpha}\cos\frac{\pi\alpha}{2} + \omega^\beta L_\beta \cos\frac{\pi\beta}{2}\right)}{\left[\begin{array}{l}2\left(R + \frac{1}{\omega^\alpha C_\alpha}\cos\frac{\pi\alpha}{2} + \omega^\beta L_\beta \cos\frac{\pi\beta}{2}\right)^2 \\ +\left(\omega^\beta L_\beta \sin\frac{\pi\beta}{2} - \frac{1}{\omega^\alpha C_\alpha}\sin\frac{\pi\alpha}{2}\right)^2\end{array}\right]} \end{aligned} \tag{7.32}$$

(3) The fractional-order impedance at the 2nd harmonic frequency is

$$\begin{aligned} Z_{\alpha\beta(2)} &= R_{(2)} + X_{L\beta(2)} + X_{C\alpha(2)} = R + (j2\omega)^\beta L_\beta + \frac{1}{(j2\omega)^\alpha C_\alpha} \\ &= \left[\begin{array}{l} R + \frac{1}{2^\alpha}\frac{1}{\omega^\alpha C_\alpha}\cos\frac{\pi\alpha}{2} + 2^\beta\omega^\beta L_\beta \cos\frac{\pi\beta}{2} \\ +j\left(2^\beta\omega^\beta L_\beta \sin\frac{\pi\beta}{2} - \frac{1}{2^\alpha}\frac{1}{\omega^\alpha C_\alpha}\sin\frac{\pi\alpha}{2}\right)\end{array}\right] \end{aligned} \tag{7.33}$$

Assuming that $\dot{U}_{(2)} = \sqrt{2}U_{(2)}\angle 0^\circ$, the current is

$$\dot{I}_{(2)} = \frac{\dot{U}_{(2)}}{Z_{\alpha\beta(2)}} = \frac{\dot{U}_{(2)}}{R + (j2\omega)^\beta L_\beta + \frac{1}{(j2\omega)^\alpha C_\alpha}}$$

$$= \frac{\sqrt{2}U_{(2)}\angle 0^\circ}{\left[\begin{array}{l} R + \dfrac{1}{2^\alpha}\dfrac{1}{\omega^\alpha C_\alpha}\cos\dfrac{\pi\alpha}{2} + 2^\beta\omega^\beta L_\beta \cos\dfrac{\pi\beta}{2} \\ +j\left(2^\beta\omega^\beta L_\beta \sin\dfrac{\pi\beta}{2} - \dfrac{1}{2^\alpha}\dfrac{1}{\omega^\alpha C_\alpha}\sin\dfrac{\pi\alpha}{2}\right)\end{array}\right]}$$
$$= I_{(2)}\angle - \varphi_2 \tag{7.34}$$

where

$$\begin{cases} I_{(2)} = \dfrac{\sqrt{2}U_{(2)}}{\sqrt{\begin{array}{l}\left(R + \frac{1}{2^\alpha}\frac{1}{\omega^\alpha C_\alpha}\cos\frac{\pi\alpha}{2} + 2^\beta\omega^\beta L_\beta\cos\frac{\pi\beta}{2}\right)^2 \\ +\left(2^\beta\omega^\beta L_\beta\sin\frac{\pi\beta}{2} - \frac{1}{2^\alpha}\frac{1}{\omega^\alpha C_\alpha}\sin\frac{\pi\alpha}{2}\right)^2\end{array}}} \\ \varphi_2 = \arctan\left(\dfrac{2^\beta\omega^\beta L_\beta\sin\frac{\pi\beta}{2} - \frac{1}{2^\alpha}\frac{1}{\omega^\alpha C_\alpha}\sin\frac{\pi\alpha}{2}}{R + \frac{1}{2^\alpha}\frac{1}{\omega^\alpha C_\alpha}\cos\frac{\pi\alpha}{2} + 2^\beta\omega^\beta L_\beta\cos\frac{\pi\beta}{2}}\right) \end{cases} \tag{7.35}$$

Similarly, the active power can be derived as

$$P_2 = \mathrm{Re}\left[\dot{U}_{(2)}\dot{I}^*_{(2)}\right] = \frac{1}{2}I^2_{(2)}\left(R + \frac{1}{2^\alpha}\frac{1}{\omega^\alpha C_\alpha}\cos\frac{\pi\alpha}{2} + 2^\beta\omega^\beta L_\beta\cos\frac{\pi\beta}{2}\right)$$
$$= \frac{U^2_{(2)}\left(R + \frac{1}{2^\alpha}\frac{1}{\omega^\alpha C_\alpha}\cos\frac{\pi\alpha}{2} + 2^\beta\omega^\beta L_\beta\cos\frac{\pi\beta}{2}\right)}{\left[\begin{array}{l} 2\left(R + \dfrac{1}{2^\alpha}\dfrac{1}{\omega^\alpha C_\alpha}\cos\dfrac{\pi\alpha}{2} + 2^\beta\omega^\beta L_\beta\cos\dfrac{\pi\beta}{2}\right)^2 \\ +\left(2^\beta\omega^\beta L_\beta\sin\dfrac{\pi\beta}{2} - \dfrac{1}{2^\alpha}\dfrac{1}{\omega^\alpha C_\alpha}\sin\dfrac{\pi\alpha}{2}\right)^2\end{array}\right]} \tag{7.36}$$

(4) Under the action of the 3rd harmonic component, the impedance can be derived as

$$Z_{\alpha\beta(3)} = R_{(3)} + X_{L\beta(3)} + X_{C\alpha(3)} = R + (j3\omega)^\beta L_\beta + \frac{1}{(j3\omega)^\alpha C_\alpha}$$
$$= \left[\begin{array}{l} R + \dfrac{1}{3^\alpha}\dfrac{1}{\omega^\alpha C_\alpha}\cos\dfrac{\pi\alpha}{2} + 3^\beta\omega^\beta L_\beta\cos\dfrac{\pi\beta}{2} \\ +j\left(3^\beta\omega^\beta L_\beta\sin\dfrac{\pi\beta}{2} - \dfrac{1}{3^\alpha}\dfrac{1}{\omega^\alpha C_\alpha}\sin\dfrac{\pi\alpha}{2}\right)\end{array}\right] \tag{7.37}$$

Supposing that $\dot{U}_{(3)} = \sqrt{2}U_{(3)}\angle 0^\circ$, the current is

$$\dot{I}_{(3)} = \frac{\dot{U}_{(3)}}{Z_{\alpha\beta(3)}} = \frac{\dot{U}_{(3)}}{R + (j3\omega)^\beta L_\beta + \frac{1}{(j3\omega)^\alpha C_\alpha}}$$

$$
\begin{aligned}
&= \frac{\sqrt{2}U_{(3)}\angle 0^\circ}{\left[\begin{array}{l} R + \dfrac{1}{3^\alpha}\dfrac{1}{\omega^\alpha C_\alpha}\cos\dfrac{\pi\alpha}{2} + 3^\beta\omega^\beta L_\beta\cos\dfrac{\pi\beta}{2} \\ +j\left(3^\beta\omega^\beta L_\beta\sin\dfrac{\pi\beta}{2} - \dfrac{1}{3^\alpha}\dfrac{1}{\omega^\alpha C_\alpha}\sin\dfrac{\pi\alpha}{2}\right)\end{array}\right]} \\
&= I_{(3)}\angle - \varphi_3
\end{aligned}
\tag{7.38}
$$

where

$$
\begin{cases}
I_{(3)} = \dfrac{\sqrt{2}U_{(3)}}{\sqrt{\begin{array}{l}\left(R + \frac{1}{3^\alpha}\frac{1}{\omega^\alpha C_\alpha}\cos\frac{\pi\alpha}{2} + 3^\beta\omega^\beta L_\beta\cos\frac{\pi\beta}{2}\right)^2 \\ +\left(3^\beta\omega^\beta L_\beta\sin\frac{\pi\beta}{2} - \frac{1}{3^\alpha}\frac{1}{\omega^\alpha C_\alpha}\sin\frac{\pi\alpha}{2}\right)^2\end{array}}} \\
\varphi_3 = \arctan\left(\dfrac{3^\beta\omega^\beta L_\beta\sin\frac{\pi\beta}{2} - \frac{1}{3^\alpha}\frac{1}{\omega^\alpha C_\alpha}\sin\frac{\pi\alpha}{2}}{R + \frac{1}{3^\alpha}\frac{1}{\omega^\alpha C_\alpha}\cos\frac{\pi\alpha}{2} + 3^\beta\omega^\beta L_\beta\cos\frac{\pi\beta}{2}}\right)
\end{cases}
\tag{7.39}
$$

The active power is

$$
\begin{aligned}
P_3 &= \mathrm{Re}\left[\dot{U}_{(3)}\dot{I}_{(3)}^*\right] = \frac{1}{2}I_{(3)}^2\left(R + \frac{1}{3^\alpha}\frac{1}{\omega^\alpha C_\alpha}\cos\frac{\pi\alpha}{2} + 3^\beta\omega^\beta L_\beta\cos\frac{\pi\beta}{2}\right) \\
&= \frac{U_{(3)}^2\left(R + \frac{1}{3^\alpha}\frac{1}{\omega^\alpha C_\alpha}\cos\frac{\pi\alpha}{2} + 3^\beta\omega^\beta L_\beta\cos\frac{\pi\beta}{2}\right)}{\left[\begin{array}{l} 2\left(R + \dfrac{1}{3^\alpha}\dfrac{1}{\omega^\alpha C_\alpha}\cos\dfrac{\pi\alpha}{2} + 3^\beta\omega^\beta L_\beta\cos\dfrac{\pi\beta}{2}\right)^2 \\ +\left(3^\beta\omega^\beta L_\beta\sin\dfrac{\pi\beta}{2} - \dfrac{1}{3^\alpha}\dfrac{1}{\omega^\alpha C_\alpha}\sin\dfrac{\pi\alpha}{2}\right)^2\end{array}\right]}
\end{aligned}
\tag{7.40}
$$

Then, the effective values of current I and active power P can be calculated as

$$
\begin{cases}
I = \sqrt{I_{(0)}^2 + I_{(1)}^2 + I_{(2)}^2 + I_{(3)}^2 + \cdots} = \sqrt{\sum\limits_{k=1}^{\infty} I_{(k)}^2} \\
P = P_0 + P_1 + P_2 + P_3 + \cdots = \sum\limits_{k=1}^{\infty} P_k
\end{cases}
\tag{7.41}
$$

7.3 Symmetrical Three-Phase Fractional-Order Circuits

Generally, the voltage generated by the generator contains some harmonic components, and the exciting current of the transformer also has some harmonic components. Therefore, in the three-phase circuit, both voltage and current may have high-order harmonic components [7, 10].

For the symmetrical three-phase fractional-order circuit, if the phase voltage of phase A is $u_A(t)$, the phase voltages of phase B and phase C can be expressed as $u_B(t) = u_A(t-2\pi/3)$ and $u_C(t) = u_A(t + 2\pi/3)$, respectively.

Assuming that $u_A(t)$ can be written as

$$\begin{aligned} u_A(t) &= \sqrt{2}U_{(1)}\sin(\omega_1 t + \varphi_1) + \sqrt{2}U_{(3)}\sin(3\omega_1 t + \varphi_3) \\ &\quad + \sqrt{2}U_{(5)}\sin(5\omega_1 t + \varphi_5) + \cdots \\ &= \sqrt{2}\sum_{k=1}^{\infty} U_{(2k+1)}\sin((2k+1)\omega_1 t + \phi_{2k+1}) \end{aligned} \tag{7.42}$$

then $u_B(t)$ and $u_C(t)$ can be expressed as

$$\begin{aligned} u_B(t) &= \sqrt{2}U_{(1)}\sin\left(\omega_1 t - \frac{2\pi}{3} + \varphi_1\right) + \sqrt{2}U_{(3)}\sin(3\omega_1 t + \varphi_3) \\ &\quad + \sqrt{2}U_{(5)}\sin\left(5\omega_1 t - \frac{4\pi}{3} + \varphi_5\right) + \cdots \\ &= \sqrt{2}\sum_{k=1}^{\infty} U_{(2k+1)}\sin\left((2k+1)\omega_1 t + \phi_{2k+1} - \frac{2k\pi}{3}\right) \end{aligned} \tag{7.43}$$

$$\begin{aligned} u_C(t) &= \sqrt{2}U_{(1)}\sin\left(\omega_1 t - \frac{4\pi}{3} + \varphi_1\right) + \sqrt{2}U_{(3)}\sin(3\omega_1 t + \varphi_3) \\ &\quad + \sqrt{2}U_{(5)}\sin\left(5\omega_1 t - \frac{2\pi}{3} + \varphi_5\right) + \cdots \\ &= \sqrt{2}\sum_{k=1}^{\infty} U_{(2k+1)}\sin\left((2k+1)\omega_1 t + \phi_{2k+1} + \frac{2k\pi}{3}\right) \end{aligned} \tag{7.44}$$

where $k = 1, 3, 5, \ldots$.

Based on the phase difference between the three-phase voltages, the symmetry of each harmonic voltage is discussed as follows:

(1) When $k = 6n + 1$ ($n = 0, 1, 2, \cdots$), the initial phases of harmonic voltages of each phase are

$$\begin{cases} \phi_A = \phi_k \\ \phi_B = \phi_k - 4n\pi - \frac{2\pi}{3} \\ \phi_C = \phi_k + 4n\pi + \frac{2\pi}{3} \end{cases} \tag{7.45}$$

Obviously, each of these harmonics constitutes a positive-sequence three-phase symmetrical voltage source.

(2) When $k = 6n + 3$ ($n = 0, 1, 2, \cdots$), the initial phase of harmonic voltages of each phase are

$$\begin{cases} \phi_A = \phi_k \\ \phi_B = \phi_k - 4n\pi - 2\pi \\ \phi_C = \phi_k + 4n\pi + 2\pi \end{cases} \tag{7.46}$$

The three-phase voltages of each harmonic frequency have the same phase, which is called the zero-sequence component in engineering.

(3) When $k = 6n + 5$ ($n = 0, 1, 2, \cdots$), the initial phases of harmonic voltages of each phase are

$$\begin{cases} \phi_A = \phi_k \\ \phi_B = \phi_k - 4n\pi - 4\pi + \frac{2\pi}{3} \\ \phi_C = \phi_k + 4n\pi + 4\pi - \frac{2\pi}{3} \end{cases} \tag{7.47}$$

It can be seen that each of these harmonics constitutes a negative-sequence three-phase symmetrical voltage source.

In conclusion, there are positive-sequence, zero-sequence, and negative-sequence components in the non-sinusoidal periodic symmetrical three-phase fractional-order circuit. The analysis of a non-sinusoidal periodic symmetric three-phase fractional-order circuit is transformed into an analysis under the action of different harmonic components.

In the non-sinusoidal periodic symmetrical three-phase fractional-order circuit, it is necessary to analyze the above three types of harmonics separately. For positive-sequence and negative-sequence components, the methods introduced in Chap. 6 can be used directly, while for the zero-sequence component, its analysis method is related to the connection method of the circuit. Here, the *Y/Y* symmetrical three-phase fractional-order circuit shown in Fig. 6.4a is taken as an example for illustration.

Each phase voltage is decomposed into the form of periodic harmonics, which is expressed as

$$\begin{cases} u_A(t) = \sqrt{2}U_{(1)}\sin(\omega t + \varphi_1) + \sqrt{2}U_{(3)}\sin(3\omega t + \varphi_3)\cdots \\ \quad = \sqrt{2}\sum\limits_{k=1}^{\infty} U_{(2k+1)}\sin((2k+1)\omega_1 t + \phi_{2k+1}) \\ u_B(t) = \sqrt{2}U_{(1)}\sin\left(\omega t + \varphi_1 - \frac{2\pi}{3}\right) + \sqrt{2}U_{(3)}\sin\left(3\omega t + \varphi_3 - \frac{2\pi}{3}\right)\cdots \\ \quad = \sqrt{2}\sum\limits_{k=1}^{\infty} U_{(2k+1)}\sin\left((2k+1)\omega_1 t + \phi_{2k+1} - \frac{2k\pi}{3}\right) \\ u_C(t) = \sqrt{2}U_{(1)}\sin\left(\omega t + \varphi_1 + \frac{2\pi}{3}\right) + \sqrt{2}U_{(3)}\sin\left(3\omega t + \varphi_3 - \frac{2\pi}{3}\right)\cdots \\ \quad = \sqrt{2}\sum\limits_{k=1}^{\infty} U_{(2k+1)}\sin\left((2k+1)\omega_1 t + \phi_{2k+1} + \frac{2k\pi}{3}\right) \end{cases} \tag{7.48}$$

Rewriting (7.48) in the form of harmonic components to obtain

$$\begin{cases} u_A = u_{A(1)} + u_{A(3)} + u_{A(5)} + \cdots = \sum_{k=1}^{\infty} u_{A(k)} \\ u_B = u_{B(1)} + u_{B(3)} + u_{B(5)} + \cdots = \sum_{k=1}^{\infty} u_{B(k)} \\ u_C = u_{C(1)} + u_{C(3)} + u_{C(5)} + \cdots = \sum_{k=1}^{\infty} u_{C(k)} \end{cases} \tag{7.49}$$

where u_{A1}, u_{A3} and u_{A5} are the positive-sequence, zero-sequence, and negative-sequence voltages, respectively. In the three-phase fractional-order circuit, the positive-sequence, negative-sequence, and zero-sequence components are dominant in harmonics. Therefore, it only needs to analyze the effect of these three types of harmonic components.

(1) For the fundamental component (positive-sequence), the equivalent circuit is shown in Fig. 7.3.

The fractional-order impedance $Z_{\alpha\beta(1)}$ is

$$Z_{\alpha\beta(1)} = R_{(1)} + X_{L\beta(1)} + X_{C\alpha(1)} = R + (j\omega)^{\beta} L_{\beta} + \frac{1}{(j\omega)^{\alpha} C_{\alpha}} \tag{7.50}$$

Due to the symmetry of the three-phase fractional-order load, the neutral-point voltage is $\dot{U}_{NN'(1)} = 0$, assuming that

$$\begin{cases} \dot{U}_{A(1)} = \sqrt{2} U_{(1)} \angle 0^{\circ} \\ \dot{U}_{B(1)} = \sqrt{2} U_{(1)} \angle -120^{\circ} \\ \dot{U}_{C(1)} = \sqrt{2} U_{(1)} \angle 120^{\circ} \end{cases} \tag{7.51}$$

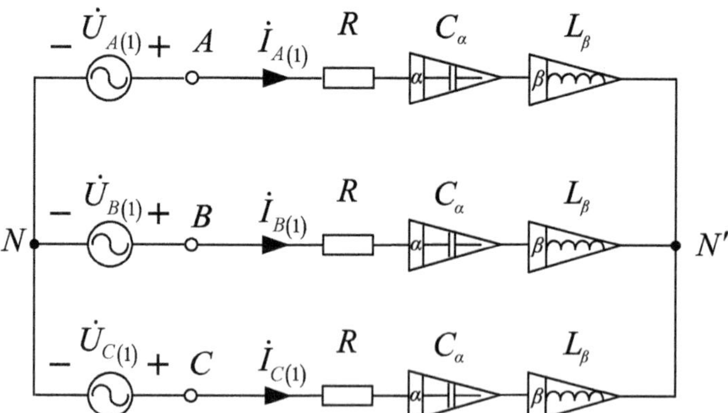

Fig. 7.3 Equivalent circuit for fundamental component

The current of phase A can be obtained as

$$\begin{aligned}\dot{I}_{A(1)} &= \frac{\dot{U}_{A(1)} - \dot{U}_{NN'(1)}}{Z_{\alpha\beta(1)}} = \frac{\dot{U}_{A(1)}}{Z_{\alpha\beta(1)}} = \frac{\dot{U}_{A(1)}}{R + \frac{1}{(j\omega)^{\alpha} C_{\alpha}} + (j\omega)^{\beta} L_{\beta}} \\ &= I_{(1)}\angle - \varphi_1 \end{aligned} \tag{7.52}$$

In the same way, the currents of phase B and C are

$$\begin{aligned}\dot{I}_{B(1)} &= \frac{\dot{U}_{B(1)} - \dot{U}_{NN'(1)}}{Z_{\alpha\beta(1)}} = \frac{\dot{U}_{B(1)}}{Z_{\alpha\beta(1)}} = \frac{\dot{U}_{B(1)}}{R + \frac{1}{(j\omega)^{\alpha} C_{\alpha}} + (j\omega)^{\beta} L_{\beta}} \\ &= I_{(1)}\angle(-\varphi_1 - 120^\circ) \end{aligned} \tag{7.53}$$

$$\begin{aligned}\dot{I}_{C(1)} &= \frac{\dot{U}_{C(1)} - \dot{U}_{NN'(1)}}{Z_{\alpha\beta(1)}} = \frac{\dot{U}_{C(1)}}{Z_{\alpha\beta(1)}} = \frac{\dot{U}_{C(1)}}{R + \frac{1}{(j\omega)^{\alpha} C_{\alpha}} + (j\omega)^{\beta} L_{\beta}} \\ &= I_{(1)}\angle(-\varphi_1 + 120^\circ) \end{aligned} \tag{7.54}$$

where

$$\begin{cases} I_{(1)} = \dfrac{\sqrt{2}U_{(1)}}{\sqrt{\left(R + \frac{1}{\omega^{\alpha} C_{\alpha}} \cos\frac{\pi\alpha}{2} + \omega^{\beta} L_{\beta} \cos\frac{\pi\beta}{2}\right)^2 + \left(\omega^{\beta} L_{\beta} \sin\frac{\pi\beta}{2} - \frac{1}{\omega^{\alpha} C_{\alpha}} \sin\frac{\pi\alpha}{2}\right)^2}} \\ \varphi_1 = \arctan\left(\dfrac{\omega^{\beta} L_{\beta} \sin\frac{\pi\beta}{2} - \frac{1}{\omega^{\alpha} C_{\alpha}} \sin\frac{\pi\alpha}{2}}{R + \frac{1}{\omega^{\alpha} C_{\alpha}} \cos\frac{\pi\alpha}{2} + \omega^{\beta} L_{\beta} \cos\frac{\pi\beta}{2}}\right) \end{cases} \tag{7.55}$$

(2) As for the 5th harmonic component (negative-sequence component), the equivalent circuit is described in Fig. 7.4.

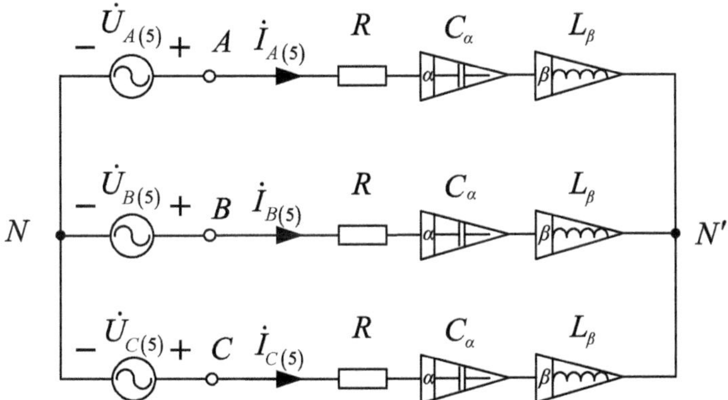

Fig. 7.4 Equivalent circuit for the 5th harmonic component

The fractional-order impedance at 5th harmonic frequency $Z_{\alpha\beta(5)}$ is

$$Z_{\alpha\beta(5)} = R_{(5)} + X_{L\beta(5)} + X_{C\alpha(5)} = R + (j5\omega)^{\beta} L_{\beta} + \frac{1}{(j5\omega)^{\alpha} C_{\alpha}}$$
$$= \begin{bmatrix} R + \frac{1}{5^{\alpha}} \frac{1}{\omega^{\alpha} C_{\alpha}} \cos\frac{\pi\alpha}{2} + 5^{\beta}\omega^{\beta} L_{\beta} \cos\frac{\pi\beta}{2} \\ +j\left(5^{\beta}\omega^{\beta} L_{\beta} \sin\frac{\pi\beta}{2} - \frac{1}{5^{\alpha}} \frac{1}{\omega^{\alpha} C_{\alpha}} \sin\frac{\pi\alpha}{2}\right) \end{bmatrix} \tag{7.56}$$

Assuming that

$$\begin{cases} \dot{U}_{A(5)} = \sqrt{2} U_{(5)} \angle 0^{\circ} \\ \dot{U}_{B(5)} = \sqrt{2} U_{(5)} \angle 120^{\circ} \\ \dot{U}_{C(5)} = \sqrt{2} U_{(5)} \angle -120^{\circ} \end{cases} \tag{7.57}$$

the current of phase A can be derived as

$$\dot{I}_{A(5)} = \frac{\dot{U}_{A(5)} - \dot{U}_{NN'(5)}}{Z_{\alpha\beta(5)}} = \frac{\dot{U}_{A(5)}}{Z_{\alpha\beta(5)}} = \frac{\dot{U}_{A(5)}}{R + (j5\omega)^{\beta} L_{\beta} + \frac{1}{(j5\omega)^{\alpha} C_{\alpha}}}$$
$$= I_{(5)} \angle -\varphi_5 \tag{7.58}$$

and the currents of phase B and C can be deduced as

$$\dot{I}_{B(5)} = \frac{\dot{U}_{B(5)} - \dot{U}_{NN'(5)}}{Z_{\alpha\beta(5)}} = \frac{\dot{U}_{B(5)}}{Z_{\alpha\beta(5)}} = \frac{\dot{U}_{B(5)}}{R + (j5\omega)^{\beta} L_{\beta} + \frac{1}{(j5\omega)^{\alpha} C_{\alpha}}}$$
$$= I_{(5)} \angle (-\varphi_5 + 120^{\circ}) \tag{7.59}$$

$$\dot{I}_{C(5)} = \frac{\dot{U}_{C(5)} - \dot{U}_{NN'(5)}}{Z_{\alpha\beta(5)}} = \frac{\dot{U}_{C(5)}}{Z_{\alpha\beta(5)}} = \frac{\dot{U}_{C(5)}}{R + (j5\omega)^{\beta} L_{\beta} + \frac{1}{(j5\omega)^{\alpha} C_{\alpha}}}$$
$$= I_{(5)} \angle (-\varphi_5 - 120^{\circ}) \tag{7.60}$$

where

$$\begin{cases} I_{(5)} = \dfrac{\sqrt{2} U_{(5)}}{\sqrt{\begin{array}{l} \left(R + \frac{1}{5^{\alpha}} \frac{1}{\omega^{\alpha} C_{\alpha}} \cos\frac{\pi\alpha}{2} + 5^{\beta}\omega^{\beta} L_{\beta} \cos\frac{\pi\beta}{2}\right)^2 \\ +\left(5^{\beta}\omega^{\beta} L_{\beta} \sin\frac{\pi\beta}{2} - \frac{1}{5^{\alpha}} \frac{1}{\omega^{\alpha} C_{\alpha}} \sin\frac{\pi\alpha}{2}\right)^2 \end{array}}} \\ \varphi_5 = \arctan\left(\dfrac{5^{\beta}\omega^{\beta} L_{\beta} \sin\frac{\pi\beta}{2} - \frac{1}{5^{\alpha}} \frac{1}{\omega^{\alpha} C_{\alpha}} \sin\frac{\pi\alpha}{2}}{R + \frac{1}{5^{\alpha}} \frac{1}{\omega^{\alpha} C_{\alpha}} \cos\frac{\pi\alpha}{2} + 5^{\beta}\omega^{\beta} L_{\beta} \cos\frac{\pi\beta}{2}}\right) \end{cases} \tag{7.61}$$

(3) As for the 3rd harmonic component (zero-sequence component), the equivalent circuit is presented in Fig. 7.5.

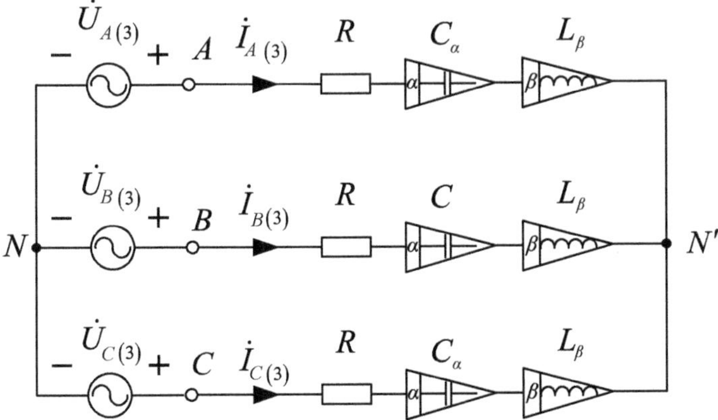

Fig. 7.5 Equivalent circuit for the 3rd harmonic component (zero-sequence component)

The fractional-order impedance is

$$Z_{\alpha\beta(3)} = R_{(3)} + X_{L\beta(3)} + X_{C\alpha(3)} = R + (j3\omega)^\beta L_\beta + \frac{1}{(j3\omega)^\alpha C_\alpha}$$
$$= \begin{bmatrix} R + \dfrac{1}{3^\alpha}\dfrac{1}{\omega^\alpha C_\alpha}\cos\dfrac{\pi\alpha}{2} + 3^\beta\omega^\beta L_\beta\cos\dfrac{\pi\beta}{2} \\ +j\left(3^\beta\omega^\beta L_\beta\sin\dfrac{\pi\beta}{2} - \dfrac{1}{3^\alpha}\dfrac{1}{\omega^\alpha C_\alpha}\sin\dfrac{\pi\alpha}{2}\right) \end{bmatrix} \tag{7.62}$$

Supposing that

$$\begin{cases} \dot{U}_{A(3)} = \sqrt{2}U_{(3)}\angle 0^\circ \\ \dot{U}_{B(3)} = \sqrt{2}U_{(3)}\angle 0^\circ \\ \dot{U}_{C(3)} = \sqrt{2}U_{(3)}\angle 0^\circ \end{cases} \tag{7.63}$$

the node voltage can be derived as

$$\left(\frac{1}{Z_{\alpha\beta}} + \frac{1}{Z_{\alpha\beta}} + \frac{1}{Z_{\alpha\beta}}\right)\dot{U}_{NN'(3)} = \frac{1}{Z_{\alpha\beta}}\left(\dot{U}_{A(3)} + \dot{U}_{B(3)} + \dot{U}_{C(3)}\right) \tag{7.64}$$

Thus, the neutral-point voltage of load is

$$\dot{U}_{NN'} = \frac{\dot{U}_{A(3)} + \dot{U}_{B(3)} + \dot{U}_{C(3)}}{3} = \frac{\sqrt{2}U_{(3)}\angle 0^\circ + \sqrt{2}U_{(3)}\angle 0^\circ + \sqrt{2}U_{(3)}\angle 0^\circ}{3}$$
$$= \sqrt{2}U_{(3)}\angle 0^\circ \tag{7.65}$$

Then, the phase current under the action of the 3rd harmonic component can be calculated as

$$\begin{cases} \dot{I}_{A(3)} = \dfrac{\dot{U}_{A(3)} - \dot{U}_{NN'(3)}}{Z_{\alpha\beta(3)}} = 0 \\ \dot{I}_{B(3)} = \dfrac{\dot{U}_{B(3)} - \dot{U}_{NN'(3)}}{Z_{\alpha\beta(3)}} = 0 \\ \dot{I}_{C(3)} = \dfrac{\dot{U}_{C(3)} - \dot{U}_{NN'(3)}}{Z_{\alpha\beta(3)}} = 0 \end{cases} \tag{7.66}$$

Based on the superposition theorem, the total phase currents can be summarized as

$$\begin{cases} i_A = i_{A(1)} + i_{A(3)} + i_{A(5)} \cdots\cdots = \sum\limits_{k=0}^{\infty} i_{A(4k+1)} \\ \quad = \sum\limits_{k=0}^{\infty} I_{(4k+1)} \sin((4k+1)\omega_1 t + \varphi_{4k+1}) \\ i_B = i_{B(1)} + i_{B(3)} + i_{B(5)} \cdots\cdots = \sum\limits_{k=0}^{\infty} i_{B(4k+1)} \\ \quad = \sum\limits_{k=0}^{\infty} I_{(4k+1)} \sin\left((4k+1)\omega_1 t + \varphi_{4k+1} - (4k+1)\frac{2\pi}{3}\right) \\ i_C = i_{C(1)} + i_{C(3)} + i_{C(5)} \cdots\cdots = \sum\limits_{k=0}^{\infty} i_{C(4k+1)} \\ \quad = \sum\limits_{k=0}^{\infty} I_{(4k+1)} \sin\left((4k+1)\omega_1 t + \varphi_{4k+1} + (4k+1)\frac{2\pi}{3}\right) \end{cases} \tag{7.67}$$

The amplitude and phase of harmonic currents are

$$\begin{cases} I_{(4k+1)} = \dfrac{\sqrt{2}U_{(4k+1)}}{\left| R + [j(4k+1)\omega]^{\beta} L_{\beta} + \frac{1}{[j(4k+1)\omega]^{\alpha} C_{\alpha}} \right|} = \dfrac{\sqrt{2}U_{(4k+1)}}{\sqrt{M^2 + N^2}} \\ \varphi_{4k+1} = \arctan\left(\dfrac{N}{M}\right) \end{cases} \tag{7.68}$$

where

$$\begin{cases} M = R + \dfrac{1}{(4k+1)^{\alpha}} \dfrac{1}{\omega^{\alpha} C_{\alpha}} \cos\dfrac{\pi\alpha}{2} + (4k+1)^{\beta} \omega^{\beta} L_{\beta} \cos\dfrac{\pi\beta}{2} \\ N = (4k+1)^{\beta} \omega^{\beta} L_{\beta} \sin\dfrac{\pi\beta}{2} - \dfrac{1}{(4k+1)^{\alpha}} \dfrac{1}{\omega^{\alpha} C_{\alpha}} \sin\dfrac{\pi\alpha}{2} \end{cases} \tag{7.69}$$

From the above analysis, it can be seen that compared with the integer-order non-sinusoidal steady-state circuit, the current of the fractional-order circuit is not only related to the impedance of the component but also can be adjusted by the order of the fractional-order component. Therefore, the fractional-order circuit has better flexibility.

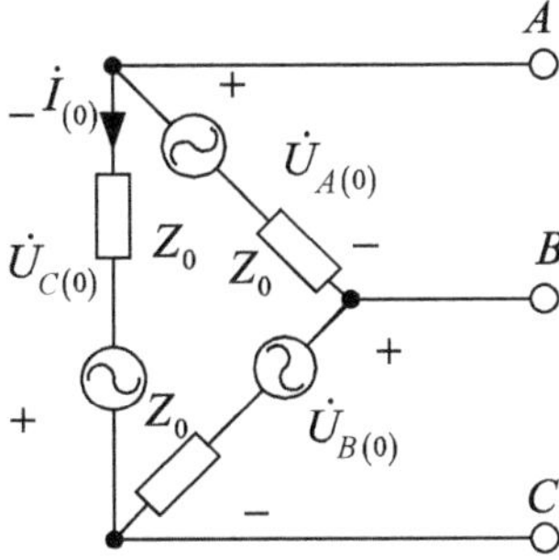

Fig. 7.6 Symmetrical Δ-connected power source

Moreover, it should be emphasized that the effect of zero-sequence harmonic components is determined by the connection type of the power supply and load. The following is a detailed analysis of the zero-sequence component.

For the zero-sequence voltages of any three-phase power supply, there are

$$\dot{U}_{A(0)} = \dot{U}_{B(0)} = \dot{U}_{C(0)} = \dot{U}_{(0)} \tag{7.70}$$

For the symmetrical Δ-connected power source shown in Fig. 7.6, the zero-sequence harmonic component produces a zero-sequence ring current, which is denoted as

$$\dot{I}_{(0)} = \frac{3\dot{U}_{(0)}}{3Z_0} = \frac{\dot{U}_{(0)}}{Z_0} \tag{7.71}$$

where $\dot{I}_{(0)}$ is the zero-sequence ring current.

The line voltage is

$$\dot{U}_{AB(0)} = \dot{U}_{BC(0)} = \dot{U}_{CA(0)} = \dot{U}_{(0)} - \dot{I}_{(0)} Z_0 = 0 \tag{7.72}$$

where Z_0 is the (one-phase) internal impedance of the power source. It is noted that the zero-sequence line voltage is zero under the effect of the ring current.

Therefore, regardless of the connection type of the load, there is a zero-sequence ring current in the power source, but other voltages and currents do not contain the zero-sequence component. In other words, the influence of the zero-sequence voltage can be eliminated by the Δ connection of the power source.

For a symmetrical three-phase circuit without the neutral line whose symmetrical three-phase power source is *Y* connected, it can always be transformed into a symmetrical *Y/Y* three-phase circuit, as shown in Fig. 7.7, in this case, the switch *S* is disconnected.

According to Fig. 7.7, the voltage of the neutral point is

$$\dot{U}_{NN'(0)} = \dot{U}_{(0)} \tag{7.73}$$

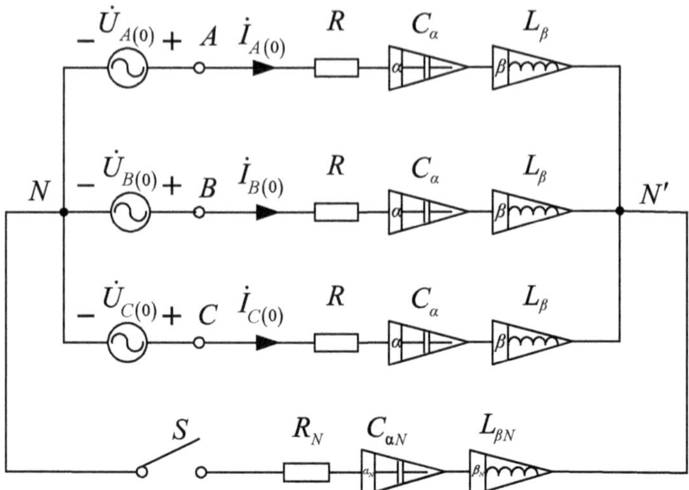

Fig. 7.7 Symmetrical power source in *Y* connection

the current can be derived as

$$\dot{I}_{A(0)} = \dot{I}_{B(0)} = \dot{I}_{C(0)} = \frac{\dot{U}_{(0)} - \dot{U}_{NN'(0)}}{Z_{\alpha\beta}} = 0 \tag{7.74}$$

and the voltages are

$$\dot{U}_{AB(0)} = \dot{U}_{BC(0)} = \dot{U}_{CA(0)} = \dot{U}_{A(0)} - \dot{U}_{B(0)} = 0 \tag{7.75}$$

It can be seen that in the symmetrical *Y*/*Y* three-phase circuit without a neutral line, only the neutral-point voltage and the source voltage have the zero-sequence component, other voltages and currents do not contain the zero-sequence components.

For the symmetrical *Y*/*Y* three-phase four-wire circuit, that is, the switch *S* in Fig. 7.7 is closed, supposing that

$$\dot{U}_{A(0)} = \dot{U}_{B(0)} = \dot{U}_{C(0)} = \dot{U}_{(0)} = \sqrt{2}U_{(0)}\angle 0^\circ \tag{7.76}$$

the impedances of the fractional-order load and the neutral line can be given by

$$\begin{cases} Z_{\alpha\beta} = R + \dfrac{1}{(j\omega)^{\alpha} C_{\alpha}} + (j\omega)^{\beta} L_{\beta} \\ Z_{\alpha\beta\_N} = R_N + \dfrac{1}{(j\omega)^{\alpha_N} C_{\alpha N}} + (j\omega)^{\beta_N} L_{\beta N} \end{cases} \tag{7.77}$$

Then, the following equations can be derived as

$$\left(\frac{1}{Z_{\alpha\beta}}+\frac{1}{Z_{\alpha\beta}}+\frac{1}{Z_{\alpha\beta}}+\frac{1}{Z_{\alpha\beta\_N}}\right)\dot{U}_{NN'(0)}=\frac{1}{Z_{\alpha\beta}}\left(\dot{U}_{A(0)}+\dot{U}_{B(0)}+\dot{U}_{C(0)}\right) \quad (7.78)$$

$$\dot{U}_{NN'(0)}=\frac{3Z_{\alpha\beta\_N}\dot{U}_{(0)}}{Z_{\alpha\beta}+3Z_{\alpha\beta\_N}} \quad (7.79)$$

Based on the symmetry of the three-phase four-wire circuit, the currents of phase A, B, and C are identical, that is

$$\begin{aligned}\dot{I}_{A(0)}=\dot{I}_{B(0)}=\dot{I}_{C(0)}=\dot{I}_{(0)}&=\frac{\dot{U}_{(0)}-\dot{U}_{NN'(0)}}{Z_{\alpha\beta}}=\frac{\dot{U}_{(0)}}{Z_{\alpha\beta}+3Z_{\alpha\beta\_N}}\\&=\frac{\sqrt{2}U_{(0)}\angle 0^\circ}{R+\frac{1}{(j\omega)^\alpha C_\alpha}+(j\omega)^\beta L_\beta+3\left(R_N+\frac{1}{(j\omega)^{\alpha_N}C_{\alpha N}}+(j\omega)^{\beta_N}L_{\beta N}\right)}\\&=\frac{\sqrt{2}U_{(0)}}{\sqrt{M^2+N^2}}\angle-\varphi_0\end{aligned} \quad (7.80)$$

Therefore, the voltage of the fractional-order load can be derived as

$$\begin{aligned}\dot{U}_{AN'(0)}=\dot{U}_{BN'(0)}=\dot{U}_{CN'(0)}&=\dot{I}_{(0)}Z_{\alpha\beta}\\&=\frac{Z_{\alpha\beta}}{Z_{\alpha\beta}+3Z_{\alpha\beta\_N}}\dot{U}_{(0)}\\&=\frac{R+\frac{1}{(j\omega)^\alpha C_\alpha}+(j\omega)^\beta L_\beta}{R+\frac{1}{(j\omega)^\alpha C_\alpha}+(j\omega)^\beta L_\beta+3\left(R_N+\frac{1}{(j\omega)^{\alpha_N}C_{\alpha N}}+(j\omega)^{\beta_N}L_{\beta N}\right)}\dot{U}_{(0)}\\&=\frac{\left[\begin{array}{l}R+\frac{1}{\omega^\alpha C_\alpha}\cos\frac{\pi\alpha}{2}+\omega^\beta L_\beta\cos\frac{\pi\beta}{2}\\+j\left(\omega^\beta L_\beta\sin\frac{\pi\beta}{2}-\frac{1}{\omega^\alpha C_\alpha}\sin\frac{\pi\alpha}{2}\right)\end{array}\right]}{M+N}\dot{U}_{(0)}\end{aligned} \quad (7.81)$$

From KVL and KCL, the zero-sequence current of the neutral line is

$$\dot{I}_{(0)}=-3\dot{I}_{A(0)}=\frac{3\sqrt{2}U_{(0)}}{\sqrt{M^2+N^2}}\angle(180^\circ-\varphi_0) \quad (7.82)$$

and the line voltages between the three-phase power source are

$$\dot{U}_{AB(0)}=\dot{U}_{BC(0)}=\dot{U}_{CA(0)}=0 \quad (7.83)$$

where

$$\begin{cases} M = \left(\begin{array}{l} R + 3R_N + \frac{1}{\omega^{\alpha} C_{\alpha}} \cos \frac{\pi\alpha}{2} + \omega^{\beta} L_{\beta} \cos \frac{\pi\beta}{2} \\ + \frac{3}{\omega^{\alpha_N} C_{\alpha N}} \cos \frac{\pi\alpha_N}{2} + 3\omega^{\beta_N} L_{\beta N} \cos \frac{\pi\beta_N}{2} \end{array} \right) \\ N = \left(\begin{array}{l} 3\omega^{\beta_N} L_{\beta N} \sin \frac{\pi\beta_N}{2} - \frac{3}{\omega^{\alpha_N} C_{\alpha N}} \sin \frac{\pi\alpha_N}{2} \\ + \omega^{\beta} L_{\beta} \sin \frac{\pi\beta}{2} - \frac{1}{\omega^{\alpha} C_{\alpha}} \sin \frac{\pi\alpha}{2} \end{array} \right) \\ \varphi_0 = \arctan\left(\frac{N}{M}\right) \end{cases} \tag{7.84}$$

The results show that, except for the line voltage of the three-phase power source, the other voltages and currents of the circuit all contain zero-sequence components.

7.4 Summary

This chapter first introduces some basic concepts, such as harmonic decomposition, harmonic impedance, average value, effective value, and several indexes of non-sinusoidal periodic signals. Then, the general analysis methods of non-sinusoidal periodic steady-state fractional-order circuits are presented. Finally, the influence of non-sinusoidal periodic harmonics on the three-phase fractional-order circuit is analyzed, especially the effect of the zero-sequence components. This chapter provides a theoretical basis for the analysis and calculation of fractional-order circuits under the action of non-sinusoidal periodic signals.

References

1. Freeborn TJ, Maundy B, Elwakil AS (2013) Measurement of supercapacitor fractional-order model parameters from voltage-excited step response. IEEE J Emerg Sel Top Circuits Syst 3(3):367–376
2. Radwan AG, Salama KN (2012) Fractional-order RC and RL circuits. Circuits Syst Signal Process 31(6):1901–1915
3. Rong C, Zhang B, Jiang Y (2020) Analysis of a Fractional-Order Wireless Power Transfer System. IEEE Trans Circuits Syst II Express Briefs 67(10):1755–1759
4. Jiang Y, Zhang B, Shu X, Wei Z (2020) Fractional-order autonomous circuits with order larger than one. J Adv Res 25:217–225
5. Radwan AG (2012) Stability analysis of the fractional-order RLC circuit. J Fract Calc Appl 3(1):1–15
6. Matsumori H, Urata K, Shimizu T et al (2018) Capacitor loss analysis method for power electronics converters. Microelectron Reliab 88–90:443–446
7. Westerlund S, Ekstam L (1994) Capacitor theory. IEEE Trans Dielectr Electr Insul 1(5):826–839
8. El-Khazali R, Tawalbeh N (2012) Realization of fractional-order capacitors and inductors. In: 5th-IFAC symposium on fractional differentiation and its applications. Nanjing, China
9. Radwan AG (2013) Resonance and quality factor of the $RL_{\beta}C_{\alpha}$ fractional circuit. IEEE J Emerg Sel Top Circuits Syst 3(3):377–385
10. Monje C, Chen Y, Vinagre B et al (2010) Fractional order systems and control—fundamentals and applications. Springer, London, pp 15–17

Chapter 8
Two-Port Fractional-Order Networks

In electronic, control, and communication systems, networks with only two pairs of terminals are often encountered, such as transmission lines, transformers, and electronic amplifiers [1]. In a network, electrical energy or signal is fed in from one pair of terminals and led out from another pair of terminals [2]. If the input and the output currents satisfy the port conditions

$$\begin{cases} i_1 = i_1' \\ i_2 = i_2' \end{cases} \tag{8.1}$$

such a network is called a two-port network [3].

In a two-port network, there is a certain relationship between the currents and voltages of the input and output ports, which can be expressed by Y-parameters, Z-parameters, H-parameters, and so on [4].

The two-port network introduced in this chapter is composed of linear resistors, fractional-order inductors, fractional-order capacitors, and linear controlled sources [5]. Such networks are called two-port fractional-order networks, as shown in Fig. 8.1.

8.1 Y-parameter

8.1.1 Y-parameter Equation

The reference directions of the currents and voltages at port $1-1'$ and port $2-2'$ are shown in Fig. 8.1. If the port voltages $\dot{U}_1$ and $\dot{U}_2$ are known, they can be regarded as independent voltage sources based on the substitution theorem [6]. According to the superposition theorem, the current of each port is equal to the sum of the current generated by each independent power supply operating alone [7]. The Y-parameter equation of the two-port fractional-order network can be expressed as

B. Zhang and X. Shu, *Fractional-Order Electrical Circuit Theory*, CPSS Power Electronics Series, https://doi.org/10.1007/978-981-16-2822-1_8

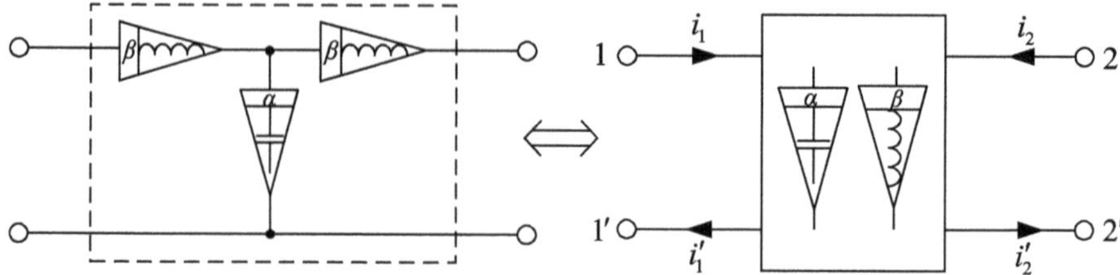

Fig. 8.1 Two-port fractional-order network

$$\begin{cases} \dot{I}_1 = Y_{\alpha\beta\_11}\dot{U}_1 + Y_{\alpha\beta\_12}\dot{U}_2 \\ \dot{I}_2 = Y_{\alpha\beta\_21}\dot{U}_1 + Y_{\alpha\beta\_22}\dot{U}_2 \end{cases} \tag{8.2}$$

where $Y_{\alpha\beta\_11}$, $Y_{\alpha\beta\_12}$, $Y_{\alpha\beta\_21}$, and $Y_{\alpha\beta\_22}$ are the Y-parameters.

Rewriting (8.2) in the matrix form as

$$\begin{bmatrix} \dot{I}_1 \\ \dot{I}_2 \end{bmatrix} = \begin{bmatrix} Y_{\alpha\beta\_11} & Y_{\alpha\beta\_12} \\ Y_{\alpha\beta\_21} & Y_{\alpha\beta\_22} \end{bmatrix} \begin{bmatrix} \dot{U}_1 \\ \dot{U}_2 \end{bmatrix} = \boldsymbol{Y}_{\alpha\beta} \begin{bmatrix} \dot{U}_1 \\ \dot{U}_2 \end{bmatrix} \tag{8.3}$$

where $\boldsymbol{Y}_{\alpha\beta}$ is called the Y-parameter matrix, and it is defined as

$$\boldsymbol{Y}_{\alpha\beta} = \begin{bmatrix} Y_{\alpha\beta\_11} & Y_{\alpha\beta\_12} \\ Y_{\alpha\beta\_21} & Y_{\alpha\beta\_22} \end{bmatrix} \tag{8.4}$$

8.1.2 Physical Meaning of Y-parameter

The physical meaning of Y-parameter can be obtained through experiments and calculations [8]. It is easily seen from (8.3) that Y-parameter has admittance characteristics.

If the port $1 - 1'$ is applied with a voltage $\dot{U}_1$ and the port $2 - 2'$ is short-circuited (i.e., $\dot{U}_2 = 0$). Substituting the conditions into (8.2), the Y-parameters can be calculated by

$$\begin{cases} Y_{\alpha\beta\_11} = \left.\dfrac{\dot{I}_1}{\dot{U}_1}\right|_{\dot{U}_2=0} \\ Y_{\alpha\beta\_21} = \left.\dfrac{\dot{I}_2}{\dot{U}_1}\right|_{\dot{U}_2=0} \end{cases} \tag{8.5}$$

The physical meanings of the Y-parameters can be easily obtained from (8.5). $Y_{\alpha\beta\_11}$ represents the input admittance of ports $1-1'$ when the port $2-2'$ is short-circuited. And $Y_{\alpha\beta\_21}$ denotes the transfer admittance from port $1-1'$ to port $2-2'$ when the port $2-2'$ is short-circuited.

Similarly, if a voltage $\dot{U}_2$ is applied to the port $2-2'$ and the port $1-1'$ is short-circuited (i.e., $\dot{U}_1=0$), then according to (8.2), the Y-parameters can be derived as

$$\begin{cases} Y_{\alpha\beta\_12} = \left.\dfrac{\dot{I}_1}{\dot{U}_2}\right|_{\dot{U}_1=0} \\ Y_{\alpha\_\beta 22} = \left.\dfrac{\dot{I}_2}{\dot{U}_2}\right|_{\dot{U}_1=0} \end{cases} \tag{8.6}$$

where $Y_{\alpha\beta\_12}$ refers to the transfer admittance from port $2-2'$ to port $1-1'$ when the port $1-1'$ is short-circuited, and $Y_{\alpha\beta\_22}$ stands for the input admittance of ports $2-2'$ when the port $1-1'$ is short-circuited.

It may be readily seen that the above Y-parameters of the two-port network are all measured when a specific port is short-circuited. Thus, the Y-parameters are also called the short-circuit parameters of the two-port network.

For the two-port fractional-order circuit shown in Fig. 8.2, its Y-parameters can be calculated by their corresponding physical meanings as

$$\begin{aligned} Y_{\alpha\beta\_11} &= \frac{1}{Z_{L\beta 1} + Z_{C\alpha}//Z_{L\beta 2}} = (j\omega)^{\beta_1} L_{\beta 1} + \frac{1}{(j\omega)^{\alpha} C_{\alpha}} //(j\omega)^{\beta_2} L_{\beta 2} \\ &= \frac{(j\omega)^{\beta_1} L_{\beta 1} + (j\omega)^{\beta_2} L_{\beta 2} + (j\omega)^{\alpha+\beta 1+\beta_2} C_{\alpha} L_{\beta 1} L_{\beta 2}}{1 + (j\omega)^{\alpha+\beta_2} C_{\alpha} L_{\beta 2}} \end{aligned} \tag{8.7}$$

$$\begin{aligned} Y_{\alpha\beta\_22} &= \frac{1}{Z_{L\beta 2} + Z_{C\alpha}//Z_{L\beta 1}} = (j\omega)^{\beta_2} L_{\beta 2} + \frac{1}{(j\omega)^{\alpha} C_{\alpha}} //(j\omega)^{\beta_1} L_{\beta 1} \\ &= \frac{(j\omega)^{\beta_1} L_{\beta 1} + (j\omega)^{\beta_2} L_{\beta 2} + (j\omega)^{\alpha+\beta 1+\beta_2} C_{\alpha} L_{\beta 1} L_{\beta 2}}{1 + (j\omega)^{\alpha+\beta_1} C_{\alpha} L_{\beta 1}} \end{aligned} \tag{8.8}$$

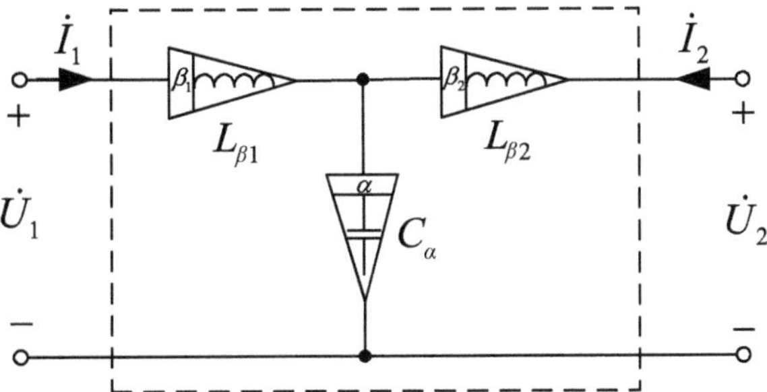

Fig. 8.2 Two-port fractional-order circuit

When the port $2-2'$ is short-circuited, applying a voltage $\dot{U}_1$ to the port $1-1'$, the current response is

$$\dot{I}_1 = \dot{U}_1 Y_{\alpha\beta\_11} \tag{8.9}$$

Then, by introducing the relationship between the current $\dot{I}_1$ and $\dot{I}_2$, that is

$$\dot{I}_2 = -\dot{I}_1 \frac{Z_{C\alpha}}{Z_{C\alpha} + Z_{L\beta2}} \tag{8.10}$$

the parameter $Y_{\alpha\beta\_21}$ can be calculated as

$$Y_{\alpha\beta\_21} = \frac{\dot{I}_2}{\dot{U}_1} = -\frac{Z_{C\alpha} Y_{\alpha\beta\_11}}{Z_{C\alpha} + Z_{L\beta2}} = -\frac{(j\omega)^{\beta_1} L_{\beta1} + (j\omega)^{\beta_2} L_{\beta2} + (j\omega)^{\alpha+\beta1+\beta_2} C_\alpha L_{\beta1} L_{\beta2}}{\left[1 + (j\omega)^{\alpha+\beta_2} C_\alpha L_{\beta2}\right]^2} \tag{8.11}$$

Similarly, when the port $1-1'$ is short-circuited, the port $2-2'$ is applied with a voltage $\dot{U}_2$, the parameter $Y_{\alpha\beta\_12}$ can be derived as

$$Y_{\alpha\beta\_12} = \frac{\dot{I}_1}{\dot{U}_2} = -\frac{Z_{C\alpha} Y_{\alpha\beta\_22}}{Z_{C\alpha} + Z_{L\beta1}} = -\frac{(j\omega)^{\beta_1} L_{\beta1} + (j\omega)^{\beta_2} L_{\beta2} + (j\omega)^{\alpha+\beta1+\beta_2} C_\alpha L_{\beta1} L_{\beta2}}{\left[1 + (j\omega)^{\alpha+\beta_1} C_\alpha L_{\beta1}\right]^2} \tag{8.12}$$

These parameters can be rewritten in the matrix form as

$$\boldsymbol{Y}_{\alpha\beta} = \begin{bmatrix} a & b \\ c & d \end{bmatrix} \tag{8.13}$$

where

$$\begin{cases} a = \dfrac{(j\omega)^{\beta_1} L_{\beta1} + (j\omega)^{\beta_2} L_{\beta2} + (j\omega)^{\alpha+\beta1+\beta_2} C_\alpha L_{\beta1} L_{\beta2}}{1 + (j\omega)^{\alpha+\beta_2} C_\alpha L_{\beta2}} \\ b = -\dfrac{(j\omega)^{\beta_1} L_{\beta1} + (j\omega)^{\beta_2} L_{\beta2} + (j\omega)^{\alpha+\beta1+\beta_2} C_\alpha L_{\beta1} L_{\beta2}}{\left[1 + (j\omega)^{\alpha+\beta_1} C_\alpha L_{\beta1}\right]^2} \\ c = -\dfrac{(j\omega)^{\beta_1} L_{\beta1} + (j\omega)^{\beta_2} L_{\beta2} + (j\omega)^{\alpha+\beta1+\beta_2} C_\alpha L_{\beta1} L_{\beta2}}{\left[1 + (j\omega)^{\alpha+\beta_2} C_\alpha L_{\beta2}\right]^2} \\ d = \dfrac{(j\omega)^{\beta_1} L_{\beta1} + (j\omega)^{\beta_2} L_{\beta2} + (j\omega)^{\alpha+\beta1+\beta_2} C_\alpha L_{\beta1} L_{\beta2}}{1 + (j\omega)^{\alpha+\beta_1} C_\alpha L_{\beta1}} \end{cases} \tag{8.14}$$

8.1.3 Equivalent Circuit of Y-parameter

1. Equivalent Circuit with Two Controlled Sources
   The *Y*-parameter Eq. (8.2) shows that the expression of the port current $\dot{I}_1$ contains two terms. One is the current generated by the admittance $Y_{\alpha\beta\_11}$ under the excitation of port voltage $\dot{U}_1$, and the other is the current excited by the port voltage $\dot{U}_2$ through the transfer admittance $Y_{\alpha\beta\_12}$. In the equivalent circuit, the second term can be expressed as a voltage-controlled current source (VCCS), which is controlled by the port voltage $\dot{U}_2$. The control coefficient is the transfer admittance $Y_{\alpha\beta\_12}$.
   Similarly, the port $2-2'$ can also be regarded as a branch composed of the admittance $Y_{\alpha\beta\_22}$ and a voltage-controlled current source in parallel, as shown in Fig. 8.3. The VCCS is controlled by port voltage $\dot{U}_1$, and the control coefficient is equal to the transfer admittance $Y_{\alpha\beta\_21}$.

2. Equivalent Circuit Including a Controlled Source
   If node 1 is regarded as the reference node, the *Y*-parameter equation shown in (8.2) can be rewritten as

$$\begin{cases} \dot{I}_1 = Y_{\alpha\beta\_11}\dot{U}_1 + Y_{\alpha\beta\_12}\dot{U}_2 \\ \dot{I}_2 = Y_{\alpha\beta\_12}\dot{U}_1 + Y_{\alpha\beta\_22}\dot{U}_2 + (Y_{\alpha\beta\_21} - Y_{\alpha\beta\_12})\dot{U}_1 \end{cases} \tag{8.15}$$

In the equivalent circuit, $Y_{\alpha\beta\_11}$ and $Y_{\alpha\beta\_22}$ represent the self-admittance of port $1-1'$ and port $2-2'$, respectively. $Y_{\alpha\beta\_12}$ is the mutual admittance between the port $1-1'$ and $2-2'$. The VCCS is controlled by the port voltage $\dot{U}_1$, and the control coefficient is $(Y_{\alpha\beta\_21} - Y_{\alpha\beta\_12})\dot{U}_1$, as shown in Fig. 8.4.

3. Passive Equivalent Circuit

If the two-port network satisfies $Y_{\alpha\beta\_12} = Y_{\alpha\beta\_21}$, it can be called a reciprocal two-port network. In this case, the equivalent circuit is composed of three fractional-order admittance, as shown in Fig. 8.5.

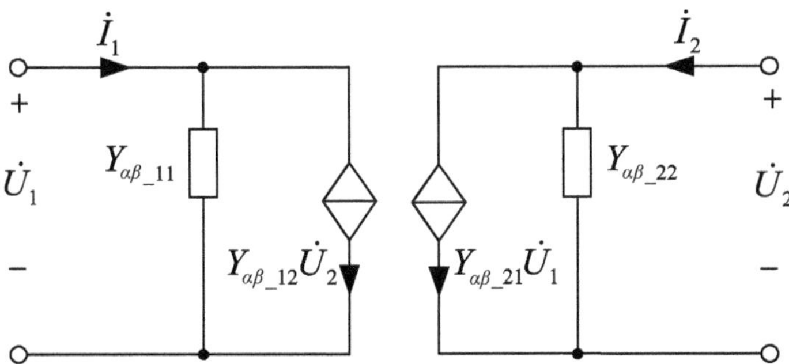

Fig. 8.3 Equivalent circuit with two controlled sources

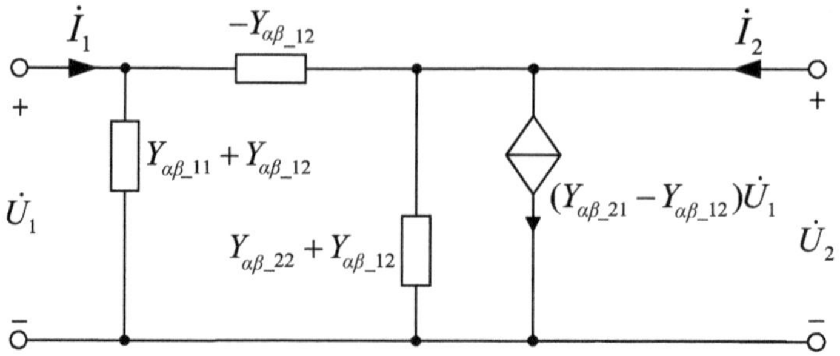

Fig. 8.4 Equivalent circuit including a controlled source

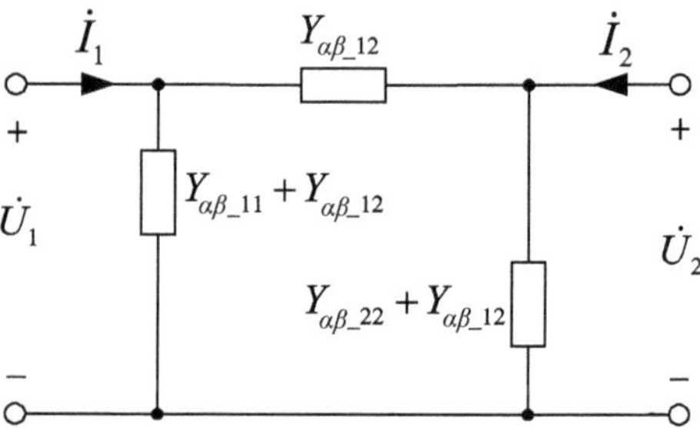

Fig. 8.5 π equivalent circuit without controlled sources

Besides, if the reciprocal two-port network satisfies

$$\begin{cases} Y_{\alpha\beta\_11} = Y_{\alpha\beta\_22} \\ Y_{\alpha\beta\_12} = Y_{\alpha\beta\_21} \end{cases} \tag{8.16}$$

it can also be called an asymmetric two-port network.

8.2 Z-parameter

8.2.1 Z-parameter Equation

Assuming that the currents are excitations and the voltages are responses, (8.3) can be rewritten as

$$\begin{bmatrix} \dot{U}_1 \\ \dot{U}_2 \end{bmatrix} = \begin{bmatrix} Y_{\alpha\beta\_11} & Y_{\alpha\beta\_12} \\ Y_{\alpha\beta\_21} & Y_{\alpha\beta\_22} \end{bmatrix}^{-1} \begin{bmatrix} \dot{I}_1 \\ \dot{I}_2 \end{bmatrix} = \boldsymbol{Y}_{\alpha\beta}^{-1} \begin{bmatrix} \dot{I}_1 \\ \dot{I}_2 \end{bmatrix} \tag{8.17}$$

and replacing the inverse Y-parameter matrix $\boldsymbol{Y}_{\alpha\beta}^{-1}$ with another matrix $\boldsymbol{Z}_{\alpha\beta}$,

$$Z_{\alpha\beta} = Y_{\alpha\beta}^{-1} \tag{8.18}$$

The Z-parameter equation of the two-port fractional-order network can be expressed as

$$\begin{bmatrix} \dot{U}_1 \\ \dot{U}_2 \end{bmatrix} = Z_{\alpha\beta} \begin{bmatrix} \dot{I}_1 \\ \dot{I}_2 \end{bmatrix} = \begin{bmatrix} Z_{\alpha\beta\_11} & Z_{\alpha\beta\_12} \\ Z_{\alpha\beta\_21} & Z_{\alpha\beta\_22} \end{bmatrix} \begin{bmatrix} \dot{I}_1 \\ \dot{I}_2 \end{bmatrix} \tag{8.19}$$

where $Z_{\alpha\beta\_11}$, $Z_{\alpha\beta\_12}$, $Z_{\alpha\beta\_21}$, and $Z_{\alpha\beta\_22}$ are Z-parameters, and they can be calculated by the Y-parameters as follows [9]:

$$Z_{\alpha\beta} = \begin{bmatrix} Z_{\alpha\beta\_11} & Z_{\alpha\beta\_12} \\ Z_{\alpha\beta\_21} & Z_{\alpha\beta\_22} \end{bmatrix} = \frac{1}{Y_{\alpha\beta\_11}Y_{\alpha\beta\_22} - Y_{\alpha\beta\_12}Y_{\alpha\beta\_21}} \begin{bmatrix} Y_{\alpha\beta\_22} & -Y_{\alpha\beta\_12} \\ -Y_{\alpha\beta\_21} & Y_{\alpha\beta\_11} \end{bmatrix} \tag{8.20}$$

$$\begin{cases} Z_{\alpha\beta\_11} = \dfrac{Y_{\alpha\beta\_22}}{Y_{\alpha\beta\_11}Y_{\alpha\beta\_22} - Y_{\alpha\beta\_12}Y_{\alpha\beta\_21}} \\ Z_{\alpha\beta\_12} = \dfrac{-Y_{\alpha\beta\_12}}{Y_{\alpha\beta\_11}Y_{\alpha\beta\_22} - Y_{\alpha\beta\_12}Y_{\alpha\beta\_21}} \\ Z_{\alpha\beta\_21} = \dfrac{-Y_{\alpha\beta\_21}}{Y_{\alpha\beta\_11}Y_{\alpha\beta\_22} - Y_{\alpha\beta\_12}Y_{\alpha\beta\_21}} \\ Z_{\alpha\beta\_22} = \dfrac{Y_{\alpha\beta\_11}}{Y_{\alpha\beta\_11}Y_{\alpha\beta\_22} - Y_{\alpha\beta\_12}Y_{\alpha\beta\_21}} \end{cases} \tag{8.21}$$

8.2.2 Physical Meaning of Z-parameter

It can be shown from (8.19) that the Z-parameters have impedance characteristics [10]. If the port $1-1'$ is applied with a current $\dot{I}_1$, and the port $2-2'$ is open-circuited (i.e., $\dot{I}_2 = 0$), and substituting these conditions into (8.17), Z-parameters can be easily obtained as

$$\begin{cases} Z_{\alpha\beta\_11} = \left.\dfrac{\dot{U}_1}{\dot{I}_1}\right|_{\dot{I}_2=0} \\ Z_{\alpha\beta\_21} = \left.\dfrac{\dot{U}_2}{\dot{I}_1}\right|_{\dot{I}_2=0} \end{cases} \tag{8.22}$$

where $Z_{\alpha\beta\_11}$ represents the input impedance of the port $1-1'$ and $Z_{\alpha\beta\_21}$ denotes the transfer impedance from port $1-1'$ to port $2-2'$ when the port $2-2'$ is open-circuited.

In the same way, if the port $2-2'$ is applied with a current $\dot{I}_2$, and the port $1-1'$ is open-circuited (i.e. $\dot{I}_1=0$), it can be deduced that

$$\begin{cases} Z_{\alpha\beta\_12} = \left.\dfrac{\dot{U}_1}{\dot{I}_2}\right|_{\dot{I}_1=0} \\ Z_{\alpha\beta\_22} = \left.\dfrac{\dot{U}_2}{\dot{I}_2}\right|_{\dot{I}_1=0} \end{cases} \tag{8.23}$$

where $Z_{\alpha\beta\_12}$ stands for the transfer impedance from port $2-2'$ to port $1-1'$ and $Z_{\alpha\beta\_22}$ refers to the input impedance of the port $2-2'$ when the port $1-1'$ is open-circuited.

Since Z-parameters are measured when a specific port is in open-circuit state, they are also called open-circuit parameters of the two-port network. As for the two-port fractional-order circuit shown in Fig. 8.2, its Z-parameters can be obtained by their physical meanings. The input impedances of port $1-1'$ and port $2-2'$ are

$$Z_{\alpha\beta\_11} = \frac{\dot{U}_1}{\dot{I}_1} = Z_{L\beta1} + Z_{C\alpha} = \frac{(j\omega)^{\alpha+\beta_1} C_\alpha L_{\beta1} + 1}{(j\omega)^\alpha C_\alpha} \tag{8.24}$$

$$Z_{\alpha\beta\_22} = \frac{\dot{U}_2}{\dot{I}_2} = Z_{L\beta2} + Z_{C\alpha} = \frac{(j\omega)^{\alpha+\beta_2} C_\alpha L_{\beta2} + 1}{(j\omega)^\alpha C_\alpha} \tag{8.25}$$

When the port $2-2'$ is open-circuited, and a current source $\dot{I}_1$ is applied to the port $1-1'$, the port voltage $\dot{U}_2$ can be calculated as

$$\dot{U}_2 = \dot{I}_1 Z_{C\alpha} = \frac{\dot{I}_1}{(j\omega)^\alpha C_\alpha} \tag{8.26}$$

from which the parameter $Z_{\alpha\beta\_21}$ can be deduced as

$$Z_{\alpha\beta\_21} = \frac{\dot{U}_2}{\dot{I}_1} = \frac{1}{(j\omega)^\alpha C_\alpha} \tag{8.27}$$

Similarly, by putting the port $1-1'$ in open-circuited state and applying a current source $\dot{I}_2$ to the port $2-2'$, the following expressions can be easily obtained as

$$\dot{U}_1 = \dot{I}_2 Z_{C\alpha} = \frac{\dot{I}_1}{(j\omega)^\alpha C_\alpha} \tag{8.28}$$

$$Z_{\alpha\beta\_12} = \frac{\dot{U}_1}{\dot{I}_2} = \frac{1}{(j\omega)^\alpha C_\alpha} \tag{8.29}$$

Hence, the Z-parameter matrix of the two-port fractional-order network can be expressed as

$$\mathbf{Z}_{\alpha\beta} = \begin{bmatrix} \frac{(j\omega)^{\alpha+\beta_1} C_\alpha L_{\beta 1}+1}{(j\omega)^\alpha C_\alpha} & \frac{1}{(j\omega)^\alpha C_\alpha} \\ \frac{1}{(j\omega)^\alpha C_\alpha} & \frac{(j\omega)^{\alpha+\beta_2} C_\alpha L_{\beta 2}+1}{(j\omega)^\alpha C_\alpha} \end{bmatrix} \tag{8.30}$$

8.2.3 Equivalent Circuit of Z-parameter

1. Equivalent Circuit with Two Controlled Sources

The Z-parameter Eq. (8.19) shows that the expression of the port voltage $\dot{U}_1$ includes two terms, one is the voltage generated by the impedance $Z_{\alpha\beta\_11}$ under the excitation of port current $\dot{I}_1$, and the other is excited by the current $\dot{I}_2$ through the transfer impedance $Z_{\alpha\beta\_12}$. Thus, in the equivalent circuit, the second term can be represented as a current-controlled voltage source (CCVS), whose control coefficient is the transfer impedance $Z_{\alpha\beta\_12}$, and it is controlled by the port current $\dot{I}_2$ [11].

In the same way, the port $2-2'$ can also be regarded as a branch composed of the series of an impedance $Z_{\alpha\beta\_22}$ and a CCVS. The CCVS is controlled by the port current $\dot{I}_1$, and its control coefficient is equal to the transfer impedance $Z_{\alpha\beta\_21}$, as shown in Fig. 8.6.

2. Equivalent Circuit with a Controlled Source

The Z-parameter equation can be regarded as mesh current equations. If node 1 is taken as the reference node, (8.19) can be rewritten as

$$\begin{cases} \dot{U}_1 = Z_{\alpha\beta\_11}\dot{I}_1 + Z_{\alpha\beta\_12}\dot{I}_2 \\ \dot{U}_2 = Z_{\alpha\beta\_12}\dot{I}_1 + Z_{\alpha\beta\_22}\dot{I}_2 + (Z_{\alpha\beta\_21} - Z_{\alpha\beta\_12})\dot{I}_1 \end{cases} \tag{8.31}$$

In the equivalent circuit, it can be easily seen that $Z_{\alpha\beta\_11}$ and $Z_{\alpha\beta\_22}$ are the self-impedances of mesh 1 and mesh 2, respectively. $Z_{\alpha\beta\_12}$ is the mutual impedance

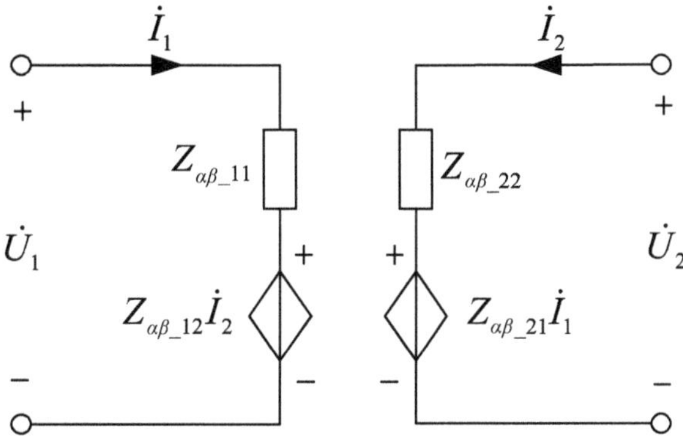

Fig. 8.6 Equivalent circuit with two controlled sources

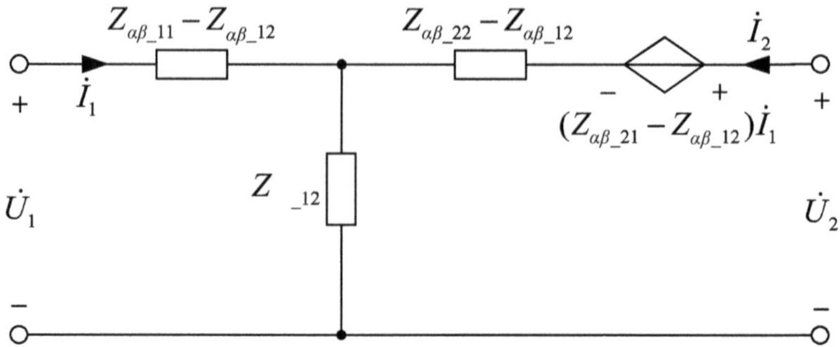

Fig. 8.7 Equivalent circuit with a controlled source

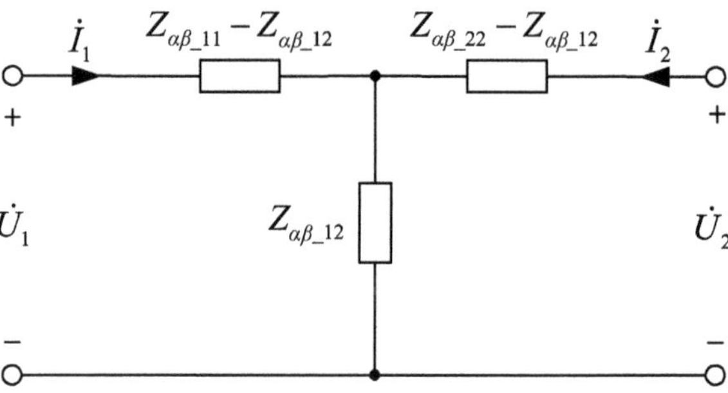

Fig. 8.8 Passive T-type equivalent circuit

between mesh 1 and mesh 2. The CCVS is connected in series to port 2 and controlled by the port current $\dot{I}_1$, as shown in Fig. 8.7.

3. Passive Equivalent Circuit

If the two-port network is reciprocal (i.e., $Z_{\alpha\beta\_12} = Z_{\alpha\beta\_21}$), the equivalent circuit can be composed of three fractional-order impedances, as shown in Fig. 8.8.

If the reciprocal two-port network satisfies

$$\begin{cases} Z_{\alpha\beta\_11} = Z_{\alpha\beta\_22} \\ Z_{\alpha\beta\_12} = Z_{\alpha\beta\_21} \end{cases} \tag{8.32}$$

it can be called a symmetric two-port network.

8.3 *H*-parameter and *G*-parameter

8.3.1 H-parameter

The H-parameter equation can be derived from the Y-parameter Eq. (8.2) when $\dot{I}_1$ and $\dot{U}_2$ are chosen as independent variables, which is

$$\begin{cases} \dot{U}_1 = \dfrac{1}{Y_{\alpha\beta\_11}}\dot{I}_1 - \dfrac{Y_{\alpha\beta\_12}}{Y_{\alpha\beta\_11}}\dot{U}_2 \\ \dot{I}_2 = \dfrac{Y_{\alpha\beta\_21}}{Y_{\alpha\beta\_11}}\dot{I}_1 + \dfrac{Y_{\alpha\beta\_22}Y_{\alpha\beta\_11} - Y_{\alpha\beta\_12}Y_{\alpha\beta\_21}}{Y_{\alpha\beta\_11}}\dot{U}_2 = \dfrac{Y_{\alpha\beta\_21}}{Y_{\alpha\beta\_11}}\dot{I}_1 + \dfrac{1}{Z_{\alpha\beta\_22}}\dot{U}_2 \end{cases} \tag{8.33}$$

(8.33) can be rewritten in the matrix form as

$$\begin{bmatrix} \dot{U}_1 \\ \dot{I}_2 \end{bmatrix} = \begin{bmatrix} H_{\alpha\beta\_11} & H_{\alpha\beta\_12} \\ H_{\alpha\beta\_21} & H_{\alpha\beta\_22} \end{bmatrix} \begin{bmatrix} \dot{I}_1 \\ \dot{U}_2 \end{bmatrix} = \boldsymbol{H}_{\alpha\beta} \begin{bmatrix} \dot{I}_1 \\ \dot{U}_2 \end{bmatrix} \tag{8.34}$$

where the H-parameters can be calculated by the Y-parameters, that is

$$\begin{cases} H_{\alpha\beta\_11} = \dfrac{1}{Y_{\alpha\beta\_11}} = \dfrac{1}{Y_{\alpha\beta\_a} + Y_{\alpha\beta\_b}} \\ H_{\alpha\beta\_12} = -\dfrac{Y_{\alpha\beta\_12}}{Y_{\alpha\beta\_11}} = \dfrac{Y_{\alpha\beta\_b}}{Y_{\alpha\beta\_a} + Y_{\alpha\beta\_b}} \\ H_{\alpha\beta\_21} = \dfrac{Y_{\alpha\beta\_21}}{Y_{\alpha\beta\_11}} = \dfrac{-Y_{\alpha\beta\_b}}{Y_{\alpha\beta\_a} + Y_{\alpha\beta\_b}} \\ H_{\alpha\beta\_22} = \dfrac{1}{Z_{\alpha\beta\_22}} = Y_{\alpha\beta\_c} + \dfrac{Y_{\alpha\beta\_a}Y_{\alpha\beta\_b}}{Y_{\alpha\beta\_a} + Y_{\alpha\beta\_b}} \end{cases} \tag{8.35}$$

Based on the relationship with Y-parameters and Z-parameters [12], that is

$$\begin{cases} H_{\alpha\beta\_11} = \left.\dfrac{\dot{U}_1}{\dot{I}_1}\right|_{\dot{U}_2=0} = \dfrac{1}{Y_{\alpha\beta\_11}} = Z_{\alpha\beta\_11} \\ H_{\alpha\beta\_21} = \left.\dfrac{\dot{I}_2}{\dot{I}_1}\right|_{\dot{U}_2=0} = \dfrac{Y_{\alpha\beta\_21}}{Y_{\alpha\beta\_11}} = K_{\text{I(short)}} \\ H_{\alpha\beta\_12} = \left.\dfrac{\dot{U}_1}{\dot{U}_2}\right|_{\dot{I}_1=0} = -\dfrac{Y_{\alpha\beta\_12}}{Y_{\alpha\beta\_11}} = \hat{K}_{\text{U(open)}} \\ H_{\alpha\beta\_22} = \left.\dfrac{\dot{I}_2}{\dot{U}_2}\right|_{\dot{I}_1=0} = \dfrac{1}{Z_{\alpha\beta\_22}} = Y_{\alpha\beta\_22} \end{cases} \tag{8.36}$$

the physical meaning of H-parameters can be obtained. $H_{\alpha\beta\_11}$ stands for the input impedance $Z_{\alpha\beta\_11}$ of the port $1-1'$, and $H_{\alpha\beta\_21}$ refers to the forward transmission ratio $K_{\mathrm{I(short)}}$ of circuit current from port $1-1'$ to port $2-2'$ when the port $2-2'$ is short-circuited. $H_{\alpha\beta\_12}$ denotes the reverse transfer ratio $\hat{K}_{\mathrm{U(open)}}$ of circuit voltage from port $2-2'$ to port $1-1'$ when the port $1-1'$ is open-circuited.

Therefore, for the two-port fractional-order circuit shown in Fig. 8.2, its H-parameters can be calculated as

$$H_{\alpha\beta\_11}=\left.\frac{\dot{U}_1}{\dot{I}_1}\right|_{\dot{U}_2=0}=\frac{1}{Y_{\alpha\beta\_11}}=\frac{1+(j\omega)^{\alpha+\beta_2}C_\alpha L_{\beta2}}{(j\omega)^{\beta_1}L_{\beta1}+(j\omega)^{\beta_2}L_{\beta2}+(j\omega)^{\alpha+\beta1+\beta_2}C_\alpha L_{\beta1}L_{\beta2}} \tag{8.37}$$

$$H_{\alpha\beta\_12}=\left.\frac{\dot{U}_1}{\dot{U}_2}\right|_{\dot{I}_1=0}=-\frac{Y_{\alpha\beta\_12}}{Y_{\alpha\beta\_11}}=\frac{1+(j\omega)^{\alpha+\beta_2}C_\alpha L_{\beta2}}{\left[1+(j\omega)^{\alpha+\beta_1}C_\alpha L_{\beta1}\right]^2} \tag{8.38}$$

$$H_{\alpha\beta\_21}=\left.\frac{\dot{I}_2}{\dot{I}_1}\right|_{\dot{U}_2=0}=\frac{Y_{\alpha\beta\_21}}{Y_{\alpha\beta\_11}}=-\frac{1}{1+(j\omega)^{\alpha+\beta_2}C_\alpha L_{\beta2}} \tag{8.39}$$

$$H_{\alpha\beta\_22}=\left.\frac{\dot{I}_2}{\dot{U}_2}\right|_{\dot{I}_1=0}=\frac{1}{Z_{\alpha\beta\_22}}=\frac{(j\omega)^{\alpha}C_\alpha}{(j\omega)^{\alpha+\beta_2}C_\alpha L_{\beta2}+1} \tag{8.40}$$

and the H-parameter matrix can be described as

$$\boldsymbol{H}_{\alpha\beta}=\begin{bmatrix}\frac{1+(j\omega)^{\alpha+\beta_2}C_\alpha L_{\beta2}}{(j\omega)^{\beta_1}L_{\beta1}+(j\omega)^{\beta_2}L_{\beta2}+(j\omega)^{\alpha+\beta1+\beta_2}C_\alpha L_{\beta1}L_{\beta2}} & \frac{1+(j\omega)^{\alpha+\beta_2}C_\alpha L_{\beta2}}{\left[1+(j\omega)^{\alpha+\beta_1}C_\alpha L_{\beta1}\right]^2}\\ -\frac{1}{1+(j\omega)^{\alpha+\beta_2}C_\alpha L_{\beta2}} & \frac{(j\omega)^{\alpha}C_\alpha}{(j\omega)^{\alpha+\beta_2}C_\alpha L_{\beta2}+1}\end{bmatrix} \tag{8.41}$$

8.3.2 G-parameter

When $\dot{U}_1$ and $\dot{I}_2$ are chosen as independent variables, the Y-parameter Eq. (8.2) becomes

$$\begin{cases}\dot{I}_1=\frac{Y_{\alpha\beta\_11}Y_{\alpha\beta\_22}-Y_{\alpha\beta\_12}Y_{\alpha\beta\_21}}{Y_{\alpha\beta\_22}}\dot{U}_1+\frac{Y_{\alpha\beta\_12}}{Y_{\alpha\beta\_22}}\dot{I}_2=\frac{1}{Z_{\alpha\beta\_11}}\dot{U}_1+\frac{Y_{\alpha\beta\_12}}{Y_{\alpha\beta\_22}}\dot{I}_2\\ \dot{U}_2=\frac{-Y_{\alpha\beta\_21}}{Y_{\alpha\beta\_22}}\dot{U}_1-\frac{1}{Y_{\alpha\beta\_22}}\dot{I}_2\end{cases} \tag{8.42}$$

the G-parameter Eq. (8.42) can be described in the following form

$$\begin{bmatrix} \dot{I}_1 \\ \dot{U}_2 \end{bmatrix} = \begin{bmatrix} G_{\alpha\beta\_11} & G_{\alpha\beta\_12} \\ G_{\alpha\beta\_21} & G_{\alpha\beta\_22} \end{bmatrix} \begin{bmatrix} \dot{U}_1 \\ \dot{I}_2 \end{bmatrix} = \boldsymbol{G}_{\alpha\beta} \begin{bmatrix} \dot{U}_1 \\ \dot{I}_2 \end{bmatrix} \tag{8.43}$$

By comparing (8.34) with (8.43), it can be found that the G-parameter matrix $\boldsymbol{G}_{\alpha\beta}$ is the inverse matrix of the H-parameter matrix $H_{\alpha\beta}$, that is

$$\begin{cases} \boldsymbol{G}_{\alpha\beta} = \boldsymbol{H}_{\alpha\beta}^{-1} \\ \boldsymbol{H}_{\alpha\beta} = \boldsymbol{G}_{\alpha\beta}^{-1} \end{cases} \tag{8.44}$$

and the G-parameters can also be derived by Y-parameters as

$$\begin{cases} G_{\alpha\beta\_11} = \left.\dfrac{\dot{I}_1}{\dot{U}_1}\right|_{\dot{I}_2=0} = \dfrac{1}{Z_{\alpha\beta\_11}} = Y_{\alpha\beta\_11} \\ G_{\alpha\beta\_21} = \left.\dfrac{\dot{U}_2}{\dot{U}_1}\right|_{\dot{I}_2=0} = -\dfrac{Y_{\alpha\beta\_21}}{Y_{\alpha\beta\_22}} = K_{U(\text{open})} \\ G_{\alpha\beta\_12} = \left.\dfrac{\dot{I}_1}{\dot{I}_2}\right|_{\dot{U}_1=0} = \dfrac{Y_{\alpha\beta\_12}}{Y_{\alpha\beta\_22}} = \hat{K}_{I(\text{short})} \\ G_{\alpha\beta\_22} = \left.\dfrac{\dot{U}_2}{\dot{I}_2}\right|_{\dot{U}_1=0} = \dfrac{1}{Y_{\alpha\beta\_22}} \end{cases} \tag{8.45}$$

where $G_{\alpha\beta\_11}$ denotes the input admittance $Y_{\alpha\beta\_11}$ of the port $1-1'$, and $G_{\alpha\beta\_21}$ represents the forward transmission ratio $K_{\text{U(open)}}$ of the circuit voltage from port $1-1'$ to port $2-2'$ when the port $2-2'$ is in open-circuited condition. $G_{\alpha\beta\_12}$ refers to the reverse transfer ratio $\hat{K}_{\text{I(short)}}$ from the port current $2-2'$ to the port $1-1'$, and $G_{\alpha\beta\_22}$ stands for the input impedance $Z_{\alpha\beta\_22}$ of the port $2-2'$ when the port $1-1'$ is short-circuited.

As for the two-port fractional-order circuit shown in Fig. 8.2, its G-parameters can be calculated by using (8.45) as

$$G_{\alpha\beta\_11} = \left.\frac{\dot{I}_1}{\dot{U}_1}\right|_{\dot{I}_2=0} = \frac{1}{Z_{\alpha\beta\_11}} = \frac{(j\omega)^{\alpha} C_{\alpha}}{(j\omega)^{\alpha+\beta_1} C_{\alpha} L_{\beta 1} + 1} \tag{8.46}$$

$$G_{\alpha\beta\_12} = \left.\frac{\dot{I}_1}{\dot{U}_1}\right|_{\dot{I}_2=0} = \frac{1}{Z_{\alpha\beta\_11}} = \frac{(j\omega)^{\alpha} C_{\alpha}}{(j\omega)^{\alpha+\beta_1} C_{\alpha} L_{\beta 1} + 1} \tag{8.47}$$

$$G_{\alpha\beta\_21} = \left.\frac{\dot{U}_2}{\dot{U}_1}\right|_{\dot{I}_2=0} = -\frac{Y_{\alpha\beta\_21}}{Y_{\alpha\beta\_22}} = \frac{1 + (j\omega)^{\alpha+\beta_1} C_{\alpha} L_{\beta 1}}{\left[1 + (j\omega)^{\alpha+\beta_2} C_{\alpha} L_{\beta 2}\right]^2} \tag{8.48}$$

$$G_{\alpha\beta\_22} = \left.\frac{\dot{U}_2}{\dot{I}_2}\right|_{\dot{U}_1=0} = \frac{1}{Y_{\alpha\beta\_22}} = \frac{1 + (j\omega)^{\alpha+\beta_1} C_{\alpha} L_{\beta 1}}{(j\omega)^{\beta_1} L_{\beta 1} + (j\omega)^{\beta_2} L_{\beta 2} + (j\omega)^{\alpha+\beta 1+\beta_2} C_{\alpha} L_{\beta 1} L_{\beta 2}} \tag{8.49}$$

Then, the G-parameter matrix of the two-port fractional-order network can be expressed as

$$G_{\alpha\beta}=\begin{bmatrix} \frac{\frac{(j\omega)^{\alpha}C_{\alpha}}{(j\omega)^{\alpha+\beta_1}C_{\alpha}L_{\beta1}+1}}{} & -\frac{1}{1+(j\omega)^{\alpha+\beta_1}C_{\alpha}L_{\beta1}} \\ \frac{1+(j\omega)^{\alpha+\beta_1}C_{\alpha}L_{\beta1}}{\left[1+(j\omega)^{\alpha+\beta_2}C_{\alpha}L_{\beta2}\right]^2} & \frac{1+(j\omega)^{\alpha+\beta_1}C_{\alpha}L_{\beta1}}{(j\omega)^{\beta_1}L_{\beta1}+(j\omega)^{\beta_2}L_{\beta2}+(j\omega)^{\alpha+\beta1+\beta_2}C_{\alpha}L_{\beta1}L_{\beta2}} \end{bmatrix} \tag{8.50}$$

8.3.3 Reciprocity and Symmetry of H-parameter and G-parameter

The H-parameters and G-parameters of the reciprocal network have the property

$$\begin{cases} H_{\alpha\beta\_12} = -H_{\alpha\beta\_21} \\ G_{\alpha\beta\_12} = -G_{\alpha\beta\_21} \end{cases} \tag{8.51}$$

Supposing that the reference directions of the voltages and currents in the H-parameter equation and the G-parameter equation are the same as those in the Y-parameter equation, it can be easily inferred that

$$\begin{cases} \hat{K}_{U(\text{open})} = -K_{I(\text{short})} \\ -\hat{K}_{I(\text{short})} = K_{U(\text{open})} \end{cases} \tag{8.52}$$

Furthermore, for the case that the network is both reciprocal and symmetric. Namely, it satisfies $Y_{\alpha\beta\_12} = Y_{\alpha\beta\_21}$ and $Y_{\alpha\beta\_11} = Y_{\alpha\beta\_22}$ at the same time. Then, the relationship between H-parameters and G-parameters can be expressed as

$$\begin{cases} H_{\alpha\beta\_11}H_{\alpha\beta\_22} - H_{\alpha\beta\_12}H_{\alpha\beta\_21} = 1 \\ G_{\alpha\beta\_11}G_{\alpha\beta\_22} - G_{\alpha\beta\_12}G_{\alpha\beta\_21} = 1 \end{cases} \tag{8.53}$$

Similarly, if the reference directions of voltages and currents in the H-parameter equation, the G-parameter equation and the Y-parameter equation are identical, it can be deduced that

$$\begin{cases} H_{\alpha\beta\_12} = G_{\alpha\beta\_12} = -H_{\alpha\beta\_21} = -G_{\alpha\beta\_21} \\ K_{U(\text{open})} = \hat{K}_{U(\text{open})} = -K_{I(\text{short})} = -\hat{K}_{I(\text{short})} \end{cases} \tag{8.54}$$

Note that only reciprocal networks have symmetry. If a two-port network satisfies (8.53) but not (8.52), it is neither reciprocal nor symmetric [13].

8.4 Transmission Parameters and Propagation Characteristics

8.4.1 Forward Transmission Parameters

When the two-port fractional-order network transfers signal from one port to another, the voltage and current at one port can be used to represent the voltage and current at the other port. If $\dot{U}_2$ and $\dot{I}_2$ are regarded as independent variables, the Y-parameter Eq. (8.2) can be rewritten as

$$\begin{cases} \dot{U}_1 = \dfrac{-Y_{\alpha\beta\_22}}{Y_{\alpha\beta\_21}}\dot{U}_2 + \dfrac{1}{Y_{\alpha\beta\_21}}\dot{I}_2 \\ \dot{I}_1 = \dfrac{Y_{\alpha\beta\_11}Y_{\alpha\beta\_22} - Y_{\alpha\beta\_12}Y_{\alpha\beta\_21}}{-Y_{\alpha\beta\_22}}\dot{U}_2 + \dfrac{Y_{\alpha\beta\_11}}{Y_{\alpha\beta\_21}}\dot{I}_2 = \dfrac{1}{Z_{\alpha\beta\_21}}\dot{U}_2 + \dfrac{Y_{\alpha\beta\_11}}{Y_{\alpha\beta\_21}}\dot{I}_2 \end{cases} \tag{8.55}$$

Assuming that the current $\dot{I}'_2$ of the port $2 - 2'$ flows out from the "+" terminal, that is, $\dot{I}'_2 = -\dot{I}_2$, as shown in Fig. 8.9.

Then, (8.55) can be further written as

$$\begin{cases} \dot{U}_1 = \dfrac{-Y_{\alpha\beta\_22}}{Y_{\alpha\beta\_21}}\dot{U}_2 + \dfrac{-1}{Y_{\alpha\beta\_21}}\dot{I}'_2 \\ \dot{I}_1 = \dfrac{1}{Z_{\alpha\beta\_21}}\dot{U}_2 + \dfrac{-Y_{\alpha\beta\_11}}{Y_{\alpha\beta\_21}}\dot{I}'_2 \end{cases} \tag{8.56}$$

Rewriting (8.56) in the form as

$$\begin{bmatrix} \dot{U}_1 \\ \dot{I}_1 \end{bmatrix} = \begin{bmatrix} A_{\alpha\beta\_11} & A_{\alpha\beta\_12} \\ A_{\alpha\beta\_21} & A_{\alpha\beta\_22} \end{bmatrix} \begin{bmatrix} \dot{U}_2 \\ \dot{I}'_2 \end{bmatrix} = \boldsymbol{A}_{\alpha\beta} \begin{bmatrix} \dot{U}_2 \\ \dot{I}'_2 \end{bmatrix} \tag{8.57}$$

which is the transmission parameter equation, and the transmission parameter is also called the T-parameter [14].

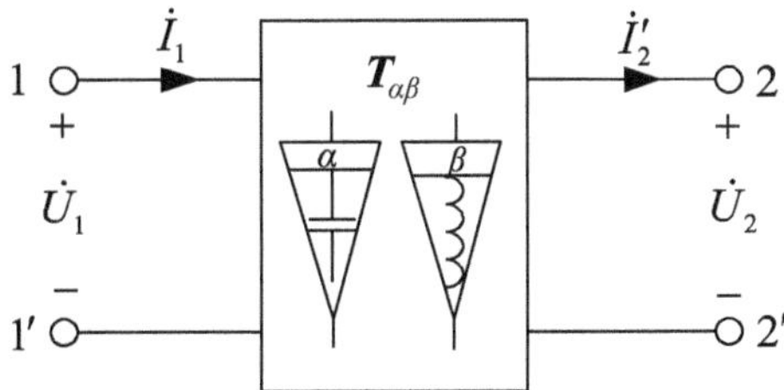

Fig. 8.9 Reference direction of currents in the transmission parameter

The physical meaning of T-parameters can be analyzed and obtained by (8.56). If voltages and currents of the port $1-1'$ are denoted as excitations, and those of the port $2-2'$ are regarded as responses.

Applying the voltage $\dot{U}_1$ and the current $\dot{I}_1$ to the port $1-1'$, respectively, and calculating the voltage response $\dot{U}_2$ of the open-circuited port $2-2'$ and the current response $\dot{I}_2$ of the shorted-circuited port $2-2'$, $A_{\alpha\beta\_11}$ and $A_{\alpha\beta\_21}$ can be obtained as

$$\begin{cases} A_{\alpha\beta\_11} = \left.\dfrac{\dot{U}_1}{\dot{U}_2}\right|_{I_2'=0} = \dfrac{Y_{\alpha\beta\_22}}{-Y_{\alpha\beta\_21}} = \dfrac{1}{K_{U(\text{open})}} \\ A_{\alpha\beta\_21} = \left.\dfrac{\dot{I}_1}{\dot{U}_2}\right|_{I_2'=0} = \dfrac{1}{Z_{\alpha\beta\_21}} \\ A_{\alpha\beta\_12} = \left.\dfrac{\dot{U}_1}{I_2'}\right|_{U_2=0} = -\dfrac{1}{Y_{\alpha\beta\_21}} \\ A_{\alpha\beta\_22} = \left.\dfrac{\dot{I}_1}{\dot{I}_2'}\right|_{U_2=0} = \dfrac{Y_{\alpha\beta\_11}}{-Y_{\alpha\beta\_21}} = -\dfrac{1}{K_{I(\text{short})}} \end{cases} \tag{8.58}$$

where $A_{\alpha\beta\_11}$ represents the reciprocal of voltage transmission ratio $K_{U(\text{open})}$ from port $1-1'$ to port $2-2'$, and $A_{\alpha\beta\_21}$ is the reciprocal of the transfer impedance $Z_{\alpha\beta\_21}$ from port $1-1'$ to port $2-2'$ when the port $2-2'$ is open-circuited. $A_{\alpha\beta\_12}$ denotes the negative reciprocal of the transfer admittance $Y_{\alpha\beta\_21}$ from the port $1-1'$ to the port $2-2'$, and $A_{\alpha\beta\_22}$ stands for the negative reciprocal of the current transfer ratio $K_{I(\text{short})}$ from the port $1-1'$ to the port $2-2'$ when the port $2-2'$ is short-circuited.

For a reciprocal two-port fractional-order network, it follows

$$A_{\alpha\beta\_11}A_{\alpha\beta\_22} - A_{\alpha\beta\_12}A_{\alpha\beta\_21} = 1 \tag{8.59}$$

If the reciprocal network is also symmetrical, which means that the Y-parameter satisfies $Y_{\alpha\beta\_12} = Y_{\alpha\beta\_21}$ and $Y_{\alpha\beta\_11} = Y_{\alpha\beta\_22}$, then there is

$$A_{\alpha\beta\_11} = A_{\alpha\beta\_22} \tag{8.60}$$

Therefore, for the two-port fractional-order circuit shown in Fig. 8.2, the T-parameters can be calculated by the physical meaning as follows:

$$A_{\alpha\beta\_11} = \left.\frac{\dot{U}_1}{\dot{U}_2}\right|_{I_2'=0} = \frac{Y_{\alpha\beta\_22}}{-Y_{\alpha\beta\_21}} = \frac{\left[1+(j\omega)^{\alpha+\beta_2}C_\alpha L_{\beta 2}\right]^2}{1+(j\omega)^{\alpha+\beta_1}C_\alpha L_{\beta 1}} \tag{8.61}$$

$$A_{\alpha\beta\_12} = \left.\frac{\dot{U}_1}{I_2'}\right|_{U_2=0} = -\frac{1}{Y_{\alpha\beta\_21}} = \frac{\left[1+(j\omega)^{\alpha+\beta_2}C_\alpha L_{\beta 2}\right]^2}{(j\omega)^{\beta_1}L_{\beta 1}+(j\omega)^{\beta_2}L_{\beta 2}+(j\omega)^{\alpha+\beta 1+\beta_2}C_\alpha L_{\beta 1}L_{\beta 2}} \tag{8.62}$$

$$A_{\alpha\beta\_21} = \left.\frac{\dot{I}_1}{\dot{U}_2}\right|_{I_2'=0} = \frac{1}{Z_{\alpha\beta\_21}} = (j\omega)^{\alpha} C_{\alpha} \tag{8.63}$$

$$A_{\alpha\beta\_22} = \left.\frac{\dot{I}_1}{\dot{I}_2'}\right|_{U_2=0} = \frac{Y_{\alpha\beta\_11}}{-Y_{\alpha\beta\_21}} = 1 + (j\omega)^{\alpha+\beta_2} C_{\alpha} L_{\beta 2} \tag{8.64}$$

Then, the T-parameter matrix of the two-port fractional-order network can be concluded as

$$\boldsymbol{A}_{\alpha\beta} = \begin{bmatrix} \frac{\left[1+(j\omega)^{\alpha+\beta_2} C_{\alpha} L_{\beta 2}\right]^2}{1+(j\omega)^{\alpha+\beta_1} C_{\alpha} L_{\beta 1}} & \frac{\left[1+(j\omega)^{\alpha+\beta_2} C_{\alpha} L_{\beta 2}\right]^2}{(j\omega)^{\beta_1} L_{\beta 1}+(j\omega)^{\beta_2} L_{\beta 2}+(j\omega)^{\alpha+\beta_1+\beta_2} C_{\alpha} L_{\beta 1} L_{\beta 2}} \\ (j\omega)^{\alpha} C_{\alpha} & 1 + (j\omega)^{\alpha+\beta_2} C_{\alpha} L_{\beta 2} \end{bmatrix} \tag{8.65}$$

8.4.2 Reverse Transmission Parameters

If $\dot{U}_1$ and $\dot{I}_1$ are taken as independent variables, the parameter matrix can be written as the inverse matrix of the T-parameter matrix, like

$$\begin{bmatrix} \dot{U}_2 \\ \dot{I}_2' \end{bmatrix} = \begin{bmatrix} A_{\alpha\beta\_11} & A_{\alpha\beta\_12} \\ A_{\alpha\beta\_21} & A_{\alpha\beta\_22} \end{bmatrix}^{-1} \begin{bmatrix} \dot{U}_1 \\ \dot{I}_1 \end{bmatrix} \tag{8.66}$$

Supposing that the current $\dot{I}_1'$ flows out from the " + " terminal of the port $1 - 1'$, there is $\dot{I}'_1 = -\dot{I}_1$, as shown in Fig. 8.10.

Then, (8.66) can be rewritten as

$$\begin{bmatrix} \dot{U}_2 \\ -\dot{I}_2 \end{bmatrix} = \frac{1}{A_{\alpha\beta\_11} A_{\alpha\beta\_22} - A_{\alpha\beta\_12} A_{\alpha\beta\_21}} \begin{bmatrix} A_{\alpha\beta\_22} & -A_{\alpha\beta\_12} \\ -A_{\alpha\beta\_21} & A_{\alpha\beta\_11} \end{bmatrix} \begin{bmatrix} \dot{U}_1 \\ \dot{I}_1 \end{bmatrix} \tag{8.67}$$

after some rearrangements, it becomes

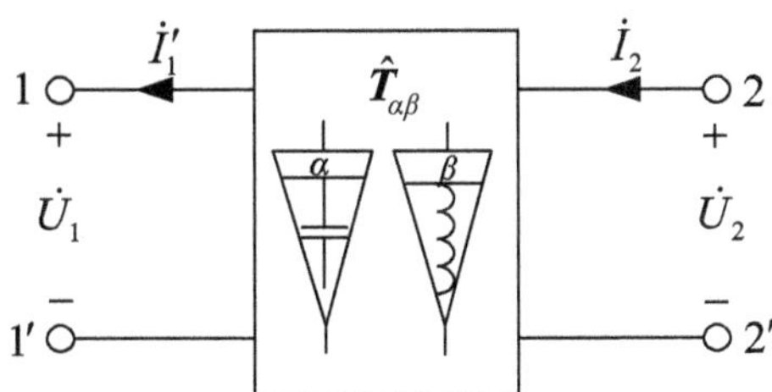

Fig. 8.10 Reference direction of currents in the reverse transmission parameter

$$\begin{bmatrix} \dot{U}_2 \\ \dot{I}_2 \end{bmatrix} = \frac{1}{A_{\alpha\beta\_11}A_{\alpha\beta\_22} - A_{\alpha\beta\_12}A_{\alpha\beta\_21}} \begin{bmatrix} A_{\alpha\beta\_22} & A_{\alpha\beta\_12} \\ A_{\alpha\beta\_21} & A_{\alpha\beta\_11} \end{bmatrix} \begin{bmatrix} \dot{U}_1 \\ \dot{I}_1 \end{bmatrix}$$
$$= \begin{bmatrix} \hat{A}_{\alpha\beta\_11} & \hat{A}_{\alpha\beta\_12} \\ \hat{A}_{\alpha\beta\_21} & \hat{A}_{\alpha\beta\_22} \end{bmatrix} \begin{bmatrix} \dot{U}_1 \\ \dot{I}_1 \end{bmatrix} = \hat{\boldsymbol{A}}_{\alpha\beta} \begin{bmatrix} \dot{U}_1 \\ \dot{I}_1 \end{bmatrix} \tag{8.68}$$

which is the reverse transmission parameter equation, also known as $\hat{T}$-parameter equation [15].

When the port $2-2'$ is the excitation and the port $1-1'$ is the response, the physical meaning of the reverse transmission parameters can be derived. Applying the voltage $\dot{U}_2$ and current $\dot{I}_2$ to the port $2-2'$ in turn, and analyzing the voltage response $\dot{U}_1$ when the port $1-1'$ is open-circuited and the current response I_1' when the port $1-1'$ is short-circuited, the physical meaning of reverse transmission parameters can be inferred that

$$\begin{cases} \hat{A}_{\alpha\beta\_11} = \left.\dfrac{\dot{U}_2}{\dot{U}_1}\right|_{I_1'=0} = \dfrac{Y_{\alpha\beta\_11}}{-Y_{\alpha\beta\_12}} = \dfrac{1}{\hat{K}_{U(\text{open})}} \\ \hat{A}_{\alpha\beta\_21} = \left.\dfrac{\dot{I}_2}{\dot{U}_1}\right|_{I_1'=0} = \dfrac{1}{Z_{\alpha\beta\_12}} \\ \hat{A}_{\alpha\beta\_12} = \left.\dfrac{\dot{U}_2}{I_1'}\right|_{U_1=0} = -\dfrac{1}{Y_{\alpha\beta\_12}} \\ \hat{A}_{\alpha\beta\_22} = \left.\dfrac{\dot{I}_2}{\dot{I}_1'}\right|_{U_1=0} = \dfrac{Y_{\alpha\beta\_22}}{-Y_{\alpha\beta\_12}} = -\dfrac{1}{\hat{K}_{I(\text{short})}} \end{cases} \tag{8.69}$$

where $\hat{A}_{\alpha\beta\_11}$ denotes the reciprocal of the reverse voltage transfer ratio $\hat{K}_{U(\text{open})}$ from port $2-2'$ to port $1-1'$, and $\hat{A}_{\alpha\beta\_21}$ represents the reciprocal of the transfer impedance $Z_{\alpha\beta\_12}$ from port $2-2'$ to port $1-1'$ when the port $1-1'$ is in open-circuited state. $\hat{A}_{\alpha\beta\_12}$ refers to the negative reciprocal of the transfer admittance $Y_{\alpha\beta\_12}$ from port $2-2'$ to port $1-1'$, and $\hat{A}_{\alpha\beta\_22}$ stands for the negative reciprocal of the reverse current transfer ratio $\hat{K}_{I(\text{short})}$ from port $2-2'$ to port $1-1'$ when the port $1-1'$ is short-circuited.

If the two-port network is reciprocal, it has the properties that

$$\hat{A}_{\alpha\beta\_11}\hat{A}_{\alpha\beta\_22} - \hat{A}_{\alpha\beta\_12}\hat{A}_{\alpha\beta\_21} = 1 \tag{8.70}$$

Comparing (8.59) and (8.70), the relationship between transmission parameter matrix and reverse transmission parameter matrix can be described as

$$\begin{bmatrix} A_{\alpha\beta\_22} & A_{\alpha\beta\_12} \\ A_{\alpha\beta\_21} & A_{\alpha\beta\_11} \end{bmatrix} = \begin{bmatrix} \hat{A}_{\alpha\beta\_11} & \hat{A}_{\alpha\beta\_12} \\ \hat{A}_{\alpha\beta\_21} & \hat{A}_{\alpha\beta\_22} \end{bmatrix} \tag{8.71}$$

If a reciprocal two-port fractional-order network is symmetrical, it will follow

$$\begin{cases} A_{\alpha\beta\_11} = A_{\alpha\beta\_22}, \hat{A}_{\alpha\beta\_11} = \hat{A}_{\alpha\beta\_22} \\ \hat{A}_{\alpha\beta\_11} = A_{\alpha\beta\_11}, \hat{A}_{\alpha\beta\_22} = A_{\alpha\beta\_22} \end{cases} \tag{8.72}$$

In this case, the transmission parameter matrix and the reverse transmission parameter matrix are the same.

For the two-port fractional-order circuit shown in Fig. 8.2, according to the corresponding physical meaning, the reverse transmission parameters can be obtained as follows:

$$\hat{A}_{\alpha\beta\_11} = \left.\frac{\dot{U}_2}{\dot{U}_1}\right|_{I_1'=0} = \frac{Y_{\alpha\beta\_11}}{-Y_{\alpha\beta\_12}} = \frac{\left[1+(j\omega)^{\alpha+\beta_1}C_\alpha L_{\beta 1}\right]^2}{1+(j\omega)^{\alpha+\beta_2}C_\alpha L_{\beta 2}} \tag{8.73}$$

$$\hat{A}_{\alpha\beta\_12} = \left.\frac{\dot{U}_2}{I_1'}\right|_{U_1=0} = -\frac{1}{Y_{\alpha\beta\_12}} = \frac{\left[1+(j\omega)^{\alpha+\beta_1}C_\alpha L_{\beta 1}\right]^2}{(j\omega)^{\beta_1}L_{\beta 1}+(j\omega)^{\beta_2}L_{\beta 2}+(j\omega)^{\alpha+\beta 1+\beta_2}C_\alpha L_{\beta 1}L_{\beta 2}} \tag{8.74}$$

$$\hat{A}_{\alpha\beta\_21} = \left.\frac{\dot{I}_2}{\dot{U}_1}\right|_{I_1'=0} = \frac{1}{Z_{\alpha\beta\_12}} = (j\omega)^\alpha C_\alpha \tag{8.75}$$

$$\hat{A}_{\alpha\beta\_22} = \left.\frac{\dot{I}_2}{\dot{I}_1'}\right|_{U_1=0} = \frac{Y_{\alpha\beta\_22}}{-Y_{\alpha\beta\_12}} = 1+(j\omega)^{\alpha+\beta_1}C_\alpha L_{\beta 1} \tag{8.76}$$

Therefore, the reverse transmission parameter matrix can be described as

$$\hat{A}_{\alpha\beta} = \begin{bmatrix} \frac{\left[1+(j\omega)^{\alpha+\beta_1}C_\alpha L_{\beta 1}\right]^2}{1+(j\omega)^{\alpha+\beta_2}C_\alpha L_{\beta 2}} & \frac{\left[1+(j\omega)^{\alpha+\beta_1}C_\alpha L_{\beta 1}\right]^2}{(j\omega)^{\beta_1}L_{\beta 1}+(j\omega)^{\beta_2}L_{\beta 2}+(j\omega)^{\alpha+\beta 1+\beta_2}C_\alpha L_{\beta 1}L_{\beta 2}} \\ (j\omega)^\alpha C_\alpha & 1+(j\omega)^{\alpha+\beta_1}C_\alpha L_{\beta 1} \end{bmatrix} \tag{8.77}$$

8.4.3 Propagation Characteristics of Two-Port Symmetric Fractional-Order Networks

When the output terminal of the two-port fractional-order network is connected with a load impedance $Z_{\alpha\beta\_L}$ [16], as shown in Fig. 8.11. Here, the impedance seen from the input port is called the input impedance, which can be calculated by T-parameters as

$$Z_{\alpha\beta\_in} = \left.\frac{\dot{U}_1}{\dot{I}_1}\right|_{\dot{U}_2=\dot{I}_2 Z_{\alpha\beta\_L}} = \frac{A_{\alpha\beta\_11}Z_{\alpha\beta\_L}+A_{\alpha\beta\_12}}{A_{\alpha\beta\_21}Z_{\alpha\beta\_L}+A_{\alpha\beta\_22}} \tag{8.78}$$

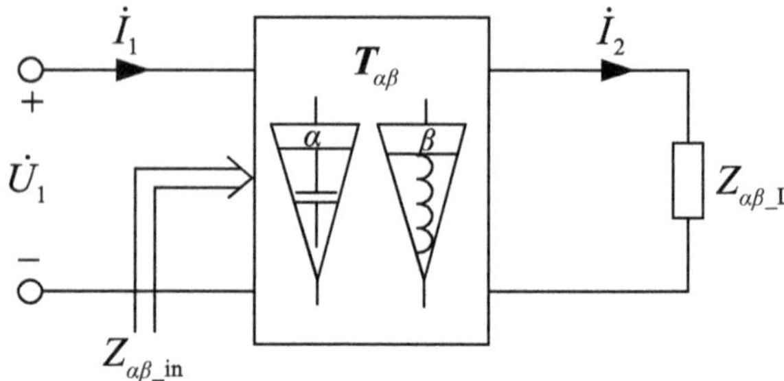

Fig. 8.11 Input impedance of two-port fractional-order network with a load

For a two-port symmetric fractional-order network, if the output terminal of the network is connected with an impedance $Z_{\alpha\beta\_c}$, and its input impedance is $Z_{\alpha\beta\_c}$ as well, then the impedance $Z_{\alpha\beta\_c}$ is called the characteristic impedance of the symmetric two-port network. The characteristic impedance can be expressed as

$$Z_{\alpha\beta\_c} = \sqrt{\frac{A_{\alpha\beta\_12}}{A_{\alpha\beta\_21}}} \tag{8.79}$$

When the output port of the symmetrical two-port network is connected to a characteristic impedance $Z_{\alpha\beta\_c}$, the voltage and the current of the port $2-2'$ can be calculated as

$$\begin{cases} \dot{U}_2 = \dot{I}_2 Z_{\alpha\beta\_c} = \dot{I}_2\sqrt{\dfrac{A_{\alpha\beta\_12}}{A_{\alpha\beta\_21}}} \\ \dot{I}_2 = \dfrac{\dot{U}_2}{Z_{\alpha\beta\_c}} = \dot{U}_2\sqrt{\dfrac{A_{\alpha\beta\_21}}{A_{\alpha\beta\_12}}} \end{cases} \tag{8.80}$$

Introducing (8.80) into (8.57), there is

$$\begin{cases} \dot{U}_1 = A_{\alpha\beta\_11}\dot{U}_2 + A_{\alpha\beta\_12}\dot{I}_2 = \left(A_{\alpha\beta\_11} + \sqrt{A_{\alpha\beta\_12}A_{\alpha\beta\_21}}\right)\dot{U}_2 \\ \dot{I}_1 = A_{\alpha\beta\_21}\dot{U}_2 + A_{\alpha\beta\_22}\dot{I}_2 = A_{\alpha\beta\_21}\dot{U}_2 + A_{\alpha\beta\_11}\dot{I}_2 \\ \quad = \left(A_{\alpha\beta\_11} + \sqrt{A_{\alpha\beta\_12}A_{\alpha\beta\_21}}\right)\dot{I}_2 \end{cases} \tag{8.81}$$

from which it is easily derived that

$$\frac{\dot{U}_1}{\dot{U}_2} = \frac{\dot{I}_1}{\dot{I}_2} = A_{\alpha\beta\_11} + \sqrt{A_{\alpha\beta\_12}A_{\alpha\beta\_21}} \tag{8.82}$$

It is noted that the voltage and current of the output port and input port have the same attenuation and phase shift when the output port is connected with a characteristic impedance $Z_{\alpha\beta\_c}$ [17]. If the propagation characteristic parameter γ, attenuation coefficient α, and phase coefficient β are introduced to express these propagation characteristics, then there are

$$\gamma = \ln\left(\frac{\dot{U}_1}{\dot{U}_2}\right) = \ln\left(\frac{\dot{I}_1}{\dot{I}_2}\right) = \ln\left(A_{\alpha\beta\_11} + \sqrt{A_{\alpha\beta\_12}A_{\alpha\beta\_21}}\right) = \alpha + j\beta \tag{8.83}$$

$$\alpha = \ln\left|\frac{\dot{U}_1}{\dot{U}_2}\right| = \ln\left|\frac{\dot{I}_1}{\dot{I}_2}\right| = \ln\left|A_{\alpha\beta\_11} + \sqrt{A_{\alpha\beta\_12}A_{\alpha\beta\_21}}\right| \tag{8.84}$$

$$\beta = \angle\dot{U}_1 - \angle\dot{U}_2 = \angle\dot{I}_1 - \angle\dot{I}_2 = \text{arctg}\frac{\text{Im}\left(A_{\alpha\beta\_11} + \sqrt{A_{\alpha\beta\_12}A_{\alpha\beta\_21}}\right)}{\text{Re}\left(A_{\alpha\beta\_11} + \sqrt{A_{\alpha\beta\_12}A_{\alpha\beta\_21}}\right)} \tag{8.85}$$

8.5 Fractional-Order Mutators

Mutator is a general term for a class of linear two-port converters, and their function is to convert circuit components from one type to another [18]. Anyone of the essential two-terminal components, such as resistor, inductor, capacitor, and memristor, can be converted into a different kind of component by the mutator. With the gradual improvement of fractional calculus theory, more fractional-order models are applied in research and engineering practice. Accordingly, it is necessary and meaningful to construct a fractional-order mutator model.

8.5.1 Basic Concepts

The two-port integer-order mutators can be generalized to two-port fractional-order mutators [19], then the relationships between the port voltages and port currents of fractional-order mutators in type I and type II can be deduced as (8.86) and (8.87), respectively.

I-type fractional-order mutator:

$$\begin{cases} \dfrac{\mathrm{d}^{a_1}}{\mathrm{d}t^{a_1}}\dot{U}_1 = \dfrac{\mathrm{d}^{a_2}}{\mathrm{d}t^{a_2}}\dot{U}_2 \\ \dfrac{\mathrm{d}^{b_1}}{\mathrm{d}t^{b_1}}\dot{I}_1 = -\dfrac{\mathrm{d}^{b_2}}{\mathrm{d}t^{b_2}}\dot{I}_2 \end{cases} \tag{8.86}$$

II-type fractional-order mutator:

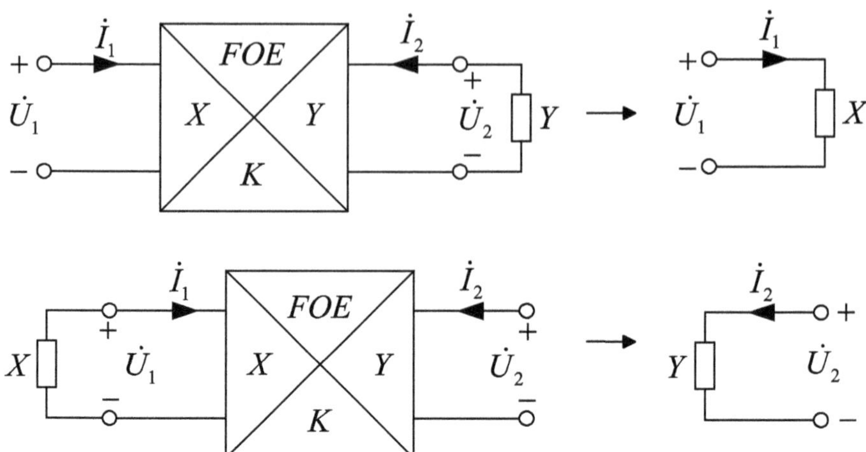

Fig. 8.12 Symbolic representation of two-port fractional-order mutators

$$\begin{cases} \frac{\mathrm{d}^{a_1}}{\mathrm{d}t^{a_1}}\dot{U}_1 = -\frac{\mathrm{d}^{b_2}}{\mathrm{d}t^{b_2}}\dot{I}_2 \\ \frac{\mathrm{d}^{b_1}}{\mathrm{d}t^{b_1}}\dot{I}_1 = \frac{\mathrm{d}^{a_2}}{\mathrm{d}t^{a_2}}\dot{U}_2 \end{cases} \tag{8.87}$$

where a_1, a_2, b_1, and b_2 are non-integer.

The universal expression of the circuit symbol of fractional-order mutators is shown in Fig. 8.12. Here, *X* and *Y* represent different types of fractional-order components, *K* represents the type of fractional-order mutator (type I or type II). *FOE* means that the mutator contains fractional-order components. The fractional-order mutator can realize the conversion between two different fractional-order components *X* and *Y*. When port 2 is connected to a component *Y*, the equivalent component seen from port 1 is *X*. Likewise, when port 1 is connected to a component *X*, the equivalent component seen from port 2 is *Y*.

According to the types of fractional-order components to be converted, the mutators can be divided into six subclasses, which are C_α-R, L_β-R, $M_{\gamma\nu}$-R, L_β-C_α, $M_{\gamma\nu}$-C_α, and $M_{\gamma\nu}$-L_β. The relationships between the port voltages and currents of mutators in different subclasses are shown in Table 8.1.

8.5.2 Fractional-Order C_α-R Mutator

Table 8.1 gives the relationship between port voltages and currents of the fractional-order C_α-R mutator in type I, which is

Table 8.1 The relationships between the port parameters of fractional-order mutators

Type	Subclass	Relationships between port parameters
I	$C_\alpha - R$	$\dot{U}_1 = \dot{U}_2, {}_0D_t^{-\alpha}\dot{I}_1 = -\dot{I}_2$
	$L_\beta - R$	${}_0D_t^{-\beta}\dot{U}_1 = \dot{U}_2, \dot{I}_1 = -\dot{I}_2$
	$L_\beta - C_\alpha$	${}_0D_t^{-\alpha}\dot{U}_1 = \dot{U}_2, \dot{I}_1 = -{}_0D_t^{-\beta}\dot{I}_2$
	$M_{\gamma,v} - R$	${}_0D_t^{-\gamma}\dot{U}_1 = \dot{U}_2, {}_0D_t^{-v}\dot{I}_1 = -\dot{I}_2$
	$M_{\gamma,v} - L_\beta$	${}_0D_t^{-\gamma}\dot{U}_1 = {}_0D_t^{-\beta}\dot{U}_2, {}_0D_t^{-v}\dot{I}_1 = -\dot{I}_2$
	$M_{\gamma,v} - C_\alpha$	${}_0D_t^{-\gamma}\dot{U}_1 = \dot{U}_2, {}_0D_t^{-v}\dot{I}_1 = -{}_0D_t^{-\alpha}\dot{I}_2$
II	$C_\alpha - R$	$\dot{U}_1 = -\dot{I}_2, {}_0D_t^{-\alpha}\dot{I}_1 = \dot{U}_2$
	$L_\beta - R$	${}_0D_t^{-\beta}\dot{U}_1 = -\dot{I}_2, \dot{I}_1 = \dot{U}_2$
	$L_\beta - C_\alpha$	${}_0D_t^{-a}\dot{U}_1 = -{}_0D_t^{-\beta}\dot{I}_2, \dot{I}_1 = \dot{U}_2$
	$M_{\gamma,v} - R$	${}_0D_t^{-\gamma}\dot{U}_1 = -\dot{I}_2, {}_0D_t^{-v}\dot{I}_1 = \dot{U}_2$
	$M_{\gamma,v} - L_\beta$	${}_0D_t^{-\gamma}\dot{U}_1 = -\dot{I}_2, {}_0D_t^{-v}\dot{I}_1 = {}_0D_t^{-\beta}\dot{U}_2$
	$M_{\gamma,v} - C_\alpha$	${}_0D_t^{-\gamma}\dot{U}_1 = -{}_0D_t^{-\alpha}\dot{I}_2, {}_0D_t^{-v}\dot{I}_1 = \dot{U}_2$

$$\begin{cases} \dot{U}_1 = \dot{U}_2 \\ {}_0D_t^{-\alpha}\dot{I}_1 = -\dot{I}_2 \end{cases} \tag{8.88}$$

The fractional-order C_α-R mutator can be realized by using a fractional-order capacitor, a controlled voltage source, and two controlled current sources, as shown in Fig. 8.13.

By using KCL and KVL, the circuit equations can be obtained as

$$\begin{cases} \dot{U}_1 = -\dot{U}_C + \dot{U}_C + \dot{U}_2 \\ \dot{I}_1 - C'_\alpha \dot{U}_C - \dot{I}_1 + \dot{I}_2 = 0 \\ C'_\alpha {}_0D_t^{\alpha}\dot{U}_C = -\dot{I}_1 \end{cases} \tag{8.89}$$

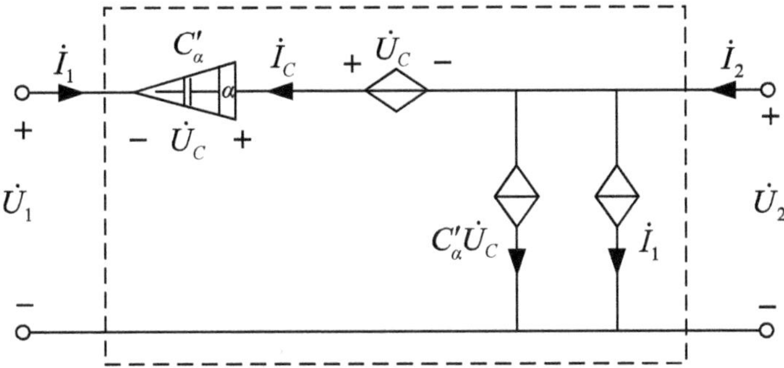

Fig. 8.13 Fractional-order C_α-R mutator realized by a fractional-order capacitor

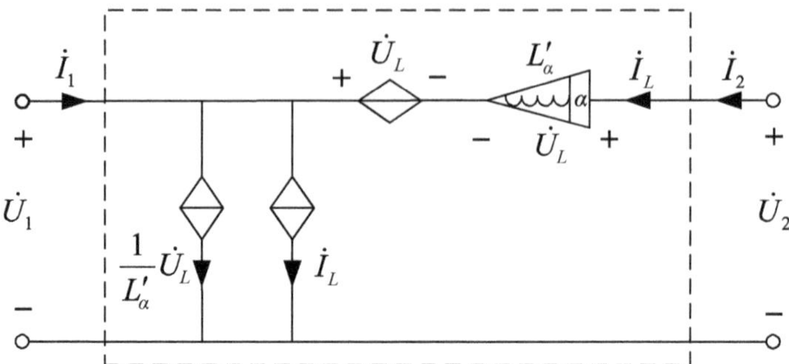

Fig. 8.14 Fractional-order C_α-R mutator realized by a fractional-order inductor

Reorganizing (8.89), it can be found that (8.89) and (8.88) are equivalent, which means that the circuit shown in Fig. 8.13 satisfies the basic property of the C_α-R mutator and can realize the mutual conversion of fractional-order capacitor and resistor.

Similarly, the C_α-R mutator can also be realized by a fractional-order inductor, a controlled voltage source, and two controlled current sources, as shown in Fig. 8.14.

Based on KCL and KVL, the circuit equations can be described as

$$\begin{cases} \dot{U}_1 = \dot{U}_L - \dot{U}_L + \dot{U}_2 \\ \dot{I}_1 + \dfrac{\dot{U}_L}{L'_\alpha} - \dot{I}_1 + \dot{I}_2 = 0 \\ L'_\alpha {}_0D_t^\alpha \dot{I}_L = \dot{U}_L \end{cases} \tag{8.90}$$

which satisfies the basic relationship (8.88). It is proved that the circuit shown in Fig. 8.14 fulfills the function of C_α-R mutator, and can realize the mutual conversion between the fractional-order capacitor and resistor.

8.5.3 Fractional-Order L_β-R Mutator

Table 8.1 provides the relationship between port voltages and currents of the fractional-order L_β-R mutator in type I, that is

$$\begin{cases} {}_0D_t^{-\beta} \dot{U}_1 = \dot{U}_2 \\ \dot{I}_1 = -\dot{I}_2 \end{cases} \tag{8.91}$$

The fractional-order L_β-R mutator can be realized by a fractional-order capacitor, two controlled voltage sources, and a controlled current source [20], as shown in Fig. 8.15.

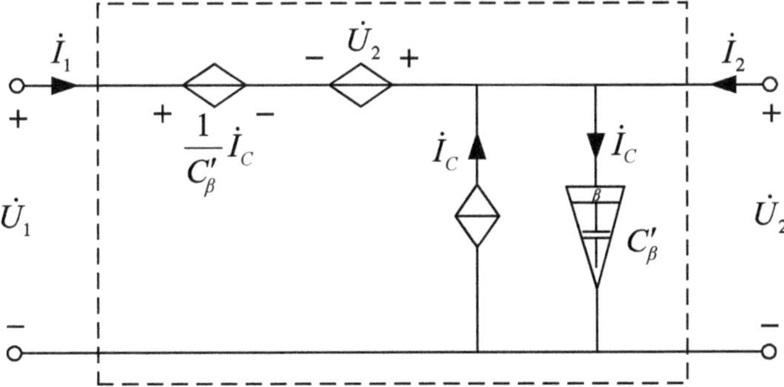

Fig. 8.15 Fractional-order L_β-R mutator realized by a fractional-order capacitor

Based on KCL and KVL, the circuit equations can be expressed as

$$\begin{cases} \dot{U}_1 = \dfrac{\dot{I}_C}{C'_\beta} - \dot{U}_2 + \dot{U}_2 \\ \dot{I}_1 + \dot{I}_C - \dot{I}_C + \dot{I}_2 = 0 \\ C'_{\beta 0} D_t^\beta \dot{U}_2 = \dot{I}_C \end{cases} \tag{8.92}$$

Through proper simplification, (8.92) is equivalent to (8.91). Therefore, the circuit shown in Fig. 8.15 satisfies the fundamental property of the L_β-R mutator and can realize the mutual conversion between fractional-order inductor and resistor.

In the same way, the L_β-R mutator can also be realized with the help of a fractional-order inductor, two controlled voltage sources, and a controlled current sources, as shown in Fig. 8.16.

By applying KCL and KVL, the circuit equations can be written as

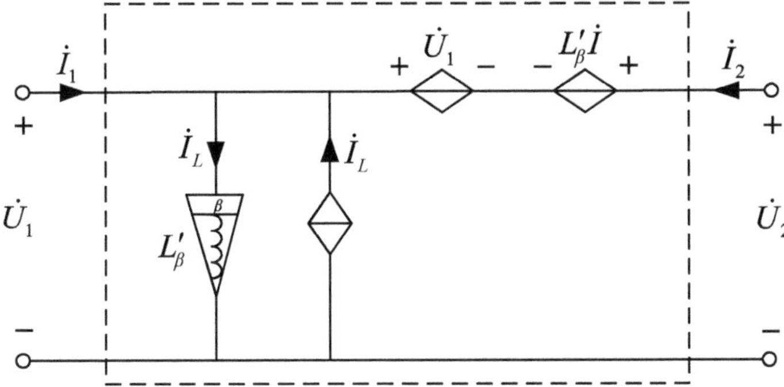

Fig. 8.16 Fractional-order L_β-R mutator realized by a fractional-order inductor

$$\begin{cases} \dot{U}_1 = \dot{U}_1 - L_\beta \dot{I}_L + \dot{U}_2 \\ \dot{I}_1 - \dot{I}_L + \dot{I}_L + \dot{I}_2 = 0 \\ L'_{\beta 0} D_t^\beta \dot{I}_L = \dot{U}_1 \end{cases} \tag{8.93}$$

which meets the essential relationship shown in (8.91). It can be inferred that the circuit shown in Fig. 8.16 can realize the mutual conversion between fractional-order inductor and resistor.

8.5.4 Fractional-Order $M_{\alpha,\beta}$-R Mutator

Table 8.1 gives the relationship between port voltages and currents of the fractional-order $M_{\alpha,\beta}$-R mutator in type I, likes

$$\begin{cases} {}_0D_t^{-\alpha} \dot{U}_1 = \dot{U}_2 \\ {}_0D_t^{-\beta} \dot{I}_1 = -\dot{I}_2 \end{cases} \tag{8.94}$$

The $M_{\alpha,\beta}$-R mutator can be realized by adopting a fractional-order inductor, a fractional-order capacitor, controlled voltage sources, and controlled current sources [21], as shown in Fig. 8.17.

On the basis of the KCL and KVL, the circuit equations are

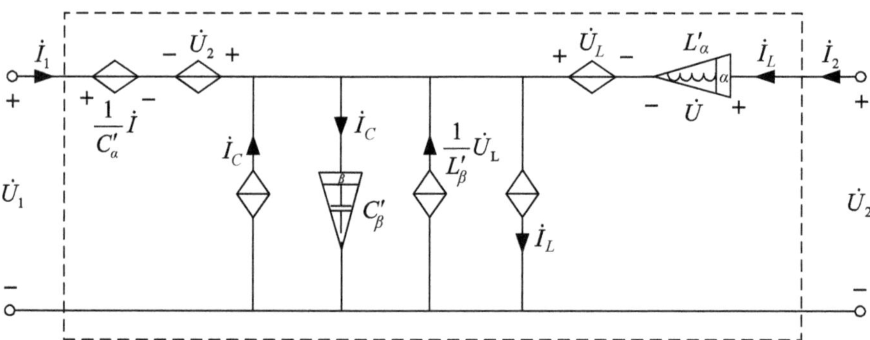

Fig. 8.17 Fractional-order $M_{\alpha,\beta}$-R mutator

$$\begin{cases} \dot{U}_1 = \dfrac{\dot{I}_C}{C'_\beta} - \dot{U}_2 + \dot{U}_L - \dot{U}_L + \dot{U}_2 \\ \dot{I}_1 + \dot{I}_C - \dot{I}_C + \dfrac{\dot{U}_L}{L'_\beta} - \dot{I}_L + \dot{I}_2 = 0 \\ C'_\alpha {}_0D_t^\alpha \dot{U}_2 = \dot{I}_C \\ L'_\beta {}_0D_t^\beta \dot{U}_2 = \dot{I}_2 \end{cases} \tag{8.95}$$

which is consistent with the fundamental relational Eq. (8.94), so the circuit shown in Fig. 8.17 has the function of the fractional-order $M_{\alpha,\beta}$-R mutator and realizes the mutual conversion between the fractional-order memristor and the resistor.

8.6 Connection Types of Multiple Two-Port Fractional-Order Networks

The connection of two-port fractional-order networks is similar to traditional two-port integer-order networks. There are five connection types: cascade, series, parallel, series–parallel, and parallel-series. It should be noted that when multiple two-port networks are combined into a composite two-port network through connections, the prerequisite for calculating the parameters of the composite two-port network is that each sub-network satisfies the port conditions (8.1) [22].

8.6.1 Cascade

Figure 8.18 shows the cascade of two networks, in which the output port of a two-port fractional-order network is directly connected to the input port of another port [23].

Assuming that the T-parameter equations of the two networks are

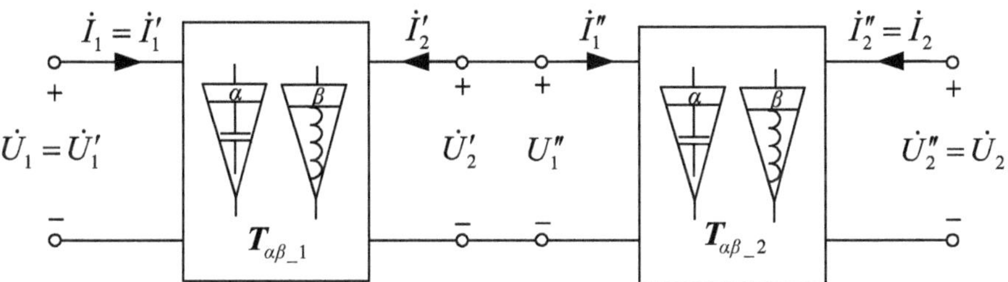

Fig. 8.18 Cascaded network

$$\begin{bmatrix} \dot{U}'_1 \\ \dot{I}'_1 \end{bmatrix} = \boldsymbol{T}_{\alpha\beta\_1} \begin{bmatrix} \dot{U}'_2 \\ -\dot{I}'_2 \end{bmatrix} \tag{8.96}$$

$$\begin{bmatrix} \dot{U}''_1 \\ \dot{I}''_1 \end{bmatrix} = \boldsymbol{T}_{\alpha\beta\_2} \begin{bmatrix} \dot{U}''_2 \\ -\dot{I}''_2 \end{bmatrix} \tag{8.97}$$

where

$$\begin{cases} \boldsymbol{T}_{\alpha\beta\_1} = \begin{bmatrix} A'_{\alpha\beta\_11} & A'_{\alpha\beta\_12} \\ A'_{\alpha\beta\_21} & A'_{\alpha\beta\_22} \end{bmatrix} \\ \boldsymbol{T}_{\alpha\beta\_2} = \begin{bmatrix} A''_{\alpha\beta\_11} & A''_{\alpha\beta\_12} \\ A''_{\alpha\beta\_21} & A''_{\alpha\beta\_22} \end{bmatrix} \end{cases} \tag{8.98}$$

and the connection relationship shown in Fig. 8.18 can be described as

$$\begin{cases} \dot{U}_1 = \dot{U}'_1 \\ \dot{I}_1 = \dot{I}'_1 \\ \dot{U}'_2 = \dot{U}''_1 \\ \dot{I}'_2 = -\dot{I}''_1 \\ \dot{U}''_2 = \dot{U}_2 \\ \dot{I}''_2 = \dot{I}_2 \end{cases} \tag{8.99}$$

Then, the T-parameter equation of the cascaded two-port fractional-order network can be written as

$$\begin{aligned} \begin{bmatrix} \dot{U}_1 \\ \dot{I}_1 \end{bmatrix} &= \boldsymbol{T}_{\alpha\beta\_1} \begin{bmatrix} \dot{U}'_2 \\ -\dot{I}'_2 \end{bmatrix} = \boldsymbol{T}_{\alpha\beta\_1} \begin{bmatrix} \dot{U}''_1 \\ \dot{I}''_1 \end{bmatrix} = \boldsymbol{T}_{\alpha\beta\_1} \boldsymbol{T}_{\alpha\beta\_2} \begin{bmatrix} \dot{U}''_2 \\ -\dot{I}''_2 \end{bmatrix} \\ &= \boldsymbol{T}_{\alpha\beta\_1} \boldsymbol{T}_{\alpha\beta\_2} \begin{bmatrix} \dot{U}_2 \\ -\dot{I}_2 \end{bmatrix} \end{aligned} \tag{8.100}$$

Replacing $\boldsymbol{T}_{\alpha\beta\_1}\boldsymbol{T}_{\alpha\beta\_2}$ with the T-parameter matrix $\boldsymbol{T}_{\alpha\beta}$ (i.e., $\boldsymbol{T}_{\alpha\beta} = \boldsymbol{T}_{\alpha\beta\_1}\boldsymbol{T}_{\alpha\beta\_2}$), (8.100) can be rewritten as

$$\begin{bmatrix} \dot{U}_1 \\ \dot{I}_1 \end{bmatrix} = \boldsymbol{T}_{\alpha\beta} \begin{bmatrix} \dot{U}_2 \\ -\dot{I}_2 \end{bmatrix} \tag{8.101}$$

which can be used to calculate the port parameters of the cascaded two-port fractional-order networks.

8.6.2 Series Connection

Figure 8.19 depicts the series connection of two fractional-order networks, in which the input ports and the output ports of two networks are connected in series [24].

As can be seen from Fig. 8.19, the relationships between the port voltages and currents can be written as

$$\begin{cases} \dot{I}_1 = \dot{I}_1' = \dot{I}_1'' \\ \dot{I}_2 = \dot{I}_2' = \dot{I}_2'' \\ \dot{U}_1 = \dot{U}_1' + \dot{U}_1'' \\ \dot{U}_2 = \dot{U}_2' + \dot{U}_2'' \end{cases} \tag{8.102}$$

and supposing that Z-parameter matrixes of two two-port networks are

$$\begin{cases} \mathbf{Z}_{\alpha\beta\_1} = \begin{bmatrix} Z'_{\alpha\beta\_11} & Z'_{\alpha\beta\_12} \\ Z'_{\alpha\beta\_21} & Z'_{\alpha\beta\_22} \end{bmatrix} \\ \mathbf{Z}_{\alpha\beta\_2} = \begin{bmatrix} Z''_{\alpha\beta\_11} & Z''_{\alpha\beta\_12} \\ Z''_{\alpha\beta\_21} & Z''_{\alpha\beta\_22} \end{bmatrix} \end{cases} \tag{8.103}$$

Then, according to (8.102) and (8.103), the Z-parameter equation of the series-connected two-port fractional-order network can be expressed as

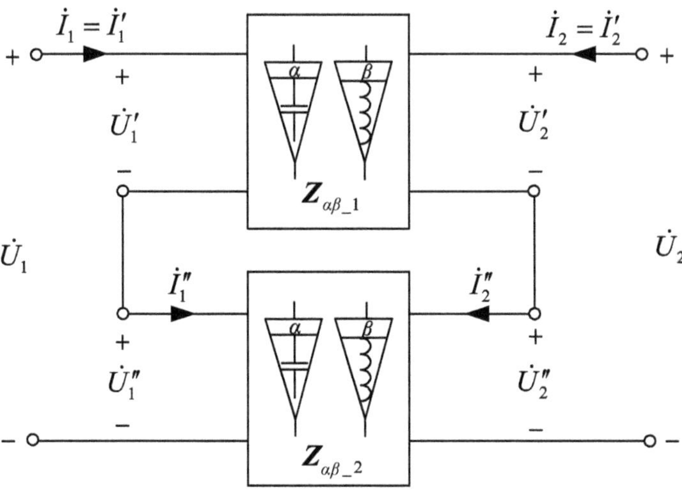

Fig. 8.19 Series-connected network

$$\begin{bmatrix} \dot{U}_1 \\ \dot{U}_2 \end{bmatrix} = \begin{bmatrix} \dot{U}'_1 \\ \dot{U}'_2 \end{bmatrix} + \begin{bmatrix} \dot{U}''_1 \\ \dot{U}''_2 \end{bmatrix} = \mathbf{Z}_{\alpha\beta\_1} \begin{bmatrix} \dot{I}_1 \\ \dot{I}_2 \end{bmatrix} + \mathbf{Z}_{\alpha\beta\_2} \begin{bmatrix} \dot{I}_1 \\ \dot{I}_2 \end{bmatrix}$$

$$= \begin{bmatrix} Z'_{\alpha\beta\_11} + Z''_{\alpha\beta\_11} & Z'_{\alpha\beta\_12} + Z''_{\alpha\beta\_12} \\ Z'_{\alpha\beta\_21} + Z''_{\alpha\beta\_21} & Z'_{\alpha\beta\_22} + Z''_{\alpha\beta\_22} \end{bmatrix} \begin{bmatrix} \dot{I}_1 \\ \dot{I}_2 \end{bmatrix} = \mathbf{Z}_{\alpha\beta} \begin{bmatrix} \dot{I}_1 \\ \dot{I}_2 \end{bmatrix} \tag{8.104}$$

where $\mathbf{Z}_{\alpha\beta} = \mathbf{Z}_{\alpha\beta\_1} + \mathbf{Z}_{\alpha\beta\_2}$.

8.6.3 Parallel Connection

Figure 8.20 presents a parallel-connected two-port fractional-order network, where the input ports and the output ports of two fractional-order networks are connected in parallel, respectively [25].

From Fig. 8.20, it is easily seen that the port voltages and currents of two networks satisfy

$$\begin{cases} \dot{U}_1 = \dot{U}'_1 = \dot{U}''_1 \\ \dot{U}_2 = \dot{U}'_2 = \dot{U}''_2 \\ \dot{I}_1 = \dot{I}'_1 + \dot{I}''_1 \\ \dot{I}_2 = \dot{I}'_2 + \dot{I}''_2 \end{cases} \tag{8.105}$$

and assuming that the Y-parameter matrixes of the two-port network are

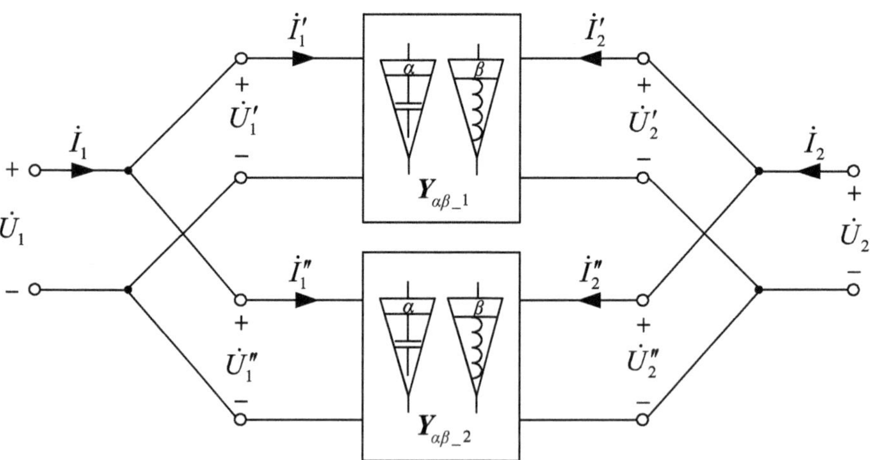

Fig. 8.20 Parallel-connected network

$$\begin{cases} \boldsymbol{Y}_{\alpha\beta\_1} = \begin{bmatrix} Y'_{\alpha\beta\_11} & Y'_{\alpha\beta\_12} \\ Y'_{\alpha\beta\_21} & Y'_{\alpha\beta\_22} \end{bmatrix} \\ \boldsymbol{Y}_{\alpha\beta\_2} = \begin{bmatrix} Y''_{\alpha\beta\_11} & Y''_{\alpha\beta\_12} \\ Y''_{\alpha\beta\_21} & Y''_{\alpha\beta\_22} \end{bmatrix} \end{cases} \tag{8.106}$$

In consequence, the Y-parameter equation of the parallel-connected two-port fractional-order network can be described as

$$\begin{aligned} \begin{bmatrix} \dot{I}_1 \\ \dot{I}_2 \end{bmatrix} &= \begin{bmatrix} \dot{I}'_1 \\ \dot{I}'_2 \end{bmatrix} + \begin{bmatrix} \dot{I}''_1 \\ \dot{I}''_2 \end{bmatrix} = \boldsymbol{Y}_{\alpha\beta\_1} \begin{bmatrix} \dot{U}_1 \\ \dot{U}_2 \end{bmatrix} + \boldsymbol{Y}_{\alpha\beta\_2} \begin{bmatrix} \dot{U}_1 \\ \dot{U}_2 \end{bmatrix} \\ &= \begin{bmatrix} Y'_{\alpha\beta\_11} + Y''_{\alpha\beta\_11} & Y'_{\alpha\beta\_12} + Y''_{\alpha\beta\_12} \\ Y'_{\alpha\beta\_21} + Y''_{\alpha\beta\_21} & Y'_{\alpha\beta\_22} + Y''_{\alpha\beta\_22} \end{bmatrix} \begin{bmatrix} \dot{U}_1 \\ \dot{U}_2 \end{bmatrix} = \boldsymbol{Y}_{\alpha\beta} \begin{bmatrix} \dot{U}_1 \\ \dot{U}_2 \end{bmatrix} \end{aligned} \tag{8.107}$$

where $\boldsymbol{Y}_{\alpha\beta} = \boldsymbol{Y}_{\alpha\beta\_1} + \boldsymbol{Y}_{\alpha\beta\_2}$.

8.6.4 Series-Parallel Connection

For the case that the input ports of two networks are connected in series, and the output ports are in parallel connection, as shown in Fig. 8.21, it is called the series–parallel connection of two fractional-order networks [26].

The H-parameters of two-port fractional-order networks are applied to analyze the network in series–parallel connection, which are expressed as

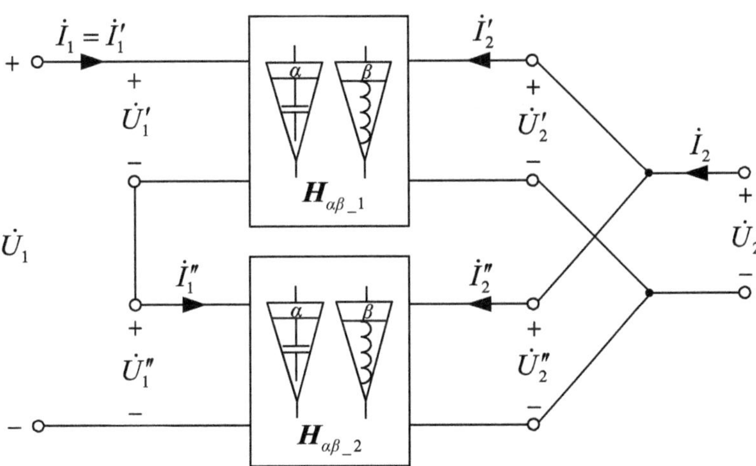

Fig. 8.21 Series–parallel network

$$\begin{cases} \boldsymbol{H}_{\alpha\beta\_1} = \begin{bmatrix} H'_{\alpha\beta\_11} & H'_{\alpha\beta\_12} \\ H'_{\alpha\beta\_21} & H'_{\alpha\beta\_22} \end{bmatrix} \\ \boldsymbol{H}_{\alpha\beta\_2} = \begin{bmatrix} H''_{\alpha\beta\_11} & H''_{\alpha\beta\_12} \\ H''_{\alpha\beta\_21} & H''_{\alpha\beta\_22} \end{bmatrix} \end{cases} \tag{8.108}$$

In the precondition that the port conditions are still followed after two sub-networks are connected in series–parallel, the H-parameter equations can be written as

$$\begin{bmatrix} \dot{U}'_1 \\ \dot{I}'_2 \end{bmatrix} = \begin{bmatrix} H'_{\alpha\beta\_11} & H'_{\alpha\beta\_12} \\ H'_{\alpha\beta\_21} & H'_{\alpha\beta\_22} \end{bmatrix} \begin{bmatrix} \dot{I}_1 \\ \dot{U}_2 \end{bmatrix} = \boldsymbol{H}_{\alpha\beta\_1} \begin{bmatrix} \dot{I}_1 \\ \dot{U}_2 \end{bmatrix} \tag{8.109}$$

$$\begin{bmatrix} \dot{U}''_1 \\ \dot{I}''_2 \end{bmatrix} = \begin{bmatrix} H''_{\alpha\beta\_11} & H''_{\alpha\beta\_12} \\ H''_{\alpha\beta\_21} & H''_{\alpha\beta\_22} \end{bmatrix} \begin{bmatrix} \dot{I}_1 \\ \dot{U}_2 \end{bmatrix} = \boldsymbol{H}_{\alpha\beta\_2} \begin{bmatrix} \dot{I}_1 \\ \dot{U}_2 \end{bmatrix} \tag{8.110}$$

Combining the relationships of the port parameters with (8.109) and (8.110), the H-parameter equation of two-port fractional-order network in series–parallel connection can be easily obtained as

$$\begin{aligned} \begin{bmatrix} \dot{U}_1 \\ \dot{I}_2 \end{bmatrix} &= \begin{bmatrix} \dot{U}''_1 \\ \dot{I}''_2 \end{bmatrix} + \begin{bmatrix} \dot{U}''_1 \\ \dot{I}''_2 \end{bmatrix} = \begin{bmatrix} H'_{\alpha\beta\_11} + H''_{\alpha\beta\_11} & H'_{\alpha\beta\_12} + H''_{\alpha\beta\_12} \\ H'_{\alpha\beta\_21} + H''_{\alpha\beta\_21} & H'_{\alpha\beta\_22} + H''_{\alpha\beta\_22} \end{bmatrix} \begin{bmatrix} \dot{I}_1 \\ \dot{U}_2 \end{bmatrix} \\ &= \boldsymbol{H}_{\alpha\beta} \begin{bmatrix} \dot{I}_1 \\ \dot{U}_2 \end{bmatrix} \end{aligned} \tag{8.111}$$

where $\boldsymbol{H}_{\alpha\beta} = \boldsymbol{H}_{\alpha\beta\_1} + \boldsymbol{H}_{\alpha\beta\_2}$.

8.6.5 Parallel-Series Connection

Figure 8.22 describes the parallel-series connection of two fractional-order networks. Here, the input ports are connected in parallel, and the output ports are in series connection [27].

The G-parameters are used to depict the parallel-series connection of two fractional-order networks [28], namely

$$\begin{cases} \boldsymbol{G}_{\alpha\beta\_1} = \begin{bmatrix} G'_{\alpha\beta\_11} & G'_{\alpha\beta\_12} \\ G'_{\alpha\beta\_21} & G'_{\alpha\beta\_22} \end{bmatrix} \\ \boldsymbol{G}_{\alpha\beta\_2} = \begin{bmatrix} G''_{\alpha\beta\_11} & G''_{\alpha\beta\_12} \\ G''_{\alpha\beta\_21} & G''_{\alpha\beta\_22} \end{bmatrix} \end{cases} \tag{8.112}$$

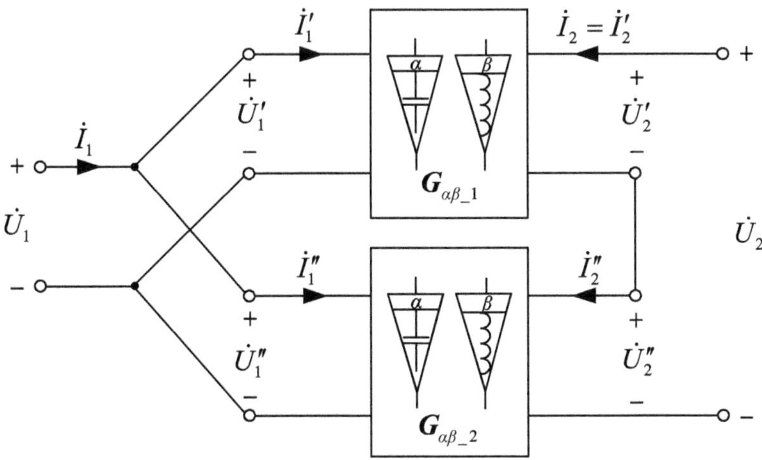

Fig. 8.22 Parallel-series network

If the port conditions of each sub-networks remain unchanged after parallel-series connection [12], the G-parameter equations can be obtained as

$$\begin{bmatrix} \dot{I}'_1 \\ \dot{U}'_2 \end{bmatrix} = \begin{bmatrix} G'_{\alpha\beta\_11} & G'_{\alpha\beta\_12} \\ G'_{\alpha\beta\_21} & G'_{\alpha\beta\_22} \end{bmatrix} \begin{bmatrix} \dot{U}_1 \\ \dot{I}_2 \end{bmatrix} = \boldsymbol{G}_{\alpha\beta\_1} \begin{bmatrix} \dot{U}_1 \\ \dot{I}_2 \end{bmatrix} \tag{8.113}$$

$$\begin{bmatrix} \dot{I}'_1 \\ \dot{U}'_2 \end{bmatrix} = \begin{bmatrix} G'_{\alpha\beta\_11} & G'_{\alpha\beta\_12} \\ G'_{\alpha\beta\_21} & G'_{\alpha\beta\_22} \end{bmatrix} \begin{bmatrix} \dot{U}_1 \\ \dot{I}_2 \end{bmatrix} = \boldsymbol{G}_{\alpha\beta\_1} \begin{bmatrix} \dot{U}_1 \\ \dot{I}_2 \end{bmatrix} \tag{8.114}$$

Substituting the relationships of the port parameters into (8.109) and (8.110), the G-parameter equation of two-port fractional-order network in parallel-series connection can be written as

$$\begin{aligned} \begin{bmatrix} \dot{I}_1 \\ \dot{U}_2 \end{bmatrix} &= \begin{bmatrix} \dot{I}'_1 \\ \dot{U}'_2 \end{bmatrix} + \begin{bmatrix} \dot{I}''_1 \\ \dot{U}''_2 \end{bmatrix} = \begin{bmatrix} G'_{\alpha\beta\_11} + G''_{\alpha\beta\_11} & G'_{\alpha\beta\_12} + G''_{\alpha\beta\_12} \\ G'_{\alpha\beta\_21} + G''_{\alpha\beta\_21} & G'_{\alpha\beta\_22} + G''_{\alpha\beta\_22} \end{bmatrix} \begin{bmatrix} \dot{U}_1 \\ \dot{I}_2 \end{bmatrix} \\ &= \boldsymbol{G}_{\alpha\beta} \begin{bmatrix} \dot{U}_1 \\ \dot{I}_2 \end{bmatrix} \end{aligned} \tag{8.115}$$

where $\boldsymbol{G}_{\alpha\beta} = \boldsymbol{G}_{\alpha\beta\_1} + \boldsymbol{G}_{\alpha\beta\_2}$.

8.7 Summary

This chapter introduces the basic concepts of the two-port fractional-order network, derives the equations, and analyzes the physical meanings of Y-parameters, Z-parameters, H-parameters, and G-parameters. The equivalent circuits of two-port fractional-order network based on Y-parameters and Z-parameters are also given. Subsequently, the equations of transmission parameters and reverse transmission parameters are deduced, and their corresponding physical meanings are explained. Besides, the propagation parameters of the symmetric two-port fractional-order network are also briefly introduced. After that, the basic concepts of fractional-order mutators are elaborate, and the circuit models of three typical fractional-order mutators are presented. Finally, the relationships between port voltages and currents of multiple two-port fractional networks in different connection ways are analyzed.

References

1. Walczak J, Swiszcz P (2005) Frequency representations of one-port networks. COMPEL: Int J Comput Math Electr Electron Eng 24(4):1142–1151
2. Adamczyk, Bogdan (2017) Circuit theorems and techniques. In: Foundations of electromagnetic compatibility: with practical applications, vol 9. Wiley, Netherlands, pp 203–241
3. Kim N, Allen JB (2013) Two-port network analysis and modeling of a balanced armature receiver. Hear Res 301(1):156–167
4. Aberle JT (2008) Two-port representation of an antenna with application to non-foster matching networks. IEEE Trans Antennas Propag 56(5):1218–1222
5. Reed J, Wheeler GJ (1957) A method of analysis of symmetrical four-port networks. IRE Trans Microw Theory Tech 4(4):246–252
6. Maundy BJ et al (2016) Fractional-order two-port networks. Math Probl Eng Theory Methods Appl 2016(5976301):1–5
7. Visser HJ (2012) Appendix E: two-port network parameters. In: Antenna theory and applications. Wiley, Netherlands, pp 245–247
8. Yang WY, Lee SC (2009) Two-port networks. In: Circuit systems with MATLAB and PSpice. John Wiley & Sons (Asia) Pte Ltd, pp 401–499
9. Johnstone GG, Deane JHB (1991) Relations between two-port parameters. Int J Electron 71(1):107–116
10. Gonzalez EA, Dorčák L, Petráš I, Terpák J (2013) On the mathematical properties of generalized fractional-order two-port networks using hybrid parameters. In: Proceedings of the 14th International Carpathian control conference. Rytro, Poland
11. Fouda ME, Elwakil AS, Radwan AG et al (2016) Fractional-order two-port networks. Math Probl Eng 2016(5976301):1–5
12. Choma J, Chen WK (2007) Two-port network models and analysis. In: Feedback networks: theory and circuit applications. World Scientific Publishing, Singapore, pp 111–223
13. Witte RA (2014) Two-port networks. In: Spectrum and network measurements. Prentice-Hall Inc., United States, pp 241–252
14. Misra DK (2004) Two-port networks. In: Radio-frequency and microwave communication circuits. Wiley
15. Hou C-L, Chang H-J, Shun-Sheng Su (2001) Equivalent circuits of a linear time-invariant two-port in terms of the transmission parameters/inverse transmission parameters and Nullator-Norator representations. J Zhongyuan Inst Technol 29(1):87–91

16. James M (1994) Input impedance derived from a transfer network. US
17. Różowicz S, Włodarczyk M, Zawadzki A (2016) Wave parameters of symmetrical two-port networks containing elements of fractional-order order. In: 17th International conference computational problems of electrical engineering (CPEE). Sandomierz
18. Valsa J (2012) Fractional-order electrical elements, networks and systems. In: Proceedings of 22nd International conference radioelektronika. Brno
19. Petrzela J (2013) Fundamental analog cells for fractional-order two-port synthesis. In: 23rd International conference radioelektronika. Pardubice
20. Rikoski RA, Choudhry A (1976) Mutator realized inductance. Proc IEEE 64(6):1012–1013
21. Soltan A, Radwan AG, Soliman AM (2016) Realizing fractional-order elements using CCII based mutators. In: Proceedings of 28th International conference on microelectronics (ICM). Giza
22. Weik MH (2001) Circuit connection. Springer, US
23. Kawaba K, Nazri W, Aun HK et al (1998) A realization of fractional power-law circuit using OTAs. In: Proceedings of IEEE Asia-Pacific conference on circuits and systems microelectronics and integrating systems. Chiang Mai, Thailand
24. Tsai MC, Gu DW (2014) Two-port networks. Springer, US
25. Witte RA (1991) Two-port networks. In: Spectrum and network measurements. Prentice-Hall Inc., United States
26. Morris NM (1993) Two-port networks. In: Electrical circuit analysis and design. Macmillan Education, UK
27. Pointon AJ, Howarth HM (1991) Two-port networks. In: AC and DC network theory. Springer, Netherlands
28. Tung LJ, Kwan BW (2015) Two-port networks. In: Circuit analysis. World Scientific, Singapore

Chapter 9
State-Variable Analysis of Fractional-Order Circuits

The state-variable method originated from modern control theory. It has been widely applied to circuit analysis [1]. The state-variable method can also be adapted to analyze fractional-order circuits.

When the state-variable method is used to analyze a system, the dynamic performance of the system is described by the differential equations composed of state variables. It can reflect the changes of all independent variables of the system so that the internal motion state of the system can be determined at the same time. Generally, the state-variable method can be put to analyze linear time-invariant, linear time-varying, and nonlinear systems, SISO (single input single output) and MIMO (multiple input multiple output) systems, continuous and discrete signal systems. Therefore, this method has wide applicability [2, 3].

The fractional-order complex circuit is analyzed by the state-variable method, the first step is to select state variables. Different circuits need different state variables. The second is to establish differential equations by the selected state variables, and the input equation and output equation of the circuit can be obtained. The circuit can be divided into a normal circuit and an ill-conditioned circuit according to whether it contains a loop composed entirely of capacitors or a cut set consisting entirely of inductors. we establish state equations of two types of circuits. Finally, the Laplace transform can solve the established state equations, and respectively write the analytical solutions of the same order and different orders of the components in the circuit.

9.1 Basic Concepts

9.1.1 State Variables

State is the information and behavior that characterize the movement of matter. The smallest set of variables that can be used to describe the state of motion of

B. Zhang and X. Shu, *Fractional-Order Electrical Circuit Theory*, CPSS Power Electronics Series, https://doi.org/10.1007/978-981-16-2822-1_9

matter is state variables [4]. The state variable is a set of variables that describe the mathematical state of a dynamical system. Intuitively, the state of a system describes enough information about a system, and it can determine its future behavior [5].

For different circuits, the choice of variables is not the same. Generally, the voltage u_C of the capacitor and the current i_L of the inductor are chosen as state variables in the circuit. For nonlinear systems and time-varying circuits, the charge q_C of the capacitor and the flux linkage ψ_L of the inductor are usually selected as the state variables.

If the circuit does not contain a cut set formed by pure inductance branches and a loop consisting of pure capacitance branches, such a circuit is called a normal" circuit. Here, the number n of state variables are equal to the number n_e of energy storage components in the normal circuit, i.e., $n = n_e$.

However, if the circuit contains a cut set of pure inductance branches and a loop of pure capacitance branches, such a circuit is an ill-conditioned circuit. The number n of state variables are equal to the number n_e of energy storage components minus the number of linearly independent cut sets n_q of pure inductance branches and linearly independent loops n_p of pure capacitance branches, which can be denoted by $n = n_e - n_q - n_p$.

The pure inductance sub-circuit is made up entirely of inductors, which is obtained by short-circuiting all capacitors, resistors, voltage sources, and disconnecting all current sources.

The pure capacitance sub-circuit is a circuit that consists entirely of capacitors, which is obtained by disconnecting all the inductors, resistors, current sources, and short-circuiting all current sources.

Figure 9.1 is an ill-conditioned circuit. The pure inductance sub-circuit and the pure capacitance sub-circuit can be obtained separately according to the above definitions. In this ill-conditioned circuit, the number of linearly independent cut sets formed by pure inductance branches is $n_q = 3$, the number of linearly independent loops consisting of pure capacitance branches is $n_p = 1$, and the number of energy storage components in the circuit is $n_e = 9$. Therefore, the number of state variables is

$$n = n_e - n_q - n_p = 9 - 3 - 1 = 5 \tag{9.1}$$

(9.1) calculates the upper bound of the number of state variables of the circuit. If the circuit contains a controlled power supply and the parameters are coincidental, the number of state variables will be further reduced, so it will not be discussed further.

9.1.2 State Space and State Vector

The state space is the set of all possible states of a system. Every state of the system corresponds to a unique point in the state space. In other words, each point in the

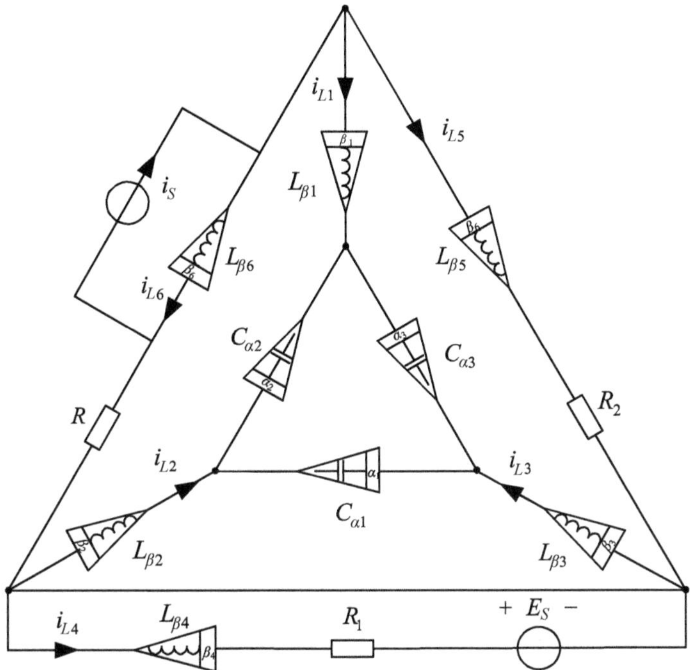

Fig. 9.1 Ill-conditioned circuit

state space corresponds to a different state of the system. The dimension of the state space is the number of state variables. Every coordinate is a state variable, and the values of the state variables completely describe the state of the system.

Taking a simple $RL_\beta C_\alpha$ circuit shown in Fig. 9.2a as an example, i_L and u_C are the state variables, the space formed by i_L, and u_C is a two-dimensional state space, as shown in Fig. 9.2b.

The state of the circuit is determined by the values of the state variables, and the state variables can be represented by an $n \times 1$ column matrix. In the state space, the

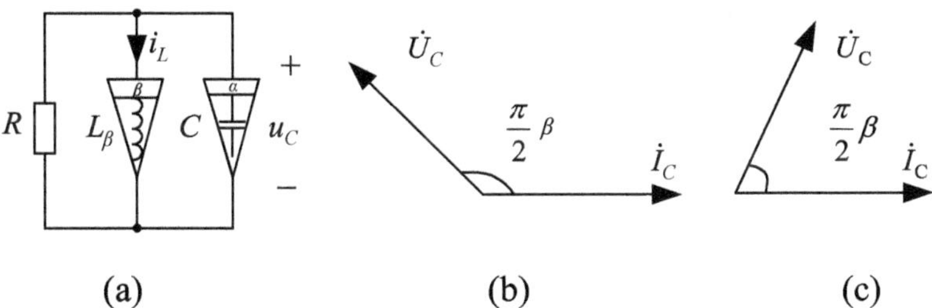

Fig. 9.2 Two-dimensional state space: **a** $RL_\beta C_\alpha$ circuit. **b** State space ($\beta > 1$). **c** State space ($0 < \beta < 1$)

vector from the origin to a certain point is called the state vector. The projection of the state vector on the base coordinates is equal to the value of each state variable, which can completely describe the state of the circuit.

When the time t changes from t_0 to ∞, the state of the circuit continuously changes, and the state vector's endpoint is moving in the state space. Thus, the state trajectory is the trajectory swept by the endpoint of the state vector moving in the state space.

9.1.3 General Form of State Equation

The state equation is a set of differential equations by the system state variables, which represents the variation of the system state caused by the input.

The state equation and the output equation are combined to form a state-space representation of a system. It is also called the state-space model. Generally, the state-space model has the following form [6–8].

$$\begin{cases} {}_{t_0}D_t^{\boldsymbol{\alpha}}\boldsymbol{x}(t) = \boldsymbol{A}\boldsymbol{x}(t) + \boldsymbol{B}\boldsymbol{u}(t) \\ \boldsymbol{y}(t) = \boldsymbol{C}\boldsymbol{x}(t) + \boldsymbol{D}\boldsymbol{u}(t) \end{cases} \tag{9.2}$$

where $\boldsymbol{\alpha} = [\alpha_1, \alpha_2, \ldots \alpha_n]^T$ are the fractional orders, $\boldsymbol{x}(t) \in \mathbb{R}^n$ is the state column vector. $\boldsymbol{u}(\mathrm{t}) \in \mathbb{R}^r$ is the input column vector. $\boldsymbol{y}(\mathrm{t}) \in \mathbb{R}^p$ is the output column vector. $\boldsymbol{A} \in \mathbb{R}^{n\times n}$ is the state matrix. $\boldsymbol{B} \in \mathbb{R}^{n\times r}$ is the input matrix. $\boldsymbol{C} \in \mathbb{R}^{p\times n}$ is the output matrix. $\boldsymbol{D} \in \mathbb{R}^{p\times r}$ is the direct transmission matrix.

9.2 State Equation of Fractional-Order Circuits

9.2.1 Normal Fractional-Order Circuit

For a fractional-order normal circuit (i.e., without the cut set composed entirely of inductors and the loop consisting entirely of capacitors), it is very convenient to write its state equation by the following method [9, 10]

Taking the fractional-order normal circuit shown in Fig. 9.3 as an example to set forth the steps to write the state equation, as follows:

Step 1: Determining the proper tree to ensure all capacitors are on the tree branches and all the inductors are on the link branches.

Step 2: Writing the cut-set equations of the single-tree branches and the loop equations of the single-link branches.

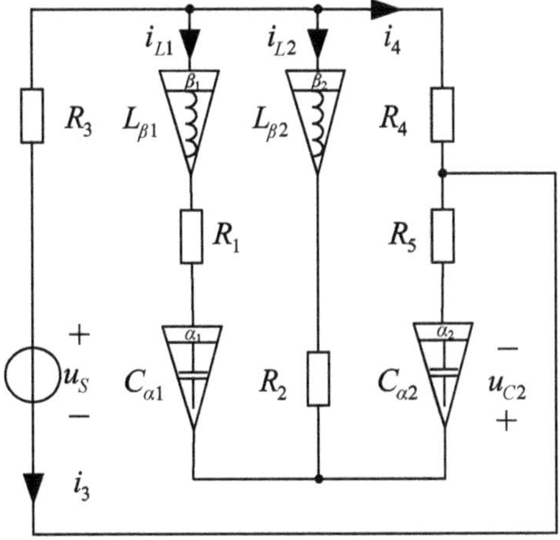

Fig. 9.3 The fractional-order normal circuit

$$\begin{cases} C_{\alpha 1}\dfrac{\mathrm{d}^{\alpha_1}u_{C1}}{\mathrm{d}t^{\alpha_1}} = i_{L1} \\ C_{\alpha 2}\dfrac{\mathrm{d}^{\alpha_2}u_{C2}}{\mathrm{d}t^{\alpha 2}} = i_{L1} + i_{L2} \\ L_{\beta 1}\dfrac{\mathrm{d}^{\beta_1}i_{L1}}{\mathrm{d}t^{\beta_1}} + i_{L1}R_1 + u_{C1} + u_{C2} + (i_{L1} + i_{L2})R_5 - i_4R_4 = 0 \\ L_{\beta 2}\dfrac{\mathrm{d}^{\beta_1}i_{L2}}{\mathrm{d}t^{\beta_2}} + i_{L2}R_2 + u_{C2} + (i_{L1} + i_{L2})R_5 - i_4R_4 = 0 \end{cases} \tag{9.3}$$

where α_1 and α_2 are the orders of the fractional-order capacitors $C_{\alpha 1}$ and $C_{\alpha 2}$, β_1 and β_2 are the orders of the fractional-order inductors $L_{\beta 1}$ and $L_{\beta 2}$, u_{C1} and u_{C2} are the voltages across $C_{\alpha 1}$ and $C_{\alpha 2}$, i_{L1} and i_{L2} are the currents flowing through $L_{\beta 1}$ and $L_{\beta 2}$, respectively. R_1, R_2, R_3, R_4 and R_5 are resistors.

Step 3: Writing the fundamental cut-set or loop equations corresponding to the non-state variables.

$$\begin{cases} i_4 + i_3 + i_{L1} + i_{L2} = 0 \\ u_S + i_3R_3 - i_4R_4 = 0 \end{cases} \tag{9.4}$$

where u_S is the voltage of the source. (9.4) can be rewritten as

$$\begin{cases} i_4 = \dfrac{u_S - R_3(i_{L1} + i_{L2})}{R_3 + R_4} \\ i_3 = -\dfrac{u_S + R_4(i_{L1} + i_{L2})}{R_3 + R_4} \end{cases} \tag{9.5}$$

Substituting (9.5) into (9.3), the state equation can be obtained as

$$\begin{bmatrix} \frac{\mathrm{d}^{\alpha_1} u_{C1}}{\mathrm{d}t^{\alpha_1}} \\ \frac{\mathrm{d}^{\alpha_2} u_{C2}}{\mathrm{d}t^{\alpha_2}} \\ \frac{\mathrm{d}^{\beta_1} i_{L1}}{\mathrm{d}t^{\beta_1}} \\ \frac{v^{\beta_2} i_{L2}}{\mathrm{d}t^{\beta_2}} \end{bmatrix} = \begin{bmatrix} 0 & 0 & \frac{1}{C_{\alpha 1}} & 0 \\ 0 & 0 & \frac{1}{C_{\alpha 2}} & \frac{1}{C_{\alpha 2}} \\ -\frac{1}{L_{\beta 1}} & -\frac{1}{L_{\beta 1}} & -\frac{R_1+R}{L_{\beta 1}} & -\frac{R}{L_{\beta 1}} \\ 0 & -\frac{1}{L_{\beta 2}} & -\frac{R}{L_{\beta 2}} & -\frac{R_2+R}{L_{\beta 2}} \end{bmatrix} \begin{bmatrix} u_{C1} \\ u_{C2} \\ i_{L1} \\ i_{L2} \end{bmatrix} + \begin{bmatrix} 0 \\ 0 \\ \frac{R_4}{L_{\beta 1}(R_3+R_4)} \\ \frac{R_4}{L_{\beta 2}(R_3+R_4)} \end{bmatrix} u_S \tag{9.6}$$

where $R = R_5 + (R_3 R_4)/(R_3 + R_4)$.

9.2.2 *Ill-Conditioned Fractional-Order Circuit*

The ill-conditioned circuit contains a loop composed entirely of capacitance branches or a cut set consisting entirely of inductance branches.

Assuming that the orders of all fractional-order components in the ill-conditioned circuit are identical, such as the fractional-order circuit shown in Fig. 9.4. Here, $C_{\alpha 1}$, $C_{\alpha 2}$, and $C_{\alpha 3}$ form a circuit, and that is composed entirely of capacitance branches. $C_{\alpha 1}$ and $C_{\alpha 2}$ can be regarded as an ill-conditioned tree. The fundamental cut-set equation corresponding to $C_{\alpha 1}$ and $C_{\alpha 2}$ can be derived as

$$\begin{cases} C_{\alpha 1} \frac{\mathrm{d}^{\alpha} u_{C1}}{\mathrm{d}t^{\alpha}} + C_{\alpha 3} \frac{\mathrm{d}^{\alpha} u_{C3}}{\mathrm{d}t^{\alpha}} + \frac{u_{C3}}{R} = i_S \\ C_{\alpha 2} \frac{\mathrm{d}^{\alpha} u_{C2}}{\mathrm{d}t^{\alpha}} - C_{\alpha 3} \frac{\mathrm{d}^{\alpha} u_{C3}}{\mathrm{d}t^{\alpha}} - \frac{u_{C3}}{R} = 0 \end{cases} \tag{9.7}$$

where i_S is the current source, R is the resistor. u_{C1}, u_{C2} and u_{C3} are the voltages across the fractional-order capacitors, respectively.

Then the voltage of the capacitor $u_{C1} = u_{C3} + u_{C2}$ is a non-state variable, can be obtained

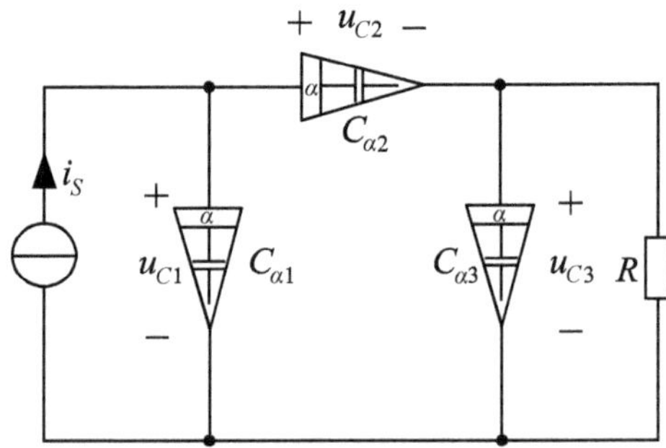

Fig. 9.4 Fractional-order ill-conditioned circuit

$$\begin{cases} (C_{\alpha1}+C_{\alpha3})\dfrac{\mathrm{d}^{\alpha}u_{C1}}{\mathrm{d}t^{\alpha}} - C_{\alpha3}\dfrac{\mathrm{d}^{\alpha}u_{C2}}{\mathrm{d}t^{\alpha}} = i_S - \dfrac{u_{C1}}{R} + \dfrac{u_{C2}}{R} \\ -C_{\alpha3}\dfrac{\mathrm{d}^{\alpha}u_{C1}}{\mathrm{d}t^{\alpha}} + (C_{\alpha2}+C_{\alpha3})\dfrac{\mathrm{d}^{\alpha}u_{C2}}{\mathrm{d}t^{\alpha}} = \dfrac{u_{C1}}{R} - \dfrac{u_{C2}}{R} \end{cases} \tag{9.8}$$

Rearranging (9.8) into a matrix form, it can be written as

$$\begin{bmatrix} C_{\alpha1}+C_{\alpha3} & -C_{\alpha3} \\ -C_{\alpha3} & C_{\alpha2}+C_{\alpha3} \end{bmatrix}\begin{bmatrix} \frac{\mathrm{d}^{\alpha}u_{C1}}{\mathrm{d}t^{\alpha}} \\ \frac{\mathrm{d}^{\alpha}u_{C2}}{\mathrm{d}t^{\alpha}} \end{bmatrix} = \frac{1}{R}\begin{bmatrix} -1 & 1 \\ 1 & -1 \end{bmatrix}\begin{bmatrix} u_{C1} \\ u_{C2} \end{bmatrix} + \begin{bmatrix} 1 \\ 0 \end{bmatrix} i_S \tag{9.9}$$

Some voltages of capacitors and currents of inductors are not state variables for the ill-conditioned circuit, which need to be represented by state variables and input excitation. The matrix form of the state equation of the ill-conditioned circuit, as shown in (9.9), is different from the normal circuit. The coefficient matrix is not diagonal, and it needs to be inverted to obtain the state equation.

Applying the matrix inverse transformation to (9.9), (9.9) can be expressed as

$$\begin{aligned} \begin{bmatrix} \frac{\mathrm{d}^{\alpha}u_{C1}}{\mathrm{d}t^{\alpha}} \\ \frac{\mathrm{d}^{\alpha}u_{C2}}{\mathrm{d}t^{\alpha}} \end{bmatrix} &= \frac{1}{R(C_{\alpha1}C_{\alpha2}+C_{\alpha1}C_{\alpha3}+C_{\alpha2}C_{\alpha3})}\begin{bmatrix} -C_{\alpha2} & C_{\alpha2} \\ C_{\alpha1} & -C_{\alpha1} \end{bmatrix}\begin{bmatrix} u_{C1} \\ u_{C2} \end{bmatrix} \\ &+ \frac{1}{C_{\alpha1}C_{\alpha2}+C_{\alpha1}C_{\alpha3}+C_{\alpha2}C_{\alpha3}}\begin{bmatrix} C_{\alpha2}+C_{\alpha3} \\ C_{\alpha3} \end{bmatrix} i_S \end{aligned} \tag{9.10}$$

9.2.3 *Normal Circuit with Fractional-Order Coupled Inductor*

The method of getting the state equation of a circuit with a fractional-order coupled inductor is the same as the normal circuit.

It is worth noting that the voltage of the fractional-order coupled inductor is determined by the current of the branch and the coupled branch, which makes the acquisition of the state equation more complicated. Here, taking the fractional-order circuit shown in Fig. 9.5 as an example. u_S is the voltage source, R_1 and R_2 are

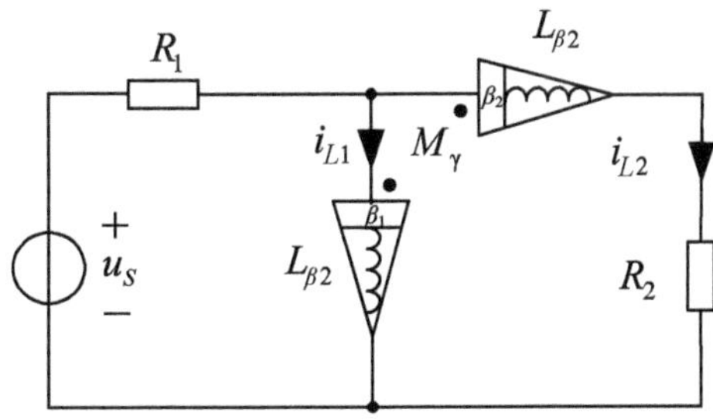

Fig. 9.5 Normal circuit with a fractional-order coupled inductor

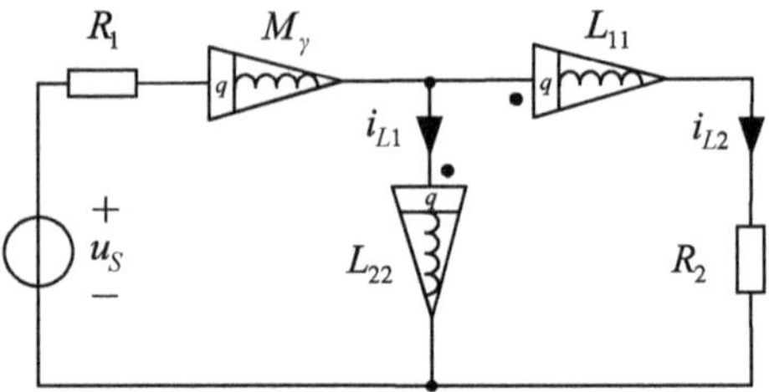

Fig. 9.6 T-type equivalent fractional-order circuit

resistors, i_{L1} and i_{L2} are the currents flowing through fractional-order inductors $L_{\beta 1}$ and $L_{\beta 2}$, respectively. M_γ is the fractional-order mutual inductor, γ is its order.

Supposing that the orders of fractional-order inductors and mutual inductors are identical, i.e., $\beta_1 = \beta_2 = \gamma = q$. The fractional-order circuit shown in Fig. 9.5 can be transformed into a T-type equivalent circuit [11], as shown in Fig. 9.6.

The T-type equivalent fractional-order circuit is an ill-conditioned circuit. The loop equations of the inductance branches are

$$\begin{cases} L_{11}\dfrac{\mathrm{d}^q i_{L1}}{\mathrm{d}t^q} + M_\gamma \dfrac{\mathrm{d}^q i_{L2}}{\mathrm{d}t^q} = u_S - R_1(i_{L1} + i_{L2}) \\ L_{22}\dfrac{\mathrm{d}^q i_{L2}}{\mathrm{d}t^q} + M_\gamma \dfrac{\mathrm{d}^q i_{L1}}{\mathrm{d}t^q} = u_S - R_1(i_{L1} + i_{L2}) - R_2 i_{L2} \end{cases} \tag{9.11}$$

where $L_{11} = L_{\beta 1} - M_\gamma$ and $L_{22} = L_{\beta 2} - M_\gamma$ are equivalent fractional-order inductors. (9.11) can be rewritten in the matrix form as

$$\begin{bmatrix} L_{11} & M_\gamma \\ M_\gamma & L_{22} \end{bmatrix} \begin{bmatrix} \frac{\mathrm{d}^q i_{L1}}{\mathrm{d}t^q} \\ \frac{\mathrm{d}^q i_{L2}}{\mathrm{d}t^q} \end{bmatrix} = -\begin{bmatrix} R_1 & R_1 \\ R_1 & R_1 + R_2 \end{bmatrix} \begin{bmatrix} i_{L1} \\ i_{L2} \end{bmatrix} + \begin{bmatrix} u_S \\ u_S \end{bmatrix} \tag{9.12}$$

By using inverse matrix of (9.12), the state equation can be obtained as

$$\begin{aligned} \begin{bmatrix} \frac{\mathrm{d}^q i_{L1}}{\mathrm{d}t^q} \\ \frac{\mathrm{d}^q i_{L2}}{\mathrm{d}t^q} \end{bmatrix} &= \begin{bmatrix} -\frac{L_{22}R_1 + M_\gamma R_1}{L_{11}L_{22} - M_\gamma^2} & -\frac{L_{22}R_2 + M_\gamma(R_1 + R_2)}{L_{11}L_{22} - M_\gamma^2} \\ -\frac{L_{11}R_1 + M_\gamma R_1}{L_{11}L_{22} - M_\gamma^2} & -\frac{L_{11}(R_1 + R_2) + M_\gamma R_1}{L_{11}L_{22} - M_\gamma^2} \end{bmatrix} \begin{bmatrix} i_{L1} \\ i_{L2} \end{bmatrix} \\ &+ \begin{bmatrix} \frac{L_{22}}{L_{11}L_{22} - M_\gamma^2} & -\frac{M_\gamma}{L_{11}L_{22} - M_\gamma^2} \\ \frac{L_{11}}{L_{11}L_{22} - M_\gamma^2} & \frac{M_\gamma}{L_{11}L_{22} - M_\gamma^2} \end{bmatrix} u_S \end{aligned} \tag{9.13}$$

9.3 Solutions to State Equation of Fractional-Order Circuits

The state equation of the fractional-order circuit can be solved by the Laplace transform method. Generally, the Laplace transform is performed on the state equation to obtain the s-domain solution. Then the time-domain solution of the state equation can be obtained by applying the inverse Laplace transform to the s-domain solution [7, 12].

9.3.1 State Equation with the Same Order

Considering that the fractional-order system can be described by the state equations, as shown in (9.2), the order of inductors and capacitors are identical, i.e., $\alpha = \beta$. Then, based on the R-L fractional differential definition, applying the Laplace transform to (9.2) yields

$$\begin{aligned} \boldsymbol{X}(s) &= \sum_{k=0}^{\infty} \boldsymbol{A}^k s^{-(k+1)\alpha} \left[\sum_{l=1}^{n} s^{l-1} \boldsymbol{x}^{(\alpha-l)}(0^+) + \boldsymbol{B}\boldsymbol{U}(s) \right] \\ &= \sum_{k=0}^{\infty} \sum_{l=1}^{n} \boldsymbol{A}^k s^{-(k+1)\alpha+l-1} \boldsymbol{x}^{(\alpha-l)}(0^+) + \sum_{k=0}^{\infty} \boldsymbol{A}^k s^{-(k+1)\alpha} \boldsymbol{B}\boldsymbol{U}(s) \end{aligned} \tag{9.14}$$

where $n\text{-}1 \le \alpha < n$, s is the Laplace transform operator, $\boldsymbol{X}(s)$ and $\boldsymbol{U}(s)$ are the Laplace transform functions of $\boldsymbol{x}(t)$ and $\boldsymbol{u}(t)$.

By using the inverse Laplace transform and the convolution theorem to (9.14), the solution of the state Eq. (9.2) is determined by

$$\boldsymbol{x}(t) = \sum_{l=1}^{n} \boldsymbol{\Phi}_l(t) \boldsymbol{x}^{(\alpha-l)}(0^+) + \int_0^t \boldsymbol{\Phi}(t-\tau) \boldsymbol{B}\boldsymbol{u}(t) \mathrm{d}\tau \tag{9.15}$$

where

$$\begin{cases} \boldsymbol{\Phi}_l(t) = \sum\limits_{k=0}^{\infty} \dfrac{\boldsymbol{A}^k t^{(k+1)\alpha-l}}{\Gamma[(k+l)\alpha - l + 1]} \\ \boldsymbol{\Phi}(t) = \sum\limits_{k=0}^{\infty} \dfrac{\boldsymbol{A}^k t^{(k+1)\alpha-1}}{\Gamma[(k+1)\alpha]} \end{cases} \tag{9.16}$$

On the basic of the Caputo fractional differential definition, the Laplace transform of (9.2) has the form as

$$\mathcal{L}\left[{}_{t_0}\mathrm{D}_t^{\alpha}\boldsymbol{x}(t)\right] = s^{\alpha}\boldsymbol{X}(s) - \sum_{k=1}^{n} s^{\alpha-k}\boldsymbol{X}^{(k-1)}(0^+) = \boldsymbol{A}\boldsymbol{X}(s) + \boldsymbol{B}\boldsymbol{U}(s) \tag{9.17}$$

(9.17) can be rewritten as

$$\begin{aligned}\boldsymbol{X}(s) &= \sum_{k=0}^{\infty} \boldsymbol{A}^k s^{-(k+1)\alpha}\left[\sum_{l=1}^{n} s^{\alpha-l}\boldsymbol{x}^{(l-1)}(0^+) + \boldsymbol{B}\boldsymbol{U}(s)\right] \\ &= \sum_{k=0}^{\infty}\sum_{l=1}^{n} \boldsymbol{A}^k s^{-(k\alpha+1)}\boldsymbol{x}^l(0^+) + \sum_{k=0}^{\infty} \boldsymbol{A}^k s^{-(k+1)\alpha}\boldsymbol{B}\boldsymbol{U}(s)\end{aligned} \tag{9.18}$$

Then, applying the inverse Laplace transform and convolution theorem to (9.18), the solution of the state Eq. (9.2) can also be denoted by

$$\boldsymbol{x}(t) = \sum_{l=1}^{n} \boldsymbol{\Phi}_l(t)\boldsymbol{x}^{(l-1)}(0^+) + \int_0^t \boldsymbol{\Phi}(t-\tau)\boldsymbol{B}\boldsymbol{u}(t)\mathrm{d}\tau \tag{9.19}$$

where

$$\begin{cases} \boldsymbol{\Phi}_l(t) = \sum\limits_{k=0}^{\infty} \frac{\boldsymbol{A}^k t^{k\alpha+l-1}}{\Gamma(k\alpha+l)} \\ \boldsymbol{\Phi}(t) = \sum\limits_{k=0}^{\infty} \frac{\boldsymbol{A}^k t^{(k+1)\alpha-1}}{\Gamma[(k+1)\alpha]} \end{cases} \tag{9.20}$$

9.3.2 State Equation with Different Orders

When a fractional-order system has the fractional-order components with different orders, its state equation is defined by [7, 12, 13]

$$\begin{bmatrix} \frac{\mathrm{d}^{\alpha_1} x_1(t)}{\mathrm{d}t^{\alpha_1}} \\ \vdots \\ \frac{\mathrm{d}^{\alpha_n} x_n(t)}{\mathrm{d}t^{\alpha_n}} \end{bmatrix} = \begin{bmatrix} A_{11} & \cdots & A_{1n} \\ \vdots & \ddots & \vdots \\ A_{n1} & \cdots & A_{nn} \end{bmatrix} \begin{bmatrix} x_1(t) \\ \vdots \\ x_n(t) \end{bmatrix} + \begin{bmatrix} B_1 \\ \vdots \\ B_n \end{bmatrix} u(t) \tag{9.21}$$

where $p_k - 1 < \alpha_k < p_k$, $p_k \in \mathbb{N}$, $x_k \in \mathbb{R}^{\bar{n}_k}$ are the state vectors, $A_{kj} \in \mathbb{R}^{\bar{n}_k \times \bar{n}_j}$, $B_k \in \mathbb{R}^{\bar{n}_k \times r}$, $u \in \mathbb{R}^r$ are the input vectors, and $k, j = 1, 2, \ldots, n$.

The initial conditions of (9.21) have the form as

$$x_k^{(j)}(0) = x_{k0}^{(j)} \in \mathbb{R}^{\bar{n}_k} \tag{9.22}$$

where $k = 1, 2, \ldots, n. j = 0, 1, 2, \ldots, p_k - 1$.

Based on the Caputo fractional differential definition and Laplace transform, (9.21) can be rewritten as

$$\begin{bmatrix} \boldsymbol{I}_{\bar{n}_1} s^{\alpha_1} - A_{11} & -A_{12} & \cdots & -A_{1n-1} & A_{1n} \\ \vdots & \vdots & \ddots & \vdots & \vdots \\ -A_{n1} & -A_{n2} & \cdots & -A_{nn-1} & \boldsymbol{I}_{\bar{n}_n} s^{\alpha_n} - A_{nn} \end{bmatrix} \begin{bmatrix} X_1(s) \\ \vdots \\ X_n(s) \end{bmatrix}$$
$$= \begin{bmatrix} B_1 \\ \vdots \\ B_n \end{bmatrix} U(s) + \begin{bmatrix} \sum\limits_{j_1=1}^{p_1} s^{\alpha_1 - j_1} x_{10}^{(j_1-1)} \\ \vdots \\ \sum\limits_{j_n=1}^{p_n} s^{\alpha_n - j_n} x_{n0}^{(j_n-1)} \end{bmatrix} \tag{9.23}$$

where $\boldsymbol{I}$ denotes the unit matrix.

By taking the inverse of the coefficient matrix of (9.23), it can be rearranged as

$$\begin{bmatrix} X_1(s) \\ \vdots \\ X_n(s) \end{bmatrix} = \begin{bmatrix} \boldsymbol{I}_{\bar{n}_1} s^{\alpha_1} - A_{11} & -A_{12} & \cdots & -A_{1n-1} & A_{1n} \\ \vdots & \vdots & \ddots & \vdots & \vdots \\ -A_{n1} & -A_{n2} & \cdots & -A_{nn-1} & \boldsymbol{I}_{\bar{n}_n} s^{\alpha_n} - A_{nn} \end{bmatrix}^{-1}$$
$$\times \left\{ \begin{bmatrix} B_1 \\ \vdots \\ B_n \end{bmatrix} U(s) + \begin{bmatrix} \sum\limits_{j_1=1}^{p_1} s^{\alpha_1 - j_1} x_{10}^{(j_1-1)} \\ \vdots \\ \sum\limits_{j_n=1}^{p_n} s^{\alpha_n - j_n} x_{n0}^{(j_n-1)} \end{bmatrix} \right\} \tag{9.24}$$

Comparing the coefficients at the same power of $s^{-\alpha_k}$, it is easy to verify that

$$\begin{bmatrix} \boldsymbol{I}_{\bar{n}_1} - A_{11} s^{-a_1} & \cdots & -A_{1n} s^{-\alpha_1} \\ \vdots & \ddots & \vdots \\ -A_{n1} s^{-\alpha_n} & \cdots & \boldsymbol{I}_{\bar{n}_n} - A_{nn} s^{-\alpha_n} \end{bmatrix} \left[\sum_{k_1=0}^{\infty} \cdots \sum_{k_n=0}^{\infty} T_{k_1 \cdots k_n} s^{-(k_1 \alpha_1 + \cdots + k_n \alpha_n)} \right] = \boldsymbol{I}_N \tag{9.25}$$

For $k_1 = \cdots = k_n = 0$

$$T_{k_1 \cdots k_n} = \boldsymbol{I}_N \tag{9.26}$$

For $k_1 = 1,\ k_2 = \ \cdots = k_n = 0$

$$T_{k_1\cdots k_n} = \begin{bmatrix} A_{11} & \ldots & A_{1n} \\ 0 & \ldots & 0 \\ \vdots & \ddots & \vdots \\ 0 & \ldots & 0 \end{bmatrix} \tag{9.27}$$

When $k_1 = \cdots = k_{i-1} = 0,\ k_i = 1,\ k_{i+1} = \cdots = k_n = 0$

$$T_{k_1\cdots k_n} = \begin{bmatrix} 0 & \ldots & 0 \\ \vdots & \ddots & \vdots \\ 0 & \ldots & 0 \\ A_{i1} & \ldots & A_{in} \\ 0 & \ldots & 0 \\ \vdots & \ddots & \vdots \\ 0 & \ldots & 0 \end{bmatrix} \tag{9.28}$$

For $k_n = 1,\ k_2 = \cdots = k_{n-1} = 0$

$$T_{k_1\cdots k_n} = \begin{bmatrix} 0 & \cdots & 0 \\ 0 & \cdots & 0 \\ \vdots & \ddots & \vdots \\ A_{n1} & \cdots & A_{nn} \end{bmatrix} \tag{9.29}$$

Else

$$T_{k_1\cdots k_n} = \begin{cases} T_{10\ldots0}T_{01\ldots1} + \cdots + T_{0\ldots01}T_{1\ldots10} & \text{for } k_1 = \cdots = k_n = 1 \\ \vdots & \\ T_{10\ldots0}T_{k_1-1,k_2,\cdots k_n} + \cdots + T_{0\ldots01}T_{k_1,k_{n-1},\cdots k_n-1} & \text{for } k_1 + \cdots + k_n > 0 \\ 0 & \text{for at least one } k_i < 0, i = 1, \ldots, n \end{cases} \tag{9.30}$$

On the basis of (9.25), (9.31) can be obtained

$$\begin{bmatrix} \boldsymbol{I}_{\bar{n}_1}s^{\alpha_1} - A_{11} & -A_{12} & \cdots & -A_{1n-1} & A_{1n} \\ \vdots & \vdots & \ddots & \vdots & \vdots \\ -A_{n1} & -A_{n2} & \cdots & -A_{nn-1} & \boldsymbol{I}_{\bar{n}_n}s^{\alpha_n} - A_{nn} \end{bmatrix}^{-1}$$

$$= \left\{ \begin{bmatrix} \boldsymbol{I}_{\bar{n}_1}s^{\alpha_1} & \cdots & 0 \\ \vdots & \ddots & \vdots \\ 0 & \cdots & \boldsymbol{I}_{\bar{n}_n}s^{\alpha_n} \end{bmatrix} \begin{bmatrix} \boldsymbol{I}_{\bar{n}_1} - A_{11}s^{-a_1} & \cdots & -A_{1n}s^{-\alpha_1} \\ \vdots & \ddots & \vdots \\ -A_{n1}s^{-\alpha_n} & \cdots & \boldsymbol{I}_{\bar{n}_n} - A_{nn}s^{-\alpha_n} \end{bmatrix} \right\}^{-1}$$

$$
= \begin{bmatrix} \boldsymbol{I}_{\bar{n}_1} - A_{11}s^{-a_1} & \cdots & -A_{1n}s^{-\alpha_1} \\ \vdots & \ddots & \vdots \\ -A_{n1}s^{-\alpha_n} & \cdots & \boldsymbol{I}_{\bar{n}_n} - A_{nn}s^{-\alpha_n} \end{bmatrix}^{-1} \begin{bmatrix} \boldsymbol{I}_{\bar{n}_1}s^{-\alpha_1} & \cdots & 0 \\ \vdots & \ddots & \vdots \\ 0 & \cdots & \boldsymbol{I}_{\bar{n}_n}s^{-\alpha_n} \end{bmatrix}
$$

$$
= \sum_{k_1=0}^{\infty} \cdots \sum_{k_n=0}^{\infty} T_{k_1 \cdots k_n} s^{-(k_1\alpha_1 + \cdots + k_n\alpha_n)} \begin{bmatrix} \boldsymbol{I}_{\bar{n}_1}s^{-\alpha_1} & \cdots & 0 \\ \vdots & \ddots & \vdots \\ 0 & \cdots & \boldsymbol{I}_{\bar{n}_n}s^{-\alpha_n} \end{bmatrix} \tag{9.31}
$$

Substituting (9.31) into (9.24) to get (9.33).

$$
\begin{bmatrix} X_1(s) \\ \vdots \\ X_n(s) \end{bmatrix} = \sum_{k_1=0}^{\infty} \cdots \sum_{k_n=0}^{\infty} T_{k_1 \cdots k_n} s^{-(k_1\alpha_1 + \cdots + k_n\alpha_n)} \begin{bmatrix} \boldsymbol{I}_{\bar{n}_1}s^{-\alpha_1} & \cdots & 0 \\ \vdots & \ddots & \vdots \\ 0 & \cdots & \boldsymbol{I}_{\bar{n}_n}s^{-\alpha_n} \end{bmatrix}
\times \left\{ \begin{bmatrix} B_1 \\ \vdots \\ B_n \end{bmatrix} U(s) + \begin{bmatrix} \sum_{j_1=1}^{p_1} s^{\alpha_1 - j_1} x_{10}^{(j_1-1)} \\ \vdots \\ \sum_{j_n=1}^{p_n} s^{\alpha_n - j_n} x_{n0}^{(j_n-1)} \end{bmatrix} \right\} \tag{9.32}
$$

Then

$$
\begin{bmatrix} X_1(s) \\ \vdots \\ X_n(s) \end{bmatrix} = \sum_{k_1=0}^{\infty} \cdots \sum_{k_n=0}^{\infty} T_{k_1 \cdots k_n} \left\{ \begin{array}{l} \begin{bmatrix} B_{10}s^{-[(k_1+1)\alpha_1 + \cdots + k_n\alpha_n]} + \cdots + \\ B_{n0}s^{-[(k_1+1)\alpha_1 + \cdots + k_{n-1}\alpha_{n-1} + (k_n+1)\alpha_n]} \end{bmatrix} U(s) \\ + s^{-(k_1\alpha_1 + \cdots + k_n\alpha_n)} \begin{bmatrix} \sum_{j_1=1}^{p_1} s^{\alpha_1 - j_1} x_{10}^{(j_1-1)} \\ \vdots \\ \sum_{j_n=1}^{p_n} s^{\alpha_n - j_n} x_{n0}^{(j_n-1)} \end{bmatrix} \end{array} \right\} \tag{9.33}
$$

Applying the inverse Laplace transform and convolution theorem to (9.33). Since $\mathcal{L}^{-1}\left(1/s^{\alpha+1}\right) = t^{\alpha}/\Gamma(\alpha + 1)$, the solution of (9.21) under the Caputo fractional differential definition can be written as

$$\begin{bmatrix} x_1(t) \\ \vdots \\ x_n(t) \end{bmatrix} = \mathcal{L}^{-1}\begin{bmatrix} X_1(s) \\ \vdots \\ X_n(s) \end{bmatrix}$$

$$= \mathcal{L}^{-1}\sum_{k_1=0}^{\infty}\cdots\sum_{k_n=0}^{\infty} T_{k_1\cdots k_n}\left\{\begin{array}{l} \begin{bmatrix} B_{10}s^{-[(k_1+1)\alpha_1+\cdots+k_n\alpha_n]}+\cdots+ \\ B_{n0}s^{-[(k_1+1)\alpha_1+\cdots+k_{n-1}\alpha_{n-1}+(k_n+1)\alpha_n]} \end{bmatrix} U(s) \\ +s^{-(k_1\alpha_1+\cdots+k_n\alpha_n)}\begin{bmatrix} \sum\limits_{j_1=1}^{p_1} s^{\alpha_1-j_1}x_{10}^{(j_1-1)} \\ \vdots \\ \sum\limits_{j_n=1}^{p_n} s^{\alpha_n-j_n}x_{n0}^{(j_n-1)} \end{bmatrix} \end{array}\right\}$$

$$= \int_0^t [\Phi_1(t-\tau)B_{10}+\cdots+\Phi_n(t-\tau)B_{n0}]u(\tau)\mathrm{d}\tau$$

$$+\sum_{k_1=0}^{\infty}\cdots\sum_{k_n=0}^{\infty} T_{k_1\cdots k_n}\begin{bmatrix} \sum\limits_{j_1=1}^{p_1}\frac{t^{k_1\alpha_1+\cdots+k_n\alpha_n+j_1-1}}{\Gamma(k_1\alpha_1+\cdots+k_n\alpha_n+j_1)}x_{10}^{j_1-1} \\ \vdots \\ \sum\limits_{j_n=1}^{p_n}\frac{t^{k_1\alpha_1+\cdots+k_n\alpha_n+j_n-1}}{\Gamma(k_1\alpha_1+\cdots+k_n\alpha_n+j_n)}x_{n0}^{j_n-1} \end{bmatrix} \tag{9.34}$$

where

$$\begin{cases} \Phi_1(t) = \sum\limits_{k_1=0}^{\infty}\cdots\sum\limits_{k_n=0}^{\infty} T_{k_1\cdots k_n}\dfrac{t^{(k_1+1)\alpha_1+\cdots+k_n\alpha_n-1}}{\Gamma[(k_1+1)\alpha_1+\cdots+k_n\alpha_n)]} \\ \vdots \\ \Phi_n(t) = \sum\limits_{k_1=0}^{\infty}\cdots\sum\limits_{k_n=0}^{\infty} T_{k_1\cdots k_n}\dfrac{t^{(k_1+1)\alpha_1+\cdots+k_{n-1}\alpha_{n-1}+(k_n+1)\alpha_n)-1}}{\Gamma\left[k_1\alpha_1+\cdots+k_{n-1}\alpha_{n-1}+(k_n+1)\alpha_n)\right]} \end{cases} \tag{9.35}$$

9.4 Summary

Compared with the circuit analysis method introduced in the previous chapter, the state-variable method is more suitable for establishing the mathematical model of the complicated fractional-order circuit. Moreover, it is essentially a matrix method, which can be widely used in fractional-order linear and nonlinear circuit simulation. This chapter mainly introduces the state-variable analysis of fractional-order circuits.

Firstly, the basic concepts of the state-variable method are introduced, including state space, state vector, state trajectory, state equation, and output equation. Then, taking ill-conditioned fractional-order circuit and normal fractional-order circuit as examples, the method of establishing state equation is briefly described. Finally, the analytical solutions of the state equations with the same order and different order circuits are given, which provides a method for the analysis of complicated fractional-order circuits.

References

1. Bubnicki Z (2005) Modern control theory. Springer, Berlin, Heideberg, Germany, pp 17–30
2. Pana A, Baloi A, Molnar-Matei F (2007) Application of the state variables method on quickly evaluation of the harmonic resonance frequencies in a transmission line. In: Proceedings of EUROCON, Warsaw, Poland, pp 1573–1578
3. Voiculescu R, Iordache M (2015) Analog circuit simulation by state variable method. UPB Sci. Bull. Ser. C 77(3):213–224
4. Kalman RE, Bertram JE (1960) Control system analysis and design via the “second method” of Lyapunov: II-discrete-time systems. J Basic Eng 82(2):394–400
5. Kuh ES, Rohrer RA (1965) The state-variable approach to network analysis. Proc IEEE 53(7):672–686
6. Monje CA, Chen YQ, Vinagre BM et al (2010) Fractional-order systems and controls. Springer, London, pp 12–19
7. Kaczorek T, Rogowski K (2015) Fractional linear systems and electrical circuits. Springer, Switzerland, pp 15–53
8. Cois O, Oustaloup A, Poinot T et al (2001) Fractional state variable filter for system identification by fractional model. In: Proceedings of European control conference (ECC), Porto, Portugal, pp 2481–2486
9. Wu C, Si G, Zhang Y, Yang N (2015) The fractional-order state-space averaging modeling of the Buck-Boost DC/DC converter in discontinuous conduction mode and the performance analysis. Nonlinear Dyn 79(1):689–703
10. Wei Z, Zhang B, Jiang Y (2019) Analysis and modeling of fractional-order buck converter based on Riemann-Liouville derivative. IEEE Access 7:162768–162777
11. Soltan A, Radwan A, Soliman AM (2015) Fractional-order mutual inductance: analysis and design. Int J Circuit Theory Appl 44:85–97
12. Kaczorek T (2008) Fractional positive continuous-time linear systems and their reachability. Appl Math Comput Sci 18:223–228
13. Kaczorek T (2011) Positive linear systems consisting of n subsystems with different fractional orders. IEEE Trans Circ Syst I Regul Pap 58(6):1203–1210

GPSR Compliance
The European Union's (EU) General Product Safety Regulation (GPSR) is a set of rules that requires consumer products to be safe and our obligations to ensure this.

If you have any concerns about our products, you can contact us on

ProductSafety@springernature.com

In case Publisher is established outside the EU, the EU authorized representative is:

Springer Nature Customer Service Center GmbH
Europaplatz 3
69115 Heidelberg, Germany

www.ingramcontent.com/pod-product-compliance
Ingram Content Group UK Ltd.
Pitfield, Milton Keynes, MK11 3LW, UK
UKHW021833270726
14058UKWH00001B/127